Transport and Chemical Transformation of Pollutants in the Troposphere

Series editors (Volumes 2 to 8 and 10): Peter Borrell, Patricia M. Borrell, Tomislav Cvitaš, Kerry Kelly and Wolfgang Seiler

Series editors (Volumes 1 and 9): Peter Borrell, Patricia M. Borrell and Pauline Midgley

Springer
Berlin
Heidelberg
New York
Barcelona
Hong Kong
London
Milan
Paris
Singapore
Tokyo

Volume 1

Transport and Chemical Transformation of Pollutants in the Troposphere

An Overview of the Work of EUROTRAC

Peter Borrell and Patricia M. Borrell
Editors

EUROTRAC-2 International Scientific Secretariat,
GSF-Forschungszentrum für Umwelt und Gesundheit, München
and
P&PMB Consultants, Garmisch-Partenkirchen

Springer

Editors
Dr. PETER BORRELL
Dr. PATRICIA BORRELL
P&PMB Consultants
Ehrwalder Strasse 9
D-82467 Garmisch-Partenkirchen

With 95 Figures and 19 Tables

The cover picture shows a scene from the 1989 GCE field campaign at S. Pietro Capofiume, near Bologna in the Po Valley. The participants were recovering after sixty hours of continuous fog measurements and it can be seen that the fog is closing in again. More details about the campaign and the results obtained can be found in chapters 3 and 8.2. *Photograph: Patricia M. Borrell*

ISBN 3-540-66775-X Springer-Verlag Berlin Heidelberg New York

Library of Congress Cataloging-in-Publication Data applied for
Transport and chemical transformation of pollutants in the troposphere: [an account of the work of EUROTRAC]. – Berlin; Heidelberg; New York; Barcelona; Hong Kong; London; Milan; Paris; Singapore; Tokyo: Springer
Vol. 1. Transport and chemical transformation of pollutants in the troposphere: an overview of the work of EUROTRAC/Peter Borrell... (ed.). – 2000
  ISBN 3-540-66775-X

Springer-Verlag is a company in the BertelsmannSpringer publishing group
© Springer-Verlag Berlin · Heidelberg 2000
Printed in Germany

Cover Design: Struve & Partner, Heidelberg
Coverpicture from Fotoverlag Huber, D-82467 Garmisch-Partenkirchen

SPIN 10514807   30/3136xz-5 4 3 2 1 0 – Printed on acid-free paper

Transport and Chemical Transformation of Pollutants in the Troposphere

Series editors (Volumes 2 to 8 and 10): Peter Borrell, Patricia M. Borrell, Tomislav Cvitaš, Kerry Kelly and Wolfgang Seiler

Series editors (Volumes 1 and 9): Peter Borrell, Patricia M. Borrell and Pauline Midgley

Foreword by the Series Editors

EUROTRAC was the European co-ordinated research project, within the EUREKA initiative, studying the transport and chemical transformation of pollutants in the troposphere. That the project achieved a remarkable scientific success since its start in 1988, contributing substantially both to the scientific progress in this field and to the improvement of the scientific basis for environmental management in Europe, was indicated by the decision of the participating countries to set up, in 1996, EUROTRAC-2 to capitalise on the achievements. EUROTRAC, which at its peak comprised some 250 research groups organised into 14 subprojects, brought together international groups of scientists to work on problems directly related to the transport and chemical transformation of trace substances in the troposphere. In doing so, it helped to harness the resources of the participating countries to gain a better understanding of the trans-boundary, interdisciplinary environmental problems which beset us in Europe.

The scientific results of EUROTRAC are summarised in this report which consists of ten volumes.

Volume 1 provides a general overview of the scientific results, together with summaries of the work of the fourteen individual subprojects prepared by the respective subproject coordinators.

Volumes 2 to 9 comprise detailed overviews of the subproject achievements, each prepared by the respective subproject coordinator and steering group, together with summaries of the work of the participating research groups prepared by the principal investigators. Each volume also includes a full list of the scientific publications from the subproject.

Volume 10 is the complete report of the Application Project, which was set up in 1993 to assimilate the scientific results from EUROTRAC and present them in a condensed form so that they are suitable for use by those responsible for environmental planning and management in Europe. It illustrates how a scientific project such as EUROTRAC can contribute practically to providing the scientific consensus necessary for the development of a coherent atmospheric environmental policy for Europe.

A multi-volume work such as this has many contributors and we, as the series editors, would like to express our thanks to all of them: to the subproject coordinators who have borne the brunt of the scientific co-ordination and who have contributed so much to the success of the project and the quality of this report; to the principal investigators who have carried out so much high-quality scientific work; to the members of the International Executive Committee (IEC) and the SSC for their enthusiastic encouragement and support of EUROTRAC; to the participating governments and, in particular, the German Government (BMBF) for funding, not only the research, but also the ISS publication activities; and finally to Mr. Christian Witschel and his colleagues at Springer-Verlag for providing the opportunity to publish the results in a way which will bring them to the notice of a large audience.

Since the first volumes in this series were published in 1997 there have been many changes. In 1998 the International Scientific Secretariat moved from its old home at the Fraunhofer Institute in Garmisch-Partenkirchen to the GSF-Forschungszentrum für Umwelt und Gesundheit in München. Dr. Pauline Midgley was appointed as Executive Secretary for EUROTRAC-2 on the retirement in 1999 of Dr. Peter Borrell, who had also been the Scientific Secretary of EUROTRAC. With these came a change of the series editors. The present editors would like to thank their former colleagues, Dr. Tomislav Cvitaš and Ms. Kerry Kelly, for their contribution to the success of the final report and also, and in particular, Dr. Wolfgang Seiler, the former Director of the International Scientific Secretariat, without whose initial efforts EUROTRAC would never have got off the ground.

Peter Borrell
Patricia M. Borrell   P&PMB Consultants, Garmisch-Partenkirchen
Pauline Midgley
    Executive Secretary   EUROTRAC-2 International Scientific Secretariat
                              GSF-Forschungszentrum für Umwelt und
                              Gesundheit, München

Preface to Volume 1

Transport and Chemical Transformation of Pollutants in the Troposphere is volume 1 of the ten volume series which constitutes the final report of EUROTRAC, the EUREKA environmental project. The volume is intended to provide an overview of the whole project which worked for eight years and at its peak involved some 250 principal investigators from 150 institutions in 25 countries in Europe.

The first chapter of Part I places EUROTRAC in its environmental context and explains how the project was organised and run. In chapters 2 to 6, members of the Scientific Steering Committee report on and attempt to synthesise the scientific contribution that EUROTRAC made to our knowledge and understanding of photo-oxidants and acidity in lower atmosphere. Chapter 7 is a summary of the work of the Application Project which was set up to extract from the EUROTRAC work the results of importance to those engaged in environmental policy development in Europe.

In Part II, chapters 8 to 14, the subproject coordinators outline the work of their respective subprojects. The chapters provide a useful overview of the scientific progress within the various fields in which EUROTRAC was active. All but two are illustrative summaries of the work that appears in the subproject volumes; for the other two we have had to use the fuller version.

In nearly every chapter there is an evaluation of what has been achieved with an indication of the uncertainties remaining and many suggestions for future work. (see *future needs* and *uncertainties* in the index). Throughout the volume their are extensive references to the work described in the other volumes and to the scientific literature so that interests can readily be followed up.

As one of the editors of this volume and as scientific coordinator of the whole project, there are so many people whom I should gratefully thank and wholeheartedly acknowledge; the principal investigators, the subproject coordinators and the members of the International Executive Committees and Scientific Steering Committee are mentioned elsewhere. However, in connection with the final report, I would like to single out four people: Professor *Øystein Hov* and Dr. *Tony Marsh* who stepped in at short notice to write chapters which others were unable to do; my co-editor, Dr. *Patricia Borrell*, without whose support, skills and sheer hard work the final report would never have been produced; and my successor as Executive Secretary of EUROTRAC-2, Dr. *Pauline Midgley* of the GSF - Forschungszentrum für Umwelt und Gesundheit in München. Her cheerful encouragement in the face of some surprising difficulties has ensured the completion of this work. May she have as much fun in her work as I had in mine and may EUROTRAC-2 have as much success as its predecessor.

Peter Borrell
P&PMB Consultants
Garmisch-Partenkirchen

Table of Contents

Part I An Overview and Synthesis of the Scientific Results

Part II   An Overview of the Scientific Work of the Subprojects

Appendix A

Appendix B

List of Contributors

Dr. Jeannette P. Beck
RIVM
National Institute of Public Health and
Environmental Protection
Antonie van Leewenhoeklaan 9
P.O.Box 1
NL-3720 BA Bilthoven
The Netherlands

Dr. Karl-Heinz Becker
Bergische Universität Wuppertal
Physikalische Chemie, Fachbereich 9
Gaußstr. 20
D-42119 Wuppertal,
Germany

Dr. Peter Borrell
P&PMB Consultants
Ehrwalder Strasse 9
D-82467 Garmisch-Partenkirchen,
Germany

Dr. Jens Bösenberg
Max-Planck-Institut für Meteorologie
Bundesstr. 55
D-20146 Hamburg,
Germany

Dr. D.J. Brassington
Imperial College
Atmospheric Chemistry Research Unit
Silwood Park, Ascot
Berkshire SL5 7PW,
Great Britain

Prof. Peter J. H. Builtjes
TNO-MEP
Department of Environmental
 Quality (MR)
P.O. Box 342
NL-7300 AH Apeldoorne
The Netherlands

Prof. Reginald Colin
Université Libre de Bruxelles
Lab. de Photophysique Moleculaire
50 avenue F.D. Roosevelt
B-1050 Bruxelles
Belgium

Prof. Dr. Adolf Ebel
Universität zu Köln, EURAD
Institut für Meteorologie und Geophysik,
Aachener Straße 201-209
D-50931 Köln,
Germany

Prof. Dr. Franz Fiedler
Universität Karlsruhe,
Forschungszentrum Karlsruhe,
Inst. für Meteorologie und Klimaforschung,
D-76128 Karlsruhe,
Germany

Dr. David Fowler
Institute of Terrestrial Ecology
Edinburgh Research Station
Bush Estate, Penicuik
Midlothian E26 0QB,
Great Britain

Dr. Rainer Friedrich
Universität Stuttgart
Institut für Energiewirtschaft und
Rationelle Energieanwendung (IER)
Heßbrühlstraße 49a
D-70565 Stuttgart,
Germany

Prof. Sandro Fuzzi
Istituto FISBAT
Area della Ricerca C.N.R.
Via Gobetti 101
I-40129 Bologna
Italy

Dr. Peringe Grennfelt
Swedish Environmental Research
Institute (IVL),
Box 47086
S-40258 Göteborg
Sweden

Dr. Øystein Hov
Norwegian Institute for Air Research
(NILU), P.O. Box 100
Instituttveien 18
N-2007 Kjeller
Norway

Dr. Niels Otto Jensen
Risö National Laboratory
Dept. of Meteorology & Wind Energy
P.O. box 49
DK-4000 Roskilde
Denmark

Dr. Dieter Kley
Forschungszentrum Jülich GmbH
Institut für Chemie und Dynamik der
Geosphäre
Postfach 1913
D-52425 Jülich,
Germany

Dr. Michael Kuhn
Kempener Straße 125,
50733 Cologne,
Germany

Dr Søren Ejling Larsen
Riso National Laboratory
Dept of Meteorology and Wind Energy
P.O. Box 49
DK-4000 Roskilde
Denmark

Dr. Georges Le Bras
CNRS - LCSR
1-c Avenue de la Recherche Scientifique
F-45071 Orléans Cedex 2
France

Dr. A. R. Marsh
Imperial College
Atmospheric Chemistry Research Unit
Silwood Park, Ascot
Berkshire SL5 7PY, UK

Dr. Gérard Mégie
CNRS
Service d'Aeronomie, B.P. 3
F-91371 Verrières Le Buisson Cedex
France

Prof. Nicolas Moussiopoulos
Aristotle University Thessaloniki
Laboratory of Heat Transfer and
Environmental Engineering
Box 483
GR-54006 Thessaloniki
Greece

Prof. Stuart A. Penkett
University of East Anglia
School of Environmental Sciences
Norwich NR4 7TJ,
Great Britain

Dr. Casimiro Pio
Universidade de Aveiro
Departamento de Ambiente e Ordenamento
P-3810 Aveiro,
Portugal

Prof. Dr. Dirk Poppe
Forschungszentrum Jülich GmbH
Institut für Chemie (ICG-3)
D-52425 Jülich,
Germany

Prof. Henning Rodhe
Stockholm University
Department of Meteorology
S-106 91 Stockholm
Sweden

Prof. Paul C. Simon
Institut d'Aeronomie Spatiale Belgique
3 Avenue Circulaire
B-1180 Bruxelles
Belgium

Dr. Sjaak Slanina
Netherlands Energy Research Foundation,
(ECN), P.O. Box 1
NL-1755 ZG Petten
The Netherlands

Dr. Roel M. van Aalst
RIVM, (EEA Topic Centre: Air Quality)
Antonie van Leeuwenhoeklaan 9
P.O.Box 1
NL-3720 BA Bilthoven
The Netherlands

Dr. Andreas Volz-Thomas
Forschungzentrum Jülich (FZJ)
Institut für Chemie und Dynamik der
Geosphäre (ICG-2)
Postfach 1913
D-52425 Jülich,
Germany

Dr. Dietmar Wagenbach
Universität Heidelberg
Institut für Umweltphysik
Im Neuenheimer Feld 366
D-69120 Heidelberg,
Germany

Prof. Dr. Peter Warneck
Max-Planck-Institut für Chemie,
Abt. Biogeochemie
Postfach 3060
Saarstraße 23
D-55020 Mainz,
Germany

Dr. Richard P. Wayne
Oxford University
Physical Chemistry Laboratory
South Parks Road
Oxford OX1 3QZ,
Great Britain

Acronyms and Abbreviations used within EUROTRAC

EUROTRAC	European Experiment on the Transport and Transformation of Environmentally Relevant Trace Constituents in the Troposphere over Europe

Subprojects

ACE	Acidity in Cloud Experiments
ALPTRAC	High Alpine Aerosol and Snow Chemistry Study
ASE	Air-Sea Exchange
BIATEX	Biosphere-Atmosphere Exchange of Pollutants
EUMAC	European Modelling of Atmospheric Constituents
GCE	Ground-based Cloud Experiments
GENEMIS	Generation of European Emission Data
GLOMAC	Global Modelling of Atmospheric Chemistry
HALIPP	Heterogeneous and Liquid Phase Processes
JETDLAG	Joint European Development of Tunable Diode Laser Absorption Spectroscopy for the Measurement of Atmospheric Trace Gases
LACTOZ	Laboratory Studies of Chemistry Related toTropospheric Ozone
TESLAS	Joint European Programme for the Tropospheric Environmental Studies by Laser Sounding
TOPAS	Tropospheric Optical Absorption Spectroscopy
TOR	Tropospheric Ozone Research
TRACT	Transport of Pollutants over Complex Terrain

Working Groups

AP	Application Project
CG	Cloud Group
CMWG	Chemical Mechanism Working Group

Organisation

IEC	International Executive Committee
SSC	Scientific Steering Committee
ISS	International Scientific Secretariat

Further acronyms and abbreviations can be found in the index, particularly under *field campaigns* and *models*.

Part I

An Overview and Synthesis of the Scientific Results

Chapter 1

EUROTRAC: Organisation Structure and Achievements

Peter Borrell[1] and Peringe Grennfelt[2]

[1]P&PMB Consultants, Ehrwalder Straße 9, D-82467 Garmisch-Partenkirchen, Germany
[2]Swedish Environmental Research Institute (IVL), Box 47086, S-40258 Göteborg, Sweden

1.1 EUROTRAC: definition; objectives

EUROTRAC, the **Euro**pean Experiment on **Tra**nsport and Transformation of Environmentally Relevant Trace Constituents in the Troposphere over Europe was a European scientific research programme within the EUREKA framework. It was accepted as a EUREKA project in 1985 and, after an two-year definition phase, began work in 1988. The project finished at the end of 1995. At its peak the eight year programme included more than 250 research groups from 24 European countries and its budget exceeded 16 million ECU per year. It was set up in the perspective of an increasing awareness of the effects of air pollution on European ecosystems. A second phase of the project was started in July 1996 (ISS, 1999).

The overall objective of EUROTRAC was to increase the scientific knowledge of the impact of human activities on the troposphere over Europe. It focused on the scientific aspects of two of the major air pollution problems occurring on the European scale:

- the chemistry and transport of photo-oxidants (in particular ozone) in the troposphere;

- the chemistry, transport and deposition of acidifying substances in the troposphere.

The principal aims were:

- to increase the basic understanding of atmospheric science;

- to promote the technological development of sensitive specific and fast response instruments for environmental research and monitoring;

- to improve the scientific basis for taking future political decisions on environmental management in the European countries.

EUROTRAC was an interdisciplinary project involving various approaches such as field experiments and campaigns, laboratory studies, comprehensive theoretical model developments and simulations, emission estimation, studies of biosphere / atmosphere exchange and the development of advanced instruments for laboratory and field measurements.

EUROTRAC was one of first environmental projects within EUREKA, which had just been started as a joint initiative of eighteen western European countries, with the purpose of increasing the competitiveness of the European science and industry in a world-wide context. Although EUROTRAC was originally a western European initiative, some research groups from the old Eastern Block countries participated from the beginning. The number of groups from eastern Europe increased appreciably after the political changes there in 1989 and 1990.

EUROTRAC was a science-driven, "bottom-up" research programme, that is: it was initiated and directed primarily by the scientists involved in the programme. As with other EUREKA projects EUROTRAC was not centralised and the scientific groups had seek their own funding, usually from their own national funding authorities. Some groups also obtained funding from the various scientific programmes of the European Commission (EC; DG-XII) which for many years had had a research programme concerned with atmospheric research. Five of the subprojects were adopted as joint projects between the EUROTRAC and the EC (DG-XII), and the EC assisted them directly by supporting their workshops and meetings and, in one case, by providing substantial support for a flight campaign.

1.2   The historical context

The problem of transboundary air pollution in Europe was recognised in the early nineteen - seventies when the transport of sulfur compounds from the continental Europe to southern Scandinavia was observed. Shortly afterwards, the long range transport of photo-oxidants from the European continent to southern England was also reported. These findings completely changed the view that air pollution was primarily a local wintertime problem giving high sulfur dioxide and soot levels threatening human health. It was realised that it is a trans-boundary problem where the effects of pollutants can be observed in pristine areas far from the emission sources. As a consequence of these observations, international scientific studies and policy actions were started in Europe, primarily within the framework of OECD. At this stage it was appreciated that there would be a need for close co-operation between scientists and policy-makers so as to establish a sound and reliable scientific basis to achieve co-ordinated international action.

The consensus from the early work was that a transboundary air pollution problem did indeed exist in Europe. These early investigations provided the basis for the Convention on Long-Range Transboundary Air Pollution (LRTAP) within the framework of the UN-ECE signed by 37 European countries in November 1989 (ECE, 1994).

In the early nineteen-eighties, after extensive evidence became available of forest decline over large parts of Europe, there was a growing awareness that the balance of many ecosystems in Europe was threatened. The symptoms and distribution of the forest decline could not easily be linked to emission sources, and it became evident that extensive scientific research would be needed to establish the knowledge necessary to underpin efficient control measures. One of the most difficult issues was the necessity for joint measures between the countries to solve the problem. It was also evident that the scientific basis needed to be strengthened, particularly because of an increasing interest in using data on the transboundary fluxes and source-receptor relationships in order to establish cost-effective abatement strategies.

There was also a realisation within the scientific community that ozone in the troposphere was increasing; as Fig. 1.1 indicates, average concentrations had in fact nearly doubled at all altitudes throughout Europe between the nineteen fifties and the eighties. While there was much speculation as to the likely causes, these were not observations that could be easily modelled. Such an increase in the general troposphere was bound to have an effect in the boundary layer which was of principal concern to the environmental agencies.

EUROTRAC had its root in two separate initiatives. In the first a group of senior European atmospheric scientists met in Stockholm in February 1985 where they identified areas of importance for scientific research in the context of the European air pollution problems. The second initiative came from the German government who wished to see environmental projects included within the EUREKA framework. A proposal was prepared, EUROTRAC was presented as a German initiative within EUREKA and it was accepted as the first environmental project at the second EUREKA ministerial conference in Hannover in November 1985.

The European Commission (EC; DG-XII) which had been running an atmospheric chemistry research programmes for a number of years, was actively involved from the beginning. Much of the then EC research programme, which was mainly concerned with the laboratory studies, formed the basis of two of EUROTRAC subprojects that were adopted as joint projects between the EC and EUROTRAC.

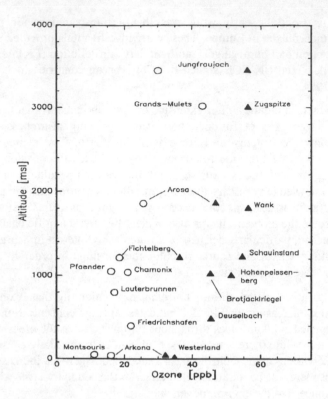

Fig. 1.1: Ozone concentrations observed in late summer at different locations in Europe. The open circles summarise data collected before and during the 1950s; the triangles are from measurements made after 1988. The data are plotted against the altitude of the different sites (Staehelin *et al.*, 1994).

The EUROTRAC proposal was welcomed by the scientific community and the participating countries. An organisational framework was established and scientists invited to propose possible subprojects to undertake the work. A formal decision initiating the project was taken by International Executive Committee (IEC) early in 1988.

1.3   The principal areas of scientific interest

The focus of work in EUROTRAC was on transboundary air pollution problems within Europe which give rise, in particular, to the effects caused by photo-chemical oxidants, and to the effects of acidification and eutrophication caused by deposition of sulfur and nitrogen compounds. In addition the research within EUROTRAC had implications for other environmental issues, such as climate change and the depletion of the stratospheric ozone layer. The scientific results are also likely to be of importance in understanding atmospheric transport and distribution of other trace constituents of the atmosphere such as persistent organic compounds, which are known to have deleterious effects on ecological systems and on health.

EUROTRAC was established to make a broad scientific approach to the most important environmental issues. The work included field studies, laboratory work, development of advanced experimental techniques and instrumentation and the development and application of advanced theoretical models. Field experiments and field campaigns were organised in several of the subprojects: co-ordinated observations and interpretations were made in order to achieve a more comprehensive picture of the phenomena under consideration, and also to compare approaches, methods and instrumentation. The issue of intercomparison and joint evaluations were also important for the laboratory projects.

The research covered the whole sequence of processes from the emission of trace substances, their transport through the air and simultaneous chemical conversion to the disappearance of the pollutants by deposition to the biosphere, the emphasis being on the scientific problems associated with the chemical and physical processes in the atmosphere.

1.4   The subprojects and working groups

The scientific work was carried out by the principal investigators and was co-ordinated within the fourteen subprojects that were organised to study particular aspects of the overall problem (for a list see Appendix A, Table 1). The aims of the subprojects were suggested initially by the scientific groups involved. They were then refined in discussion with the Scientific Steering Committee (SSC) to meet the aims of EUROTRAC as a whole. Each subproject was run by a project coordinator assisted by a deputy co-ordinator and a small steering committee. Workshops were' organised on an annual basis and each principal investigator within the subproject was expected to produce an annual report on which his contribution was assessed by the steering group. Each year too, an overview of the subproject work was produced and this was assessed by the SSC.

The SSC, which bore the overall responsibility for the scientific programme, regularly reviewed the progress of the subprojects and intervened if it was felt that the progress was not good enough. Of the fourteen projects approved at the start of EUROTRAC, eleven continued to the end of the project; two were reorganised and one was closed, the principal investigators being transferred to another subproject with similar interests. A further subproject, GENEMIS, was started part way through the project when it was realised that there was a need for evaluated emission data by the various atmospheric models.

It is fair to say that the undoubted success that EUROTRAC enjoyed was largely due to the dedication of the subproject coordinators and steering groups. Their work can be judged from the remainder of this chapter, the later chapters in this volume and from volumes 2 to 9 of this series (Borrell *et al.*, 1996).

As the project progressed it was realised that certain areas in between the existing subprojects needed attention. Two working groups, the Chemical Mechanism Working Group (CMWG) and the Cloud Group (CG) were set up with members selected from the appropriate subprojects. These tackled particular problems and their success may be judged from the EUROTRAC special publications which resulted from their work (Wirtz *et al.*, 1994; Flossmann and Cvitaš, 1995; Poppe *et al.*, 1996).

A further special subproject, the Application Project (AP), is dealt with in section 1.8.

1.5   The direction and management of EUROTRAC

The overall responsibility of EUROTRAC was held by the International Executive Committee (IEC), on which there was one representative from each of the EUREKA countries involved together with a representative of the European Commission (see Appendix A, Table 2). Seventeen countries were directly involved in the IEC, but there were also principal investigators from a further nine countries participating in the project(see Appendix A, Table 3). The representatives came mainly from the agencies responsible for funding atmospheric science in the participating countries and their principal interest was in ensuring that the project reached its stated goals and did indeed provide the necessary support for policy development.

The scientific direction of EUROTRAC was provided by the Scientific Steering Committee (SSC), to which fifteen distinguished scientists were appointed by IEC (see Appendix A, Table 4). The task of the SSC was to set the scientific focus of the project, evaluate and recommend project proposals, evaluate and assess the performance of the various subprojects and ensure that the overall outcome of the project fulfilled the original objectives.

EUROTRAC was co-ordinated by the International Scientific Secretariat (ISS) situated at the Fraunhofer Institute for Environmental Research (IFU) in Garmisch-Partenkirchen (see Appendix A, Table 5). The ISS was responsible for implementing the decisions of the IEC and SSC, for maintaining regular contacts between the various parts in the project (between the different subprojects, between subproject coordinators and principal investigators etc.) and the relations between EUROTRAC and other national and international organisations. The ISS was also responsible for publishing the various reports and proceedings described below.

In order to encourage scientific contacts within the EUROTRAC community, and between it and other scientists in the field within Europe, EUROTRAC Symposia were organised on a biennial basis (1990, 1992, 1994 and 1996). The Symposia were attended by several hundred scientists from all over Europe as well as some key people from outside Europe, primarily the USA. Efforts were made to publish the Proceedings of the Symposia within the year that the Symposia were held and these volumes provide lively snapshots of the activities and concerns of the community at that time, as well as a handy inter-disciplinary source of information for people in the field (Borrell *et al.*, 1997). The Symposia became a main focal point for scientific presentations and discussions on atmospheric chemistry within Europe (Borrell, 1998).

In order to keep all the participants informed on what was happening within EUROTRAC, and to present EUROTRAC to the interested community of scientists and administrators outside the project, the EUROTRAC Newsletter was published twice a year. Its circulation figure in 1996 was 1900 issues.

The scientific progress within EUROTRAC was presented in a series of multi-volume Annual Reports. Each subproject produced the material for its own volume or section. This consisted of a an overview prepared by the subproject coordinator and steering group, together with short reports from each of contributing research groups. In the general report an attempt was made to present a scientific overview of the whole project and this was combined with a shorter report from each subproject. Internally, the Annual Reports provided a useful management tool both within the individual subprojects and for the project as a whole. Externally they provided a useful source of information about the project and, in time, an excellent advertisement for the project. The quality of the work presented in several of the subproject volumes was such that they became a reliable way for research workers both inside and outside the project to keep up to date with the latest news from the particular field.

An innovation within the project was the series of EUROTRAC Special Publications. These met the need for the presentation of material not suitable for the normal literature but necessary for the appreciation of the work. They included detailed descriptions of observation stations for ALPTRAC (Kromp-Kolb, 1993) and TOR (Cvitaš and Kley, 1993), results of model evaluation campaigns and exercises and a field phase handbook for TRACT.

A list of the official EUROTRAC publications is given in Appendix B.

The main vehicle of publication for the scientific work of EUROTRAC is the refereed scientific literature. Lists of the publications which have resulted from the work of the project are provided in volumes 2 to 9 of this report (Borrell *et al.*, 1996).

1.6   The funding of EUROTRAC

As with all EUREKA projects, each country was expected to fund the participation of its own scientists in the project. There were no central funds and Principal Investigators were expected to seek their own support from their national funding agencies. In some countries just one agency was responsible; in others, several.

A number of investigators, particularly in the laboratory subprojects, competed successfully for the funds made available by the European Commission (DG-XII) for work in atmospheric chemistry. The subprojects within EUROTRAC provided an useful forum for recruiting groups to propose international joint work on problems or urgent scientific interest.

The EC also provide support for the co-ordination of five of the subprojects, which were designated as joint projects between the EC and EUROTRAC. In addition the EC provided generous funding for one of the flight campaigns in the subproject TRACT.

The Application Project (section 1.8) was an exception to the general funding rules in that the participating countries funded the project centrally through the ISS. The status reflects the "top-down" nature of a project, defined by the SSC and IEC. All the other subprojects were "bottom-up" in that the ideas and personnel were proposed by the scientists themselves, and then considered and agreed with the SSC.

Germany, as the lead country in this EUREKA project, funded the International Scientific Secretariat (ISS), which was established at Fraunhofer Institut für Atmosphärische Umweltforschung (IFU) in Garmisch-Partenkirchen. They supported two scientists with secretarial and support staff and provided co-ordination costs. As the project progressed it was realised that the co-ordination costs granted would not suffice to meet the needs of the growing project. Fourteen of the other countries together with the EC then offered to contribute to the special expenses incurred in the production of the annual reports, the biennial symposia, the review of EUROTRAC and the Application Project.

1.7   Evaluation of progress

The regular evaluation of the progress of the subproject by the SSC has already been alluded to. The first part of such an evaluation was normally made by one or two scientists from the SSC who provided a written and an oral report to the committee. The SSC then discussed the progress with the subproject coordinator and deputy before formulating their conclusions and recommendations. These were then communicated to the members of the subproject and to the IEC before being published in the Newsletter.

The review gave a chance for the SSC to express any concerns they had about the subproject, for them to hear about any particular problems faced by the subproject coordinator, and usually to make constructive suggestions for improvement, most of which were heeded. During the eight years it was necessary to re-organise two subproject and to close one, the principal investigators being transferred to another with similar scientific interests.

In 1991 the IEC decided to seek an outside evaluation of the project before approving the continuation into the second half. After inviting tenders, the IEC appointed a UK consultancy, Serco Space Ltd. The firm itself invited seven scientific experts from North America to advise it on the scientific aspects of the programme, while it concentrated on the organisational aspects and the profile of EUROTRAC, both within the project and in the community outside. Serco worked through the second half of 1991 and made its first report at a joint meeting of the IEC, SSC and subproject coordinators early in 1992. A revised version was presented at Symposium '92 and the final report was published in the early summer. (Serco Space, 1992).

The overall tenor of the report was very favourable to EUROTRAC. One strong recommendation, however, was that a special effort be made to bring together the scientific results of the project and make them available in a form suitable for those concerned with environmental management and policy development in Europe. The IEC was pleased with the report. It implemented promptly the various small recommendations and, in order to answer the main recommendation, set in train the mechanism which led to the formation of the Application Project, described in the following section.

1.8   The Application Project

The Application Project (AP) was set up in 1993 in order to fulfil the policy-oriented objective of EUROTRAC. The objective of the Application Project itself was to "assimilate the scientific results from EUROTRAC and present them in a condensed form, together with recommendations where appropriate, so that they are suitable for use by those responsible for environmental planning and management in Europe". Within the AP three themes were addressed: Photo-oxidants, Acidification and Tools. Within the first two, Photo-oxidants and Acidification, the potential scientific contributions from EUROTRAC to the policy process for each of the environmental problems were evaluated and discussed. The Tools theme was a recognition that much of the work within EUROTRAC was laying the foundations for the future policy related scientific work that will be required as the abatement measures develop. Among the tools considered by the AP were the development of models for tropospheric processes, advanced instrumentation for monitoring of atmospheric trace constituents and the provision of the accurate chemical data required for the models through laboratory work.

The AP consisted of 10 scientists divided into three groups, each covering one of the themes. As part of the project each of the 250 Principal Investigators (PIs) was requested to say which of their results they believed to be of potential importance, directly or indirectly to policy development. Some 120 principal investigators responded and thus made a direct contribution to the policy-oriented evaluation of the project. The major part of the evaluation was made by reviewing the published work and reports of the principal investigators and subprojects. The draft report was reviewed by the IEC, SSC, the subproject coordinators and the principal investigators and the many comments taken into account in preparing the final draft which was finally accepted by the IEC. In this way the AP report represented an overall consensus of the whole project on the issues discussed.

The executive summary of the results of the Application Project is presented in chapter 7 of this volume and the full account is given in volume 10 of this series (Borrell *et al.*, 1997).

1.9   Links to environmental policy development

While the AP provided the EUROTRAC results to those responsible for policy development in Europe, there were numerous other contacts which facilitated the transfer of knowledge understanding from science to policy development. As already mentioned, a number of national and international policy actions were taken in parallel with the EUROTRAC activities, in order to alleviate the environmental problems. Most important internationally was the Convention on Long-Range Transboundary Air Pollution (LRTAP). Agreements have been signed on nitrogen oxides (protocol in 1988), volatile organic compounds (protocol in 1991) and sulfur dioxide (second protocol signed in 1994). The continuing interest in developing cost-effective control strategies has put an even stronger pressure on the scientific community to develop valid source-receptor relationships for the major pollutants. In order to ensure the links to the LRTAP convention and in particular its body for monitoring and modelling (EMEP), a formal link was established and the Scientific Secretary was appointed observer at the EMEP Steering Body meetings. A number of the project participants were already actively involved as scientific experts in EMEP and in the negotiation processes. Contributions were made at EMEP workshops and joint work between EMEP and EUROTRAC scientists was undertaken in model development and quality assurance.

A formal link was also established between EUROTRAC and the EC (DG-XI) Expert Group on Ozone. The group advised on the forthcoming directive on ozone and many of the results obtained in EUROTRAC will certainly underpin this. A recent development in Europe is the establishment of the European Environmental Agency (EEA). Two of their topic centres (Air Quality and Emissions) are concerned with the atmospheric environment and a number of their scientists are involved in EUROTRAC work.

There are also numerous informal links with policy development since many EUROTRAC scientists are advisers to their own national governments and carry with them the knowledge and experience gained in EUROTRAC. It is the combination of these links with the more formal ones which give a practical realisation to the general scientific consensus on the scientific issues that is required for the development of effective international policy on air pollution.

For the marine conventions, the links have been less obvious, probably due to the lesser importance of atmospheric transport as a source of the pollution load, but also because of a less developed approach in terms of the establishment of quantitative source-receptor relationships.

1.10   Some other achievements of EUROTRAC

In addition to the scientific success and the application of the results to environmental policy development by EUROTRAC, there were a number of other worthwhile by-products of the project.

The establishment of an international scientific community in Europe

The subprojects brought together groups of European scientists to work on various aspects of an inter-disciplinary problem and so created networks of individuals with a common interest. These have served not only the interests of EUROTRAC and the groups themselves by providing a platform for future common endeavours in the field. The groups also serve environmental policy development in Europe, in that they provide the common understanding and agreement on the scientific issues which is necessary to alleviate complex environmental problems (this volume chapter 7; Borrell *et al.*, 1997; chapter 5).

Education of experts to deal with environmental issues

The common understanding of the scientific issues required for international environmental policy development can only be among experts in the field. The experience of international work that individuals gained within EUROTRAC certainly contributed to their own development and knowledge of these matters. The Application Project itself was an educational tool for the majority of the principal investigators many of whom were encouraged for the first time to consider the implications of their results for the policy process.

The education of a large number of students in atmospheric chemistry

The work within EUROTRAC has provided a practical training for both Masters and Doctoral students in all of the participating countries. The degrees and qualifications were obtained in both institutions of higher education and in the research institutes associated with them. Undoubtedly the liveliness of the science, and of many of the workshops and meetings was increased appreciably by the presence of students who brought new ideas and a fresh outlook on old problems. Lists of degrees obtained through EUROTRAC work are included as part of the publication lists, published in the other volumes of this series (Borrell *et al.*, 1996).

Development and manufacture of instrumentation for monitoring and research

The growth in scientific understanding in a particular area is paralleled by improvements in instrumentation and the development of new instruments. EUROTRAC had three subprojects devoted specifically to instrument development. In addition many instruments were developed and improved as part of the work in other subprojects and a variety of them have been produced commercially (This volume, chapters 6 and 11; Bösenberg *et al.*, 1997; Borrell *et al.*, 1997).

1.11   Some lessons learned from EUROTRAC

The previous section emphasised the educational aspects of EUROTRAC for the scientific community and the countries in Europe. Here a number of direct lessons concerning the running of an international environmental project are indicated.

The "bottom-up" approach strengthens the motivation and innovation

Encouraging the scientists themselves to define the scientific problems to be studied and to set the practical goals to be achieved proved to be a powerful way to motivate the whole project. In this way everyone is a true participant and responsible for working out their own ideas.

Flexibility of approach

The possible disadvantage in a long-term project is the potential lack of flexibility in responding to new issues. However the formation of working groups to tackle particular problems and the Application Project itself demonstrated how flexibility could be achieved.

Success depends upon enthusiastic co-ordination

An important ingredient for the successful conclusion of the project is the quality of the subproject coordinators and steering group members, on whom the principal burden of work falls. Most of the co-ordination was done as part of the subproject coordinator's normal scientific and administrative work and most received little extra payment, either cash or kind, for what they did.

An active central secretariat helps

The central secretariat was necessary to ensure that the work flowed smoothly and on time. It is essential that everyone within the project is clear about their role and their responsibility, and so feels part of the project. It is important that everyone sees that his or her contribution is appreciated and evaluated; the annual and final reports were particularly important in this context. This is particularly true in a project where each principal investigator obtains his or her own funds; the whole could so easily have dissipated.

Success also depends on realistic goals and regular evaluation

The founders of the project set much store in defining aims for the project as a whole, and general scientific objectives. These ensured that the project did not become diffuse, a particular danger in this area where there are so many issues which are related and which might be included.

Realistic goals and regular reviews are necessary

They also insisted that the subprojects set realistic goals and instituted a review procedure to ensure that attention was paid to them. The annual reports from the principal investigators, reviewed by the subproject steering groups, and the annual reports from the subprojects, reviewed by the Scientific Steering Committee, together with the regular SSC reviews of progress kept everyone's concentration on fulfilling the promises made in the aims. The fixed time span of the project, long enough to achieve much, but not too long to allow everything to fossilise, was a help in itself.

Distributed funding can be used to advantage

As already explained each principal investigator was expected to obtain funds for his or her own work and each country funded their own participants, apart from the few who obtaining direct support from the EC. The lack of synchronisation in the funding from some twenty countries certainly created appreciable difficulties for the detailed co-ordination of the work since the work of the subproject was strongly influenced by the vicissitudes in funding (timing and policy) in the participating countries. However it was noticeable that when a principal investigator was firmly committed to a project, he or she generally found a way to contribute, even in the face of adverse funding conditions, and so the subprojects prospered and met their goals.

However distributed funding had some marked advantages too. The funding in most countries is short-term, and the priorities change within a country as time progresses. In a long term project the effects, provided they are not too great, average out and it is still possible to achieve the goals. With a single funding source a change of priorities could kill a subproject, or indeed a project.

The principal advantage of distributed funding is that far more resources can be brought to bear on a problem that could ever be provided by one country alone. This includes the EC as well, which does not have limitless resources and has many claims upon them. EUROTRAC provided a model of how to harness the resources, manpower, expertise and finance, for a group of European countries and apply them in a cost-effective way so as to provide a scientific basis for the solution of a difficult international environmental problem.

1.12   Future needs - EUROTRAC-2

Clearly work such as that undertaken in EUROTRAC is never fully completed. Even when there is a clear understanding of the scientific situation many uncertainties, some of them major, remain. It was therefore hardly surprising that in 1995 there was a demand both from the scientific community and from the participating countries that there should be a further project to follow EUROTRAC.

The nature of the project was elaborated by the scientific groups and a detailed proposal approved by the International Executive Committee. The proposal was prepared on behalf of institutions in 26 countries by the ISS (ISS, 1995; Borrell, 1997) and submitted to EUREKA for consideration in December 1995. The new project, EUROTRAC-2, was approved by the EUREKA High Level Group at their Liége meeting in February 1996 and the new project began work with an inaugural meeting of the new IEC in June 1996.

The overall objective of EUROTRAC-2 (The Transport and Chemical Transformation of Environmentally Relevant Trace Constituents in the Troposphere over Europe; Second Phase) is (ISS, 1999):

"to support the further development of abatement strategies within Europe by providing an improved scientific basis for the quantification of source-receptor relationships for photo-oxidants and acidifying substances".

The project will provide an integrated scientific evaluation of photo-oxidants and acidifying substances in the troposphere over Europe, using the extended knowledge base and improved methodologies gained in the first phase of EUROTRAC. In particular advanced techniques in linking numerical atmospheric models and observational data will be used to provide an unprecedented capability for understanding the complex interactions between the sources, transformations and deposition which determine the concentrations of trace substances appropriate to environmental impact. The work will thus provide scientific support for future measures needed to control photo-oxidants and the deposition of acidifying substances and nutrients.

The specific objectives are:

1:   *Quantification of atmospheric interactions*

To quantify the anthropogenic and natural contributions of relevant emissions and atmospheric processes to the abundance and the long term changes of photo-oxidants and acidifying substances in the planetary boundary layer and free troposphere.

2:   *Evaluation of feedback mechanisms*

To evaluate the consequences of feed-back mechanisms, for example: the feed-back between the concentrations of tropospheric photo-oxidants and biogenic emissions; the feed-back between the concentrations of photo-oxidants and those of climatically relevant atmospheric constituents; and the feed-back between the changing intensity of ultraviolet radiation and photo-oxidant production.

3:   *Contribution to the formulation of abatement strategies and future air quality*

To contribute to the formulation and improvement of strategies for reducing the anthropogenic contribution to the abundance of photo-oxidants and acidifying substances and to the prediction of future air quality on shorter and longer time scales.

In these objectives there is the recognition that, although much was achieved in the first phase, there is still much to be done before the scientific understanding is good enough for policy development. The authorities responsible for implementing environmental policy must have confidence in the predictions of the models before they will be prepared to enact the necessary but expensive abatement measures that are likely to be needed to deal with the present environmental situation.

The present directions in policy development themselves provide a further driving force for better science. The latest protocols (ECE, 1994) to the Geneva Convention on Long-Range Transboundary Air Pollution (LRTAP) are based on the concept of critical loads to eco-systems (Nilsson and Grennfelt, 1988): the landscape is divided into ecologically coherent areas and an attempt is made to specify the maximum load of a particular pollutant which the area will tolerate. In much of Europe these areas are relatively small and so, in the future, the models used to determine deposition or exposure will have to cope with additional detail on smaller scales. This necessarily requires a better and more detailed scientific understanding.

It is also the case that pollutants cannot be dealt with separately; there are undoubtedly synergistic effects between pollutants and the reduction of one pollutant can lead to changes in another. Thus a reduction in ammonia emissions is likely to increase the transport distance of sulfur dioxide (Borrell *et al.*, 1997).

Another complication is the potential feedbacks and interactions between systems which, for convenience, are often regarded as independent. When plants are exposed to pollutants they do not act as simple passive receivers; they can be expected to respond actively and in doing so change the environment in which they find themselves. These interactions are little studied and less well understood, but as more detailed forecasting is required it will not be possible to neglect them.

The new project, while still firmly scientific, will be more closely linked to policy development. This should ensure that the results can be more rapidly utilised and that the scientific community is more aware of the current scientific needs of those responsible for policy development and can perhaps adapt its work to met them.

1.13   References

Borrell, P., P.M. Borrell, T. Cvitaš, K. Kelly, W. Seiler (eds); 1996, *Transport and Chemical Transformation of Pollutants in the Troposphere*, Springer Verlag, Heidelberg, Volumes 1 to 10 (in press).

Borrell, P., Ø. Hov, P. Grennfelt and P. Builtjes (eds); 1997; *Photo-oxidants, Acidification and Tools: Policy Applications of EUROTRAC Results*, Springer Verlag, Heidelberg; Vol. 10 of the EUROTRAC final report.

Borrell, P.; 1997, *www.gsf.de/eurotrac*, formerly *www.eurotrac.fhg.de*.

Borrell, P.; 1998, Symposium '98: reflections, EUROTRAC Newsletter **20**, 24–27

Borrell, P.M., P. Borrell, T Cvitaš, K. Kelly, W. Seiler (eds); 1996, *Proc. EUROTRAC Symp. '96*, Computational Mechanics, Southampton, Vol. 1 1057 pp., Vol. 2 822 pp.

Bösenberg, J., D. Brassington, P. Simon (eds); 1997; *Instrument Development for Atmospheric Research and Monitoring*, Springer Verlag, Heidelberg; Vol. 8 of the EUROTRAC final report.

ECE; 1994: The Economic Commission for Europe (ECE), Convention on Long-Range Transboundary Air Pollution (LRTAP), United Nations, Geneva.

Flossmann, A., T. Cvitaš, (eds); 1995, *Clouds: Models and Mechanisms*, EUROTRAC ISS, Garmisch-Partenkirchen.

ISS; 1995; EUROTRAC Phase 2, A proposal for a new EUREKA Project to follow EUROTRAC, *EUROTRAC Newsletter* **15**, 11–19;

ISS; 1999; *EUROTRAC-2: Project Description and Handbook*, EUROTRAC-2 ISS, München, 1998. See also: http://www.gsf.de/eurotrac

Cvitaš, T., D. Kley (eds); 1994, *The TOR Network*, EUROTRAC ISS, Garmisch-Partenkirchen.

Kromp-Kolb, H., W. Schoener, P. Seibert; 1993, *ALPTRAC Data Catalogue*, EUROTRAC ISS, Garmisch-Partenkirchen.

Nilsson, J., P. Grennfelt; 1988: Critical loads for sulfur and nitrogen; report from the Stockholm workshop (March 1988). *NORD Miljørapport*, **15**, Nordic Council of Ministers, Copenhagen.

Poppe, D., Y. Andersson-Skoeld, A. Baart, P.H.J. Builtjes, M. Das, F. Fiedler, Ø Hov, F. Kirchner, M. Kuhn, P.A. Makar, J.B. Milford, M.G.M. Roemer, R.Ruhnke, D. Simpson, W.R. Stockwell, A. Strand, B. Vogel, H. Vogel; 1996, *Gas-phase Reactions in Atmospheric Chemistry and Transport Models: a model intercomparison*, EUROTRAC ISS, Garmisch-Partenkirchen

Serco Space, 1992, *Review of EUROTRAC*, Serco Space Ltd, Sunbury on Thames.

Staehelin, J., J. Thudium, R. Buechler, A. Volz-Thomas, W. Graber; 1994, Trends in surface ozone concentrations at Arosa (Switzerland), *Atmos Environ.* **28**, 75–87.

Wirtz, K., C. Roehl, G.D. Hayman and M.E. Jenkin, 1994, *LACTOZ Re-evaluation of the EMEP MSC-W Photo-oxidant Model*, EUROTRAC ISS, Garmisch-Partenkirchen

Zimmermann, H.; 1995, *The TRACT Field Measurement Campaign (Field Phase Report)* EUROTRAC ISS, Garmisch-Partenkirchen.

Chapter 2

Photo-oxidants

Anthony R. Marsh[1] and Peter Borrell[2]

[1]Imperial College Atmospheric Chemistry Research Unit, Silwood Park, Berkshire SL5 7PY, UK
[2]P&PMB Consultants, Ehrwalder Straße 9, D-82467 Garmisch-Partenkirchen, Germany

2.1 Introduction: photo-oxidants, tropospheric ozone and EUROTRAC

The decision that EUROTRAC should study photo-oxidants stemmed directly from the environmental concerns about ozone: the apparent doubling of the ambient concentrations since the beginning of the century to levels close to those which inhibit crop growth, and the frequent summer smog episodes which are thought to be hazardous to health.

The decision also reflected the scientific complexity of the problem which would require an appreciable effort from a large number of research groups in order to reach the common understanding necessary to develop cost-effective abatement measures. Environmental policy development requires access to reliable computer models which encompass the scientific understanding and allow one to simulate atmospheric behaviour. Such models need detailed data about the meteorological conditions, the emissions of primary pollutants over the area concerned and an understanding of the dynamics of the atmosphere together with a comprehensive scheme for the chemical reactions occurring. A vast amount of computer power is also needed to cope not only with the simulation itself but also with the presentation and interpretation of the results.

Each of the items mentioned is itself complex. A recent computer simulation of scheme for the atmospheric chemistry, for example, includes several hundred species and several thousand reactions. Not all of these are of equal importance but a surprising fraction are necessary to reproduce the details of the transformation that takes place as the primary chemical pollutants, such as volatile organic compounds (volatile organic compounds (VOC)) and the nitrogen oxides, are converted to secondary pollutants such as ozone and PAN (peroxyacetyl nitrate). The complexity of this mathematically non-linear system makes simple analysis of the behaviour of any individual chemical species within the atmosphere difficult and prone to error.

The photo-oxidants considered within EUROTRAC were ozone, O_3, nitrogen dioxide, NO_2, peroxyacetyl nitrate, PAN, and hydrogen peroxide, H_2O_2. The project concentrated on the secondary pollutant ozone, part of which is produced in the atmosphere by photochemical reactions involving the primary emissions of the oxides of nitrogen, NO_x, mainly as nitric oxide, NO, and VOCs. NO_2 is mainly a secondary pollutant which itself can cause health problems at high concentrations. Some vegetative damage has been attributed to hydrogen peroxide and it plays a significant role in the formation of acid rain. PAN is a strong lachrymator and was identified as an indicator of photochemical smog formation many years ago in Los Angeles. It plays a role in the long range transport of photo-oxidants from the conurbations where they are principally emitted to "pristine" background areas far from the sources.

At the start of EUROTRAC there were many uncertainties in quantifying the relationship between these primary emissions, the precursors, and the formation of the secondary pollutants, the photo-oxidants. There was a need for a better understanding of the relationship between these ozone precursors and the spatial and temporal formation of both ozone itself and the other important photo-oxidants. At the end of the project, EUROTRAC had advanced considerably both the fundamental scientific knowledge and the characterisation of photo-oxidants in Europe. The project contributed substantially to the determination of the kinetics of reactions of important atmospheric processes, to the improvement of techniques for measuring photo-oxidants in the atmosphere, to the characterisation of the behaviour of ozone and its precursors on a European-wide scale, to the improvement of emission estimates, to the study of deposition processes to the biosphere (the ultimate sink for all atmospheric pollutants) and to the development of numerical simulation models in which the scientific understanding of photo-oxidants is encapsulated.

2.2   Photo-oxidant formation and its relationship to policy development

Thanks to the emissions of trace substances from the biosphere and pollutants from human activities, the atmosphere constitutes a vast chemical reactor with, as already mentioned, hundreds of chemical species reacting with each other in thousands of reactions. The purpose of this section is to outline some of the processes involved and to indicate some of the advances which have been made.

a.   The chemistry of photo-oxidation

The key to understanding the chemical transformation processes in the atmosphere lies in the behaviour of free radicals. The most important of these are the hydroxyl radical, OH, the hydroperoxy radical, HO_2, and the large number of organic peroxy radicals, denoted generally as RO_2. These radicals control the oxidation of most pollutants.

The scientific interest in these problems centres on elucidating the mechanisms by which these radicals react with the more stable chemical species and on the chemical fate of the products, and also, if possible, determining the rates of reaction at temperatures of atmospheric interest. Fig. 2.1 shows diagrammatically the cycle of chain reactions involving the principal radicals and the primary pollutants. As the whole process is photochemically driven by sunlight, the concentrations of these free radicals follow a diurnal cycle.

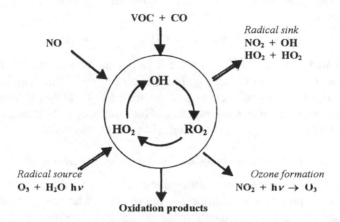

Fig. 2.1:   A schematic presentation of the mechanism used in the photo-oxidant model.

There are many feedback processes possible between these derived species, one of which is illustrated by the interaction of ozone with the oxides of nitrogen which was one of the first of the photolytic processes to be identified in the atmosphere.

$$NO_2 + h\nu \, (\lambda \approx 400 \, nm) \; \rightarrow \quad NO + O \tag{1}$$

$$O \; + \; O_2 \qquad\qquad\qquad \rightarrow \quad O_3 \tag{2}$$

Reactions (1) and (2) can be reversed by the rapid reaction of NO with ozone:

$$NO + O_3 \qquad\qquad\qquad \rightarrow \quad NO_2 + O_2 \tag{3}$$

This process establishes a photochemical steady state between O_3, NO and NO_2, the concentrations depending on the insolation. The oxidation of NO to NO_2 by processes other reaction (3) leads to the formation of ozone.

In summer, away from the immediate proximity to pollution sources, significant ozone formation follows from the conversion of nitric oxide to nitrogen dioxide by RO_2 radicals derived in turn from OH radical reactions with VOCs. OH reacts with methane for example, to form the methyl radical and the methylperoxy radical:

$$CH_4 \quad + \quad OH \qquad \rightarrow \quad CH_3 \; + \; H_2O \tag{4}$$

$$CH_3 \quad + \quad O_2 \qquad\quad \rightarrow \quad CH_3O_2 \tag{5}$$

And the methylperoxy radical can then react with NO to form NO_2, and hence ozone by reactions (1) and (2), and re-forms an OH radical:

$$CH_3O_2 + NO \rightarrow CH_3O + NO_2 \tag{6}$$

$$CH_3O + O_2 \rightarrow HCHO + HO_2 \tag{7}$$

$$HO_2 + NO \rightarrow OH + NO_2 \tag{8}$$

Thus much of the ozone is produced locally by positive photochemical feedback cycles. The general result, found by summing equations (2) to (8) for the hydrocarbon RH is:

$$RH + 4O_2 + h\nu \rightarrow R'CHO + H_2O + 2O_3 \tag{9}$$

The carbonyl species, R'CHO, is itself usually reactive and can be photolysed to yield more radicals and further O_3. The number of cycles, the chain length, of the radical chain reactions producing ozone before the NO_x catalyst is finally removed by a chain termination step is a measure of the efficiency of NO_x as a catalyst. It is the number of molecules of ozone produced per NO_x molecule.

b.   *Photo-oxidant formation under varying NO_x concentrations*

The effect of the interactions in the atmosphere on photo-oxidant formation can perhaps be appreciated by considering the situation where NO_x is produced at a source and moves away gradually being diluted and dispersed. It is illustrated in Fig. 2.2, which indicates the relative behaviour of the OH and HO_2 radicals, the chain length the net ozone production rate as a function of the NO_x concentration. The pollutants become more dilute as one moves from right to left on the diagram.

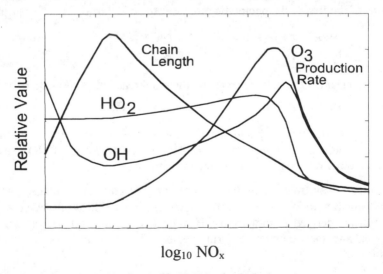

Fig. 2.2:   Variation of radical behaviour with NO_x concentrations

Perhaps the first feature to note is that none of the values shown is a linear function of NO_x concentration and it for this reason that a proper understanding of the fundamental processes creating ozone are required before reliable control strategies can be developed.

At high NO_x concentrations, found close to its sources, Fig. 2.2 indicates that NO is a major scavenger of both ozone and the radical pool, resulting in low radical and ozone concentrations. The reaction of OH with NO_2 is then appreciable

$$OH + NO_2 \rightarrow HNO_3 \tag{10}$$

removing the NO_x from the chain so inhibiting further ozone formation.

As the plume moves away the NO_x concentration falls and O_3 concentrations can rise by entrainment from the surroundings to a value close to that dictated by the photochemical steady state between NO_x and O_3, reactions (1) to (3). The OH concentration increases resulting in the oxidation of VOC (reactions (4) to (8)) and O_3 formation.

Eventually the NO_x concentration falls far enough so that the O_3 production rate peaks and starts to fall, limited now by the availability of NO_x.

As the air becomes cleaner the importance of longer lived species such as CH_4 and carbon monoxide, CO, which are present at relatively high concentrations, and slowly reacting non-methane hydrocarbons (NMHC), such as ethane, C_2H_6, and propane, C_3H_8, increases. The crossover in the relative ratio of HO_2/OH shown in the figure is due to the change in the CO concentration which is also falling from higher values near a traffic source, say, to lower ambient values on the left hand side of the figure. High CO concentrations reduce the concentration of the OH radical:

$$OH + CO \rightarrow H + CO_2 \tag{11}$$

$$H + O_2 \rightarrow HO_2 \tag{12}$$

but because the HO_2 radical concentration is always much greater than that of OH, reaction (11) does not appreciably alter the HO_2 concentration. The ratio of radicals falls because of the increase in OH; so does the chain length for ozone production. Under these low NO_x concentrations loss of HO_2 by recombination to H_2O_2 is becomes appreciable.

Finally at extremely low NO_x concentrations, normally encountered only in remote locations, the competition for reaction with HO_2 between reaction (3) with NO and reaction (13) with O_3

$$O_3 + HO_2 \rightarrow OH + 2O_2 \tag{13}$$

followed by reactions (11) and (12) determines whether ozone is produced or destroyed. The cross-over point depends on pressure, humidity and temperature which vary with altitude so that the behaviour will vary at different levels in the troposphere.

These features shown in Fig. 2.2 indicate the problems facing policy makers. Cities are major NO_x sources and are represented by the extreme right hand side of the diagram. Within them ozone will be either titrated out directly by NO emissions, reaction (3), or the ozone formation rate will be low since the radical pool is reduced. Small reductions in NO_x will thus lead to an *increase* in ozone in cities. The practical effect is indicated in Fig. 2.3 which shows some results for the summer months in Belgium. At weekends, despite the decrease in traffic and industrial activity and the fall in NO_x emissions, the ozone concentrations actually increase (Dumollin *et al.*, 1999). In Austria the opposite effect is observed with ozone levels remaining approximately constant or decreasing at weekends (Schneider, 1999).

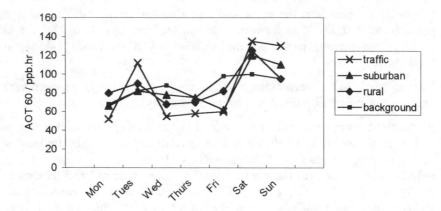

Fig. 2.3:   Daily AOT60 values for Belgium averaged between 1987 and 1996. The "weekend effect" can readily be seen (Dumollin *et al.*, 1999)

The effect can only be avoided by a dramatic cut in emissions which is large enough to reduce NO_x concentrations to the lower side of the ozone production peak shown in Fig. 2.2.

Thus city emissions usually produce ozone in the surrounding area. Downwind dispersion reduces the NO_x concentration creating an area of high ozone production; the peak in Fig. 2.2. The ratio $(HO_2+RO_2)/OH$ rises and the chain length increases. Since the ozone production rate exhibits quite a wide peak (Fig. 2.2), O_3 is formed over a wide range of NO_x concentrations and, in spatial and temporal terms, large areas of excess ozone can be produced and persist under the right atmospheric conditions. This simple picture of the consequences of city emissions indicates why the ozone problem is often divided into local and regional scales.

Fig. 2.2 also indicates why NO_x emissions have such a dramatic effect on the free troposphere, above the boundary layer. A clean free troposphere would be represented by the left hand side of the diagram with O_3 being destroyed. As the NO_x concentration rises the O_3 production rate changes sign and ozone is produced. Further the free tropospheric ozone lacks a real sink for ozone destruction except for transport back to the boundary layer by subsidence. The loss of NO_x fortunately provides a natural restriction on the process of catalytic production; its lifetime ranging from a few hours to a few days depending on the altitude. Thus one might expect the effects of a city on the free troposphere to be localised. However, products of the NO_x removal, such as PAN, can act as reservoirs for the transport of NO_x within the troposphere, and later decompose slowly liberating NO_x and creating ozone in pristine areas far from the sources.

Another threat to the upper troposphere are aircraft emissions (Brasseur *et al.*, 1998). For example the flight corridors over the North Atlantic represent a concentrated linear source of NO_x emissions and are accompanied by ozone formation.

The foregoing discussion illustrates the need to have good kinetic data on many atmospheric reactions to provide a reliable basis for the interpretation of observations of photo-oxidant behaviour. The discussion has concentrated on summer O_3 episodes of very high concentrations but the concentration of O_3 throughout the year is important especially in the growing season. Furthermore the build up of ozone in the free troposphere may be a year round phenomenon. Thus the formation of photo-oxidants throughout the year under a wide range of conditions is important.

2.3   Reactions and mechanisms involved in photo-oxidant formation

Within EUROTRAC the chemical processes for the homogeneous formation of ozone were largely studied in the subproject LACTOZ, *Laboratory studies of chemistry related to tropospheric ozone*. A number of areas of uncertainty were identified at the outset and specific sets of reactions that required a firm kinetic data base were targeted.

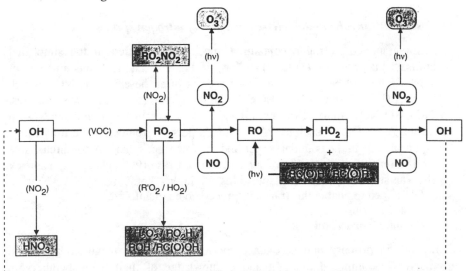

Fig. 2.4:   The mechanism of VOC oxidation during the day (Le Bras, 1996).

Although the rate constants for reactions equivalent to reaction (4), the initial attack of OH on many hydrocarbons, were well known the subsequent behaviour of the resultant hydrocarbon radicals were not well characterised under atmospheric conditions. Fig. 2.4, which outlines the daytime oxidation of VOC, indicates the number of reactions involved for each hydrocarbon.

The principal work concentrated on determining the rate constants for reactions of free radicals with stable species and on inter-free radical reactions. Although the techniques for studying these were on the whole well established, new techniques were developed and many improvements made to existing techniques (Borrell *et al.*, 1997).

The following sections outline the major groups of reactions studied within LACTOZ (Le Bras, 1997).

a.   Reactions of alkyl peroxy ($\mathbf{RO_2}$) *and alkoxy* ($\mathbf{RO}$) *radicals*

Alkyl peroxy and alkoxy radicals are directly involved in the reaction chain, producing ozone, reactions (5) and (6), but there are numerous side reactions which can compete and must also be included within the reaction scheme. Detailed results can be found in the LACTOZ volume (Le Bras, 1997; chapter 2, Tables 1 to 8).

Since so many hydrocarbons are involved, the question naturally arises as to whether it is necessary to measure the rate constants for all of them. Much effort has been devoted to studying the trends in the rate constants determined and their correlations with various fundamental parameters in order to determine *structure - reactivity relationships* for groups of radicals. In this way one hopes to develop a predictive capacity for unmeasured species and avoid time-consuming measurements. Some progress was made towards this goal within the subproject.

b.   Reactions involving the inter-conversion of nitrogen species

The importance of nitrogen containing species is indicated in the simplified reaction scheme, reaction (1) to (8) and also in Fig. 2.3. The reactions involve not only NO_x (NO and NO_2) but also other species often described as NO_y, which includes N_2O_5, HNO_3, HO_2NO_2, HONO, and the alkyl nitrates and peroxy nitrates, $RONO_2$, RO_2NO_2. Some of these such as HNO_3 represent loss routes for NO_x in the atmosphere. Others such as PAN, provide possible routes for the long range transport of NO_x in the boundary layer and the free troposphere. A comprehensive set of data for the nitrogen species are given in Le Bras (1997), chapter 2, Tables 9 to 14. The stability/structure relationships for higher homologues of PAN were also determined to enable this transport process to be quantified.

c.   Photolysis processes

Some of the primary and secondary species involved in photo-oxidation are themselves decomposed by light and a knowledge of their photochemistry is necessary for a quantitative description of oxidation. In the laboratory the painstaking and meticulous work is required to determine the absorption cross-sections and quantum yields necessary to calculate photolysis rates and only a few groups in the world are prepared to undertake it. A whole range of results are given by Le Bras (1997), section. 2.6 and Table 15.

d.   Reactions of ozone with alkenes

Alkenes, which are part of the hydrocarbon mix in petroleum, are reactive enough to react directly with O_3. The reaction produces an adduct which rapidly decomposes to oxygenated radicals which, in turn, can eventually form OH radicals or participate in the photo-oxidant reaction sequence in other ways. Alkenes can thus affect ozone production in a complex manner, attaining a particular importance in radical generation under low light conditions such as in winter, at low sun angles or at night-time. Much work was done on the "Criegee intermediate" which is thought to play a key role in the breakdown and the many reactions studied are listed in Le Bras (1997), Tables 5a, 5b, 6a, 6b and 7.

e.   Chemistry of the NO_3 radical

The nitrate radical is formed by the reaction of NO_2 with O_3.

$$NO_2 + O_3 \qquad \rightarrow \qquad NO_3 + O_2 \qquad\qquad (14)$$

It is easily photolysed during the day and so is only present at night. However, being a strong oxidising agent and able to react directly with hydrocarbons, it provides a path for oxidation of hydrocarbons when photo-oxidation is not possible. It also provides a loss process for NO_x from the atmosphere via the formation of N_2O_5, so that its night-time behaviour can have appreciable effects on the overall formation of ozone. A interesting discovery was that NO_3 reacts with RO_2 radicals allowing chain reactions to be propagated at night, which may increase the rate of night-time oxidation. Members of the subproject conducted a substantial review of NO_3 chemistry for the EC (Wayne et al., 1991). Many of the results obtained within LACTOZ are summarised in Le Bras (1997), section 2.2.

f.   Mechanism of oxidation of aromatic and biogenic hydrocarbons

Aromatic hydrocarbons, such as benzene, toluene and the xylenes, are produced from a range of anthropogenic sources and are themselves rated as health hazards, particularly in the urban environment. They too react with the OH radical as in reaction (4) but the subsequent processes are far from clear. The reactions are difficult to study in the laboratory since the initial reaction produces one or more reactive radicals which initiate a sequence of reactions in which virtually none of the intermediates are known. Even the final products are not characterised fully and it is still not possible to obtain an adequate carbon balance between the reactant consumed and the products formed.

The situation for biogenic hydrocarbons, such as isoprene and the terpenes, which can be important VOCs in rural locations, was not much better at the outset of the project.

The principal investigators in LACTOZ have made substantial progress in elucidating the mechanisms involved and have produced detailed schemes for the OH- and NO_3- initiated oxidation of isoprene and also for its ozonisation, as well elucidating the likely mechanism for the initial stages of the OH attack on benzene and toluene. (Le Bras, 1997; sections 2.7 and 2.8)

2.4   Observations in the atmosphere

Atmospheric observations make a crucial contribution to our understanding of the transport and transformation of pollutants in the troposphere: they serve to determine the distribution of photo-oxidants and their precursors over the region, to study detailed chemical processes where the details of the emissions and the meteorology are sufficiently well known at a particular site and they serve as the yardstick by which progress in atmospheric modelling can be judged. As measurements are made over time they serve to quantify the human influence on the atmospheric environment in space and time and, when measurements have been made for a long period, they indicate the trends in pollutant concentrations.

The principal observational subproject within EUROTRAC was TOR, Tropospheric Ozone Research. More than thirty observation stations were established throughout Europe, most equipped to measure the principal chemical components and the meteorological parameters and some capable of measuring radiation fluxes and a range of VOCs as well (Cvitaš et al., 1994). Much emphasis was placed on the standardisation of measurements and calibration procedures and a comprehensive quality control exercise was carried out for VOCs.

One of the side results was the development of new instruments some of which has been commercialised. These include a sensitive NO_x, NO, and NO_2 instrument, instruments for HCHO and H_2O_2, and radiation monitors for $J(NO_2)$ and $J(O_3)$. In addition an H_2O_2 sonde has been developed. (Hov, 1997; Borrell et al., 1997; chapter 5)

At the outset the TOR principal investigators had decided on a number of questions which might be answered with the results from the observational network and during the second half of the project task groups were established to do this (Hov, 1997). The questions and the answers are discussed in some detail in section 2.4b.

a.   Trends

The earliest quantitative ozone measurements were made at Montsouris near Paris between 1876 and 1911 using the Schönbein method. A 24 hr average concentration of O_3 of about 10 ppb was found which is a factor of three to four less than that found in similar locations today. Furthermore, this data did not show any day with concentrations greater than 40 ppb. (Hov, 1997; Borrell et al., 1997). Similar values were found on re-analysis of similar data from the Pic du Midi (Marenco et al., 1994). Although there is some question about the validity of these data when compared to those from Montsouris, they certainly show that the O_3 concentrations in the free troposphere at the beginning of the twentieth century were appreciably less than today.

The ozone increase in the second half of the century is illustrated by the measurements made a variety of sites throughout Europe; see Fig. 1.1. The data is from stations at differing heights and the measurements made in the nineteen fifties all fall at smaller concentrations than those for the corresponding height made in the eighties. There appears to have been a doubling in O_3 concentrations during the period probably corresponding to the enormous increases in traffic volumes and in energy utilisation during the period (Staehelin et al., 1994).

World wide, there is a statistically significant increase in ozone in the Northern Hemisphere. However, ozone has fallen at the South Pole, a decline which may be due to the reduced transfer from the stratosphere to the troposphere as a result of the sharp reductions in ozone in the stratosphere in that region. (Hov, 1997)

There are no long term direct observations of precursor concentrations from which trends might be deduced. Within the TOR, measurements of VOCs started at some stations in the summer of 1987 and do provide nearly a decade of data in which there appears to be a statistically significant increase the sum of C_2–C_5 hydrocarbons. A decrease was found for the alkenes. (Hov, 1997)

A longer trend record for NO_x can be gleaned from the nitrate ion records in ice cores which were examined, together with those for sulfate and heavy metals, in the subproject ALPTRAC. These correlate well with each other and show a marked increase between 1950 and 1980. The implicit conclusion is that the increase in the nitrate ion results from the dramatic increase in automotive exhaust emissions of NO_x; since this increase correlates very well with the increase in ozone concentrations over Europe it seems that automotive exhaust emissions are a major cause of the ozone increase. (Fuzzi and Wagenbach, 1997; Borrell et al., 1997)

b.   Current observations

Any attempt to interpret current measurements of photo-oxidants has to consider a variety of temporal and spatial scales. Summer smog episodes of elevated ozone concentrations, usually associated with anticyclonic conditions, can occur for a periods of few days over large areas of Europe, and pose appreciable health problems for the population affected. There are also seasonal trends that differ in detail in different parts of Europe and give a complex overall picture to the regional behaviour. On smaller scales, urban photo-oxidant concentrations differ in their diurnal behaviour, because of local emissions, from their rural surroundings and both are affected by the regional situation. Concentrations measured at high altitude sites arise from a mixture of boundary layer and free tropospheric air, and the effects have to be disentangled in any interpretation of the measurements. The diurnal cycling of the boundary layer height must also be taken into account in such complex terrain. Finally the European scale observations are superimposed upon, and contribute to, the northern hemisphere and global background concentrations.

It is possible, therefore, to divide any discussion of photo-oxidant observations in many ways. The TOR investigators identified six topics that might be addressed using the data from the TOR stations (Hov, 1997).

(i)   Spatial and temporal variations of ozone.

(ii)   Distribution of ozone precursors.

(iii)   Photochemical production rates.

(iv)   Exchange between the boundary layer and the free troposphere.

(v)   Exchange between the stratosphere and troposphere.

(vi)   Transport across regional boundaries.

(i) Spatial and temporal observations of ozone

The measurements show that there is a general gradient in O_3 concentrations in the boundary layer across Europe in summer. The average summer concentrations in the North West have the lowest values, (30 to 40 ppb). The highest concentrations are in south eastern Europe, (60 to 70 ppb).

Fig. 2.5 shows contour maps for O_3 concentrations for the winter and summer. The excesses and deficits of ozone are relative to the four remoter sites at the western edge of Europe. These are taken to represent the marine boundary layer normally upwind of Europe and they show small annual variations with an average concentration just above 30 ppb. Relative to these sites there is a winter deficit in ozone concentrations in the North West of between 0 and 5 ppb. The deficit rises to about 10 ppb in the South East. However, the maximum deficit in winter ozone, about 20 ppb, is observed in central Europe where NO_x concentrations are high in winter. The average deficit is about 10 % in winter. In summer the excess is about 30–40 %.

Sites which show a seasonal variability usually have a peak in early summer, the amplitude of the cycle apparently increasing as one moves from west to east and north to south. Seasonal variations also appear at higher altitude and these were examined at mountain top sites, and with sondes and Lidar sounding from lower sites within the summer boundary layer below *ca.* 3 km. At heights of 3, 5 and 7 km there is a summer maximum in ozone concentrations. At 3 km the horizontal gradient of ozone is small across Europe suggesting that local pollution is not normally a major effect at this altitude, although observations at the Zugspitz, (2937 m), appear to contradict this.

An attempt was made to deduce short term trends from the TOR data. In the boundary layer there is no systematic trend within the time span of the measurements, a result which probably reflects the short time span well as changes to emission patterns within the period. The concentrations in the incoming air at particular sites was also examined in terms by analysing back trajectories. While there might be a slight decrease in O_3 in air from polluted regions; unpolluted air shows no trend.

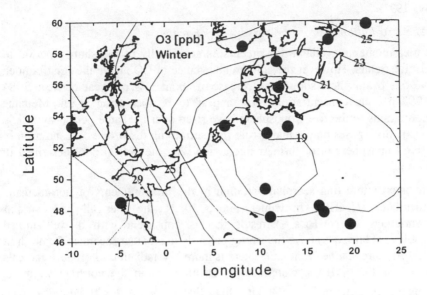

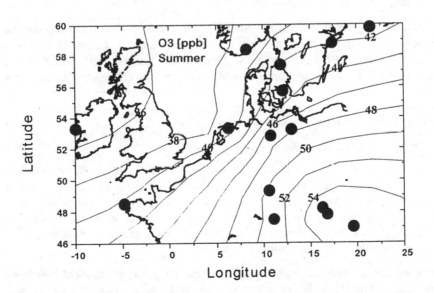

Fig. 2.5:   Isoline representation of the surface grid results for ozone in winter and summer, on maps showing the locations of the boundary layer measurement sites.

Ozone in the free troposphere appears to have increased by between 1 and 2 % per year with growth rates being greater in the 1970s than in the 1980s. One mountain top site reported a decline in O_3 since 1994 which would be consistent both with less injection of ozone from the stratosphere and increased UV radiation in the troposphere, both of these resulting from decreasing O_3 in the stratosphere. Only further measurements over an extended time scale will resolve such speculation (Hov, 1997).

(ii) The distribution of ozone precursors over Europe

The measurements available for precursors are usually of too short a period to determine trends. However the column abundance of CO above the Jungfraujoch in the free troposphere showed a decrease of about 1 % per year between 1984 to 1993 but the concentrations now appear to be increasing again. Methane concentrations in the free troposphere have risen over the same period by 0.5 % per year. Both gases play a major role in determining the ozone concentration in the free troposphere through their interaction with the radical pool described in section 2.2.

There is an interesting seasonal variation in the measurements of non-methane hydrocarbons (NMHC). In winter, above 50° N, the sites all have similar concentrations of NMHCs, a result that is consistent with a well mixed atmosphere at latitudes above the strong polar air stream that separates the north in winter. At more southerly latitudes there is more UV radiation to produce radicals and process the NMHCs. Consequently an air mass in the north in winter is "younger", in terms of its $\int[OH]dt$, than that of one at lower latitudes, even though the northern air mass may have passed over emission sources at an earlier time (Hov, 1997).

In summer the measurements of NMHCs are some 60–70 % lower due to photochemical removal. One anomaly is propene, the concentration of which is higher than that expected from anthropogenic sources. Since it reacts with OH in a matter of a few hours in summer it should have very low concentrations far from anthropogenic sources. The observation appears to point to biogenic emissions of alkenes from both land and sea. Another compound that appears to be anomalous with respect to the emission inventories is propane.

(iii) The photochemical production rate of ozone.

As indicated in Fig. 2.2, the photochemical production rate of O_3 is strongly dependent on the NO_x concentration and, although difficult to measure, it is a useful parameter to know in order to characterise the prevailing conditions. At a single site the change in ozone is controlled not only by the chemical production and loss rates but also by meteorological advection and turbulent diffusion, with advection usually being the dominant process.

Two methods of estimating the photochemical chemical production rate have been used. The first is to measure the NO, NO_2 and peroxy radical concentrations, the last being difficult under field conditions. In the second method, the value of $J(NO_2)$, the effective rate of photolysis of NO_2, is measured together with the concentrations of NO, NO_2 and O_3, and the Leighton photostationary state ratio, PSS = $[O_3][NO]/[NO_2]$, is estimated. The ratio is based on the conversion of NO to NO_2 only by O_3, reactions (1) to (3), and, in the presence of photochemical O_3 production, the ratio changes from that expected. The difference between the photolysis loss of NO_2 and the production rate of NO_2 from NO and O_3 is a measure of the local ozone production rate, but the difference is generally very small. The measurements must thus be accurate, which again is hard to achieve in the field.

Ozone production rates were measured at several sites within the TOR network. The PSS has a maximum value at local noon when the sunlight is at a maximum and such diurnal behaviour was observed at all sites. The influence of the active precursors was indicated by the local O_3 production rate which was higher at polluted sites (typically about 20 ppb/h in summer but sometimes 50 ppb/h) compared with more remote locations (5 to 10 ppb/h). These higher rates were associated with conditions of high NO_x concentrations (Fig. 2.2) and strong sunlight.

Peroxy radical concentrations have been measured at two sites, Schauinsland in south western Germany and Izaña in the Canary Islands. There was good agreement between the peroxy radical method and the PSS method for measuring local ozone production rates at the remote island site where the chemistry is dominated by CO and CH_4 oxidation and low NO_x concentrations.

At Schauinsland, where the air is more polluted, the agreement between the two methods was poor with peroxy radical technique giving a lower value by a factor of 3 for the local ozone production rate compared to the PSS. The reasons for the differences between the PSS method and the peroxy radical techniques at the Schauinsland site are not fully understood but it is believed that it may be due to the radical measurements missing the complex radicals formed from the degradation of biogenic emissions at the forested site, or to the then unknown effect of humidity on the peroxy radical amplifier.

The observations from the well-instrumented Schauinsland site have been used to test model predictions of behaviour. In summer a correlation between ($[O_3] + [NO_2]$) and the products of NO_x oxidation, ($[NO_y] - [NO_x]$), suggests that four to five molecules of O_3 are produced per NO_x molecule. The site is influenced by the urban plume from the city of Freiburg and it is thought that the production is achieved during the time it takes for the urban plume to reach the site, a distance of some 10kms. In winter there is a negative correlation between ($[O_3] + [NO_2]$) and ($[NO_y] - [NO_x]$) showing that ozone is consumed as NO_x is lost, a conclusion that is consistent with the Europe-wide ozone deficits in winter, mentioned above, Fig. 2.5 (Hov, 1997).

(iv) The exchange between the boundary layer and the free troposphere

There is exchange between these two regions of the troposphere: both photo-oxidants and their precursors are transported upwards from the boundary layer, where they are produced, into the free troposphere. In the downwards direction ozone can be transferred from the free troposphere to initiate the photo-oxidation processes which remove primary pollutants from the boundary layer.

The problem was encapsulated in TOR as two questions:

What is the seasonal, latitudinal and vertical variation of ozone within the boundary layer and the free troposphere?

How much excess ozone in the boundary layer passes to the free troposphere?

Possible meteorological mechanisms for exchange are subsidence, convective growth, convective cloud formation, and large scale updraft in cyclones and frontal systems. Other mechanisms are possible but were not considered as appropriate to the data available in TOR. However as these processes are highly correlated with each other, it is difficult to distinguish clearly between them.

By definition subsiding air cannot be injected into the boundary layer. However, the process is important in interpreting measurements at a fixed altitude such as at a high elevation site. Convection causes boundary layer growth by entrainment. It can be induced by topography or thermal buoyancy with velocities as high as 5 ms^{-1}, although more commonly they are in the range 1 to 2 ms^{-1}.

Fig. 2.6 shows the diurnal behaviour of the boundary layer. The temporal variation in the boundary layer height causes repeated pumping of air between the boundary layer and the free troposphere so that the net transport between the two layers, averaged over longer times, is smaller than might be estimated from a single event. For a height of 1200 m, modelling suggests that, over a several day period of subsidence, transport of ozone occurs in both directions. Rates of 0.75 ppb/hr to –0.25 ppb/hr have been estimated from observations in summer photochemical episodes.

When convection occurs it can make an appreciable contribution. The exchange is dominated by the diurnal cycle: in early morning convection brings down O_3 from aloft into a boundary layer, which has been depleted in O_3 overnight by loss at the surface sink and reaction with NO_x. Transfer rates as high as 5 ppb/hr can occur under these conditions. In the afternoon the convective flux of O_3 changes sign. The lower boundary layer loses some of its photochemically produced O_3 with rates of the order of –2.5 ppb/hr. At the same time the upper layer is enhanced by rates up to 0.75 ppb/hr. Average rate of loss of O_3 from the boundary layer over the three summer months were ca. –0.2 ppb/hr under cloud free conditions. Cloud enhanced the loss rate by only a small amount. The results indicate that the net effect is to decrease boundary layer O_3 concentrations and increase them aloft during the summer over polluted areas. Over less polluted areas the opposite occurs.

Frontal systems are important in are important in mixing boundary layer air with that of the free troposphere because they can have lengths of hundreds of kilometres and sweep over large regions of the continent. They often terminate photochemical summer smog episodes. Cold fronts cause horizontal convergence

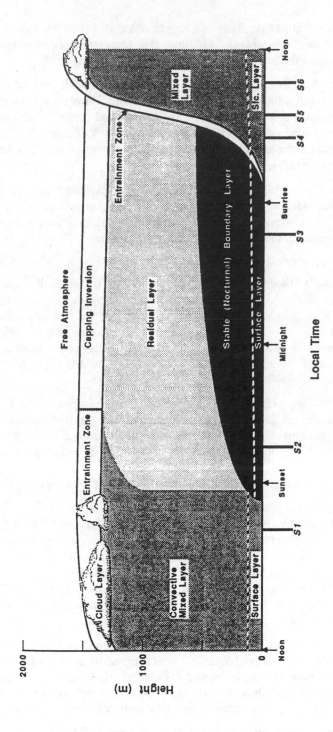

Fig. 2.6: The boundary layer in high pressure regions over land consists of three major parts; a very turbulent mixed layer, a less turbulent residual layer and a nocturnal stable layer of sporadic turbulence.

and produce vertical motion. They can also be associated with tropopause folding that brings down stratospheric O_3 into the free troposphere although no contributions from the stratosphere have been observed at sea level.

Unfortunately there is not enough data over a wide enough area in the TOR database to generalise the behaviour of frontal systems. However it is thought that each frontal passage at a site averages about 4 hours and causes a decrease of *ca.* −2.5 ppb/hr in boundary layer O_3. As there are about 80 fronts a year over the European continent the annual rate for the loss of O_3 from the boundary layer rate is *ca.* −0.09 ppb/hr (Hov, 1997).

Some progress has been made on modelling this exchange process and the calculated rates indicate of the magnitude of the transport between the boundary layer and the free troposphere. However, a full evaluation awaits a validated climate and chemistry model of the European atmosphere.

(v) Exchange across the tropopause

It was originally thought, before the importance of photochemical ozone formation was realised, that the ozone in the troposphere was all transported down from the stratosphere across the tropopause into the free troposphere and ultimately down into the boundary layer. It is now known that both photochemistry and downward transport from the stratosphere have a role to play and much of work in the last thirty years has been devoted to trying to estimate the contributions of each.

Tropopause folding which is often associated with vigorous frontal systems provides one way of injecting stratospheric air into the troposphere. Within TOR, two algorithms were developed to detect tropopause folds in the vertical O_3 sounding data set. Both were based on peaks in ozone below the tropopause and correspondingly low humidity both of which are characteristics of stratospheric air. Ozone concentrations can reach 100 to 400 ppb within a fold region.

The potential vorticity (PV) is a conserved parameter for stratospheric air since its value is an order of magnitude higher in the stratosphere than in the troposphere below the tropopause. The first algorithm, developed by the Köln group for cases for which a PV analysis is available, used a criterion of PV >1 PVU for tropopause folding in data obtained from the ECMWF (European Centre for Medium term Weather Forecasting), analyses. The algorithm also included criteria based on the $[O_3]$/PV ratio.

The second, the KMI/CNRS algorithm, was developed for cases where PV analyses are not available in the data. It uses a criterion of high wind speeds to indicate the presence of a jetstream and the vertical wind gradient to indicate the presence of a front, both these features being associated with tropopause folding. This algorithm was tested for periods when PV analyses were available.

The algorithms were applied to the data from several TOR sites. The second identified 4.5 % of the profiles as indicating folding on an annual basis. When applied to the results from the UCCLE and OHP sites a frequency of about 8 %

was found. This is much larger than the normally accepted frequency of folds. However, it is consistent with how the fold disperses over a large area as it descends further into the troposphere to form what is known as a stratospheric intrusion of ozone.

To compare results between the algorithms, the EURAD model (see next section), has been used to estimate the 'detectable' area for the two algorithms. The KMI/CNRS algorithm appears to detect a fold within 100 to 150 kms of the jetstream axis. The PV algorithm can detect a fold up to 500kms on the cyclonic side of the jetstream and to about 100 to 150 kms the other side. Thus a typical fold appears to extend some 1000 to 1500 kms along the jet axis. This relative detection rate is consistent with their application to a data set of 146 soundings from UCCLE. The first algorithm identified 13 events in the data, the second found 6 events.

For observations at mountain top sites the tracer ^7Be, which is formed in the upper atmosphere from the action of cosmic rays, was used as an indicator of stratospheric air when it exceeded a threshold value. The use of a ^7Be threshold at the two mountain sites of the Zugspitz and Wank showed 85 events at the higher Zugspitz and 58 at the Wank for the period 1984 to 1989. These events showed a maximum in autumn while the analysis of the vertical soundings showed no significant seasonal pattern.

The EURAD model results for the northern hemisphere show a seasonal trend in fold occurrence with a late spring/summer minimum. Above the tropopause, in the lower stratosphere, the soundings show a pronounced ozone concentration maximum in the spring and a minimum in the autumn. This higher O_3 concentration in the source region with the longer lifetimes of the enriched O_3 layers in the troposphere in the late spring/summer period could explain the differences in seasonal behaviour between the observations and the global modelling.

The objective of the study was to estimate ozone injection rates into the troposphere from the stratosphere. The results appear to show that folds are more frequent than previously assumed for the northern hemisphere. The amount transferred per fold is uncertain but the earlier estimates of 100–120 ppb/PVU now appear to be too high and a value of ca. 50 ppb/PVU seems more likely. Using this lower figure together with the increased frequency of folding coincidentally yields a value for the flux of ozone not very different from earlier estimates. The final figure of 6 $(3.7–8.5) \times 10^{10}$ mol cm^{-2} s^{-1} for the northern hemisphere also agrees with global circulation model estimates of trace gases $N_2O/NO_y/O_3$ in the atmosphere. This agreement is obtained considering only tropopause folding as the mechanism of transfer. It ignores any contribution from other potential sources such as cut-off lows or rising of the tropopause (Hov, 1997; Ebel et al., 1997).

2.5   Modelling studies

The large EUROTRAC efforts in model development and in the applications of the simulation models are described in a subsequent chapter. Here we concentrate on the some of the results concerning photo-oxidants.

a.   The global scale

The role of global modelling within a regional project is to ascertain how Europe is influenced by photo-oxidants from the global background and what Europe contributes to the global atmosphere.

The problem of the relative contributions of transport of O_3 down from the stratosphere and photochemical production of O_3 has already been mentioned (2.4.b (v)). The global budget of O_3 itself is uncertain not least because of the complexity of NMHC emissions, but it is estimated that photochemical reaction involving NMHCs, from both anthropogenic and biogenic sources, contribute about half of the net production of O_3 in the troposphere.

The MOGUNTIA model was used within the subproject GLOMAC to look at the historical trends in global O_3 concentrations and confirmed that the increase from the earlier O_3 concentrations is due to anthropogenic photochemically-produced O_3. The highest O_3 production occurs between $30° N$ and $60° N$ with O_3 destruction occurring over the remote tropical oceans. The model indicates an average lifetime of tropospheric for O_3 in the northern hemisphere to be about a month compared to 1 to 2 days for NO_x. The lifetime of ozone actually varies from a few minutes in urban areas within the boundary layer to several days in a clean boundary layer, and to many months in the upper free troposphere that has low temperatures and low water vapour concentrations (Ebel et al., 1997).

Such a long lifetime for O_3 has implications for its transport and distribution in the northern hemisphere. North America contributes about 30% of man-made NO_x emissions to the northern hemisphere. This NO_x together, with direct export of O_3, is estimated to add about 30 Tg of O_3 per year to the northern hemisphere which raises the question of a transatlantic influence on the level of photo-oxidants in Europe (Ebel et al., 1997). Measurements, made at Izaña in the Canary Islands within TOR, show high O_3 concentrations persist until August in air masses from the northern part of the North Atlantic. At most remote sites in the northern hemisphere O_3 concentrations have usually fallen from the highest values by this time of year so these high values are probably the result of the export of both O_3 and its precursors from the North American continent into the free troposphere over the Atlantic (Hov, 1997).

b.   The regional scale

The results from many of the modelling activities on a regional scale have been already been mentioned in section 2.4 or will dealt with in chapter 5. Here we give the results from an analysis of a summer smog episode which was the subject of an experimental study within TOR and then analysed in some detail with the EURAD model. The episode was for the period 31st July to 4th August 1990, occurring during a large anticyclone with high temperatures, ranging from 25 to 30 °C. It was terminated by a cold front moving across Europe from the west. (Ebel *et al.*, 1997).

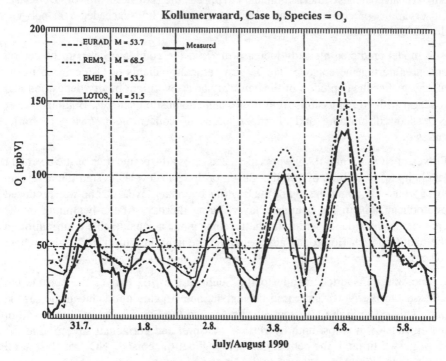

Fig. 2.7:   Ozone time series for the Kollumerwaard TOR station in the Netherlands. The measured ozone concentrations (ppb) are shown together with results from four long-range transport models used in Europe, for the episode from 31st July to 5th August 1990.

The first task for the EURAD model was to reproduce the observed concentration fields for the major species for the whole of Europe. Fig. 2.7 shows the predictions from the model of ozone concentrations for a single site in the Netherlands. Subsequently the episode was analysed in seven categories to identify the contribution each made to the ozone concentration and how that contribution varied during the episode. The categories were:

(i)   Horizontal large scale transport.

(ii)   Vertical large scale transport.

(iii) Small scale vertical turbulence.

(iv)   Losses due to dry deposition.

(v)   Chemical production and loss.

(vi)   Processes due to clouds *e.g.* transport, aqueous chemistry.

(vii) Emissions.

To examine fluxes across horizontal boundaries an inner region of central Europe was defined that contained the highest concentrations of O_3. This region had the west coast of the UK as its western boundary and the eastern border of Poland formed its eastern edge. The southern boundary crossed Switzerland and Austria and the Northern boundary clipped the northern tip of Denmark. Observations of concentrations within parts of this box exceeded 100 ppb O_3, during the episode.

The model predicted a steady increase of O_3 in the boundary layer in the central box area during the episode. The largest vertical gradients of O_3 occurred in the morning after the depletion of the boundary layer overnight by dry deposition and reaction with NO emissions from traffic and combustion sources. Boundary layer NO_x showed morning and afternoon peaks of concentration owing to traffic emissions.

Fluxes of NO_2, O_3 and PAN were calculated across the boundaries of the box at 11 levels up to an altitude of 100hPa. Seven of these 11 levels covered the height up to 1550 m - *i.e.* about the top of the daytime boundary layer. Ozone was produced photochemically in the boundary layer during the day. At the beginning of the episode the subsidence within the anticyclone made a considerable contribution to the ozone budget. Horizontal advection was the main loss term for O_3 throughout the episode.

Ozone was transported mainly to the South in the first part of the episode but transport to the North increased throughout and became dominant in the second part. This led to high O_3 in southern France and the Bay of Biscay. At the end of the episode O_3 was lost both by vertical transport and horizontal advection as the front passed through the central box. Throughout the episode NO_2 was lost at all levels up to 1800m by transport to the North and South.

The O_3 production rate over the whole central European box area averaged more than 10 ppb/hour and peaked at 16 ppb/hour. The maximum ozone production occurred in the second level of the model at heights between 75–150 m.

All the budgets for NO_2, O_3 and PAN were determined for the duration of the episode. For example the total O_3 produced photochemically within the central box area over all levels was nearly half a million tonnes. The O_3 budget for the whole European area is shown in Fig. 2.8.

The budget for Europe as a whole showed that O_3 increases at all altitude levels in the model throughout the episode. The overall picture of the O_3 budget for Europe during the episode is that the main sources of ozone are photochemical production in the boundary layer and transport from the upper atmosphere to the lower atmosphere by large scale motions. The loss processes are dry deposition and large scale horizontal advection.

In the central European box about 40 % of the O_3 produced chemically is deposited and about 40 % is lost by vertical transport, even though advection acts as a source. For the whole region of Europe, 70 % of the photochemically produced O_3 is deposited in the area and some 23 % is lost by transport.

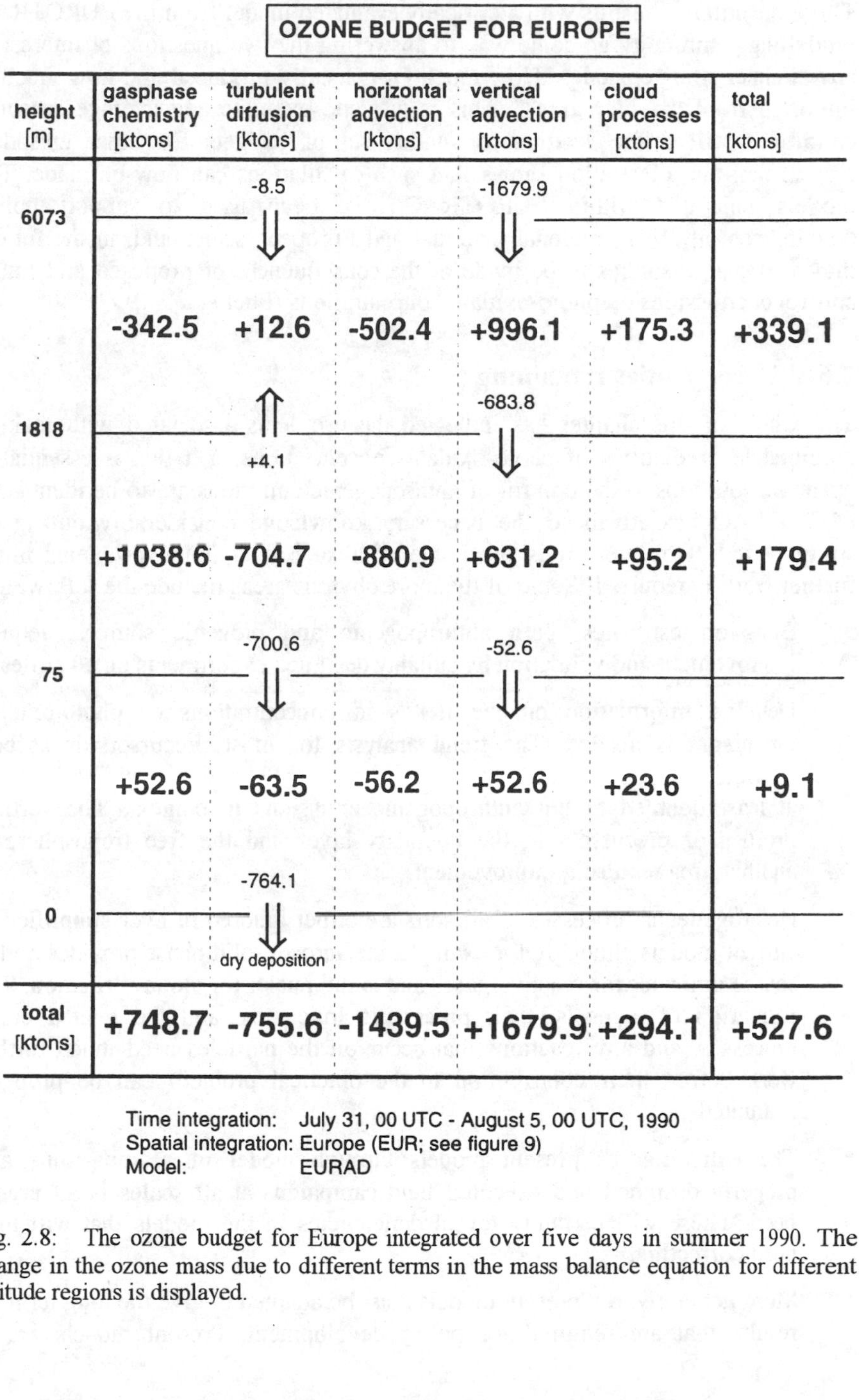

OZONE BUDGET FOR EUROPE					

height [m]	gasphase chemistry [ktons]	turbulent diffusion [ktons]	horizontal advection [ktons]	vertical advection [ktons]	cloud processes [ktons]	total [ktons]
		−8.5 ⇓		−1679.9 ⇓		
6073						
	−342.5	+12.6	−502.4	+996.1	+175.3	+339.1
		+4.1 ⇑		−683.8 ⇓		
1818						
	+1038.6	−704.7	−880.9	+631.2	+95.2	+179.4
		−700.6 ⇓		−52.6 ⇓		
75						
	+52.6	−63.5	−56.2	+52.6	+23.6	+9.1
		−764.1 ⇓ dry deposition				
0						
total [ktons]	+748.7	−755.6	−1439.5	+1679.9	+294.1	+527.6

Time integration: July 31, 00 UTC - August 5, 00 UTC, 1990
Spatial integration: Europe (EUR; see figure 9)
Model: EURAD

Fig. 2.8:   The ozone budget for Europe integrated over five days in summer 1990. The change in the ozone mass due to different terms in the mass balance equation for different altitude regions is displayed.

Most of the NO_x emissions, about 70 to 85 %, were lost by chemical conversion. PAN is produced at all levels and 80 to 95 % was lost by transport out of the region.

These quantitative results with a critically evaluated model from the EUROTRAC modelling community go some way to answering the two questions of interest in any summer smog episode: "How much O_3 is locally produced and how much is imported from a wider area?" This is just one example and average seasonal conditions will differ, leading to another set of budgets for other episodes. Nonetheless the illustration shows that such calculations can now be made. The models generated within EUROTRAC have been used to support policy development at urban, regional, national and European scales and, in the future, they will enable studies to be made of the consequences of proposed and future control of emissions on photo-oxidant concentrations (Ebel *et al.*, 1997).

2.6   Uncertainties remaining

The whole of this chapter has indicated the problems associated with making quantitative predictions of photo-oxidant concentrations. Yet this is essential if optimum solutions to the control of anthropogenic emissions are to be identified. EUROTRAC has advanced the necessary knowledge considerably during its study period. However, it is inevitable with such a complex problems much further work is required. Some of the more obvious areas include the following.

*   Emission estimates, both anthropogenic and biogenic sources, require improvement and validation by suitably designed experiments on all scales.

*   Detailed information on the trends in concentrations of photo-oxidant precursors is needed. The trend analysis for most precursors is at best restricted to a decade of modern measurements. This time scale needs to be at least doubled to put anthropogenic emissions in context. The vertical profiles of precursors in the boundary layer and the free troposphere is another area requiring improvement.

*   Heterogeneous processes of all sorts are either ignored or over simplified in current models although, for example, the aerosol solid phase provides both a source and sink for reactive gases and is the nuclei for cloud formation. The formation of aerosols, their processing in clouds, and the related cloud processes, and any reactions that occur on the particles need much further work before their contribution to the practical problem can be properly evaluated.

*   The validation of present models through model intercomparisons and properly designed and executed field campaigns at all scales is an urgent need. These will certainly reveal deficiencies in the models that will then need correcting.

*   More generally, the present models must be adapted to give the long term results that are required for policy development. Present models are

largely "episodic", capable of dealing with periods of a just few days or so. Most of the present efforts, and the vast increases in computing power, have gone and are going into improving the details. However policy makers require models capable of giving averages for a year or several years so that likely effects of new policy measures for a foreseeable period can be evaluated. New methods will have to be developed to obtain such results from the present detailed models.

EUROTRAC made a considerable contribution to the photo-oxidant problem both scientifically and in terms of policy development. Much has been achieved but many uncertainties still remain. A major contribution to the future that the project has made is the identification of a number of key areas that require further attention. Some of these are being addressed in EUROTRAC-2.

2.7   References

Borrell, P., Ø. Hov, P. Grennfelt, P. Builtjes (eds); 1997; *Photo-oxidants, Acidification and Tools: Policy Applications of EUROTRAC Results,* Springer Verlag, Heidelberg; Vol. 10 of the EUROTRAC final report.

Brasseur, G.P., R.A. Cox, D. Hauglustaine, I. Isaksen, J. Lelieveld, D.H. Lister, R. Sausen, U. Schmann, A. Wahner, P. Wiesen; 1998, European Scientific Assessment of the Atmospheric Effects of Aircraft Emissions, *Atmos. Environ.* **32**, 2329–2418.

Cvitaš, T., D. Kley (eds); 1994, The TOR Network, EUROTRAC ISS, Garmisch-Partenkirchen.

Dumollin, J., A. Derouane, G. Dumont; 1999, *Assessing ozone in ambient air in Belgium,* International Conference on Air Quality in Europe: Challenges for the 2000s, Venice, May 1999.

Ebel, A., R. Friedrich, H. Rodhe (eds); 1997; *Tropospheric Modelling and Emission Estimation,* Springer Verlag, Heidelberg; Vol. 7 of the EUROTRAC final report.

Fuzzi, S., D. Wagenbach (eds); 1997, *Cloud Multi-phase Processes and High Alpine Air and Snow Chemistry,* Springer Verlag, Heidelberg; Vol. 5 of the EUROTRAC final report.

Hov, Ø. (ed); 1997; *Tropospheric Ozone Research,* Springer Verlag, Heidelberg; Vol. 6 of the EUROTRAC final report.

Le Bras, G. (ed); 1997; *Chemical Processes in Atmospheric Oxidation,* Springer Verlag, Heidelberg; Vol. 3 of the EUROTRAC final report.

Marenco, A., N. Phillipe, G. Hervé; 1994, Ozone measurements at the Pic du Midi observatory, *EUROTRAC Annual Report for 1993, part 9,* EUROTRAC ISS, Garmisch-Partenkirchen, pp. 121–130.

Schneider, J; 1999, An assessment of Austrian Ozone Data based on different indicators, *Bericht des Umweltbundesamts,* Wien, BE 106.

Slanina, J. (ed); 1997, *Biosphere-Atmosphere Exchange of Pollutants and Trace Substances,* Springer Verlag, Heidelberg; Vol. 4 of the EUROTRAC final report.

Staehelin, J., J. Thudium, R. Buehler, A. Volz-Thomas, W. Graber, 1994, Trends in surface ozone concentrations at Arosa (Switzerland), *Atmos. Environ.* **28**, 75–87.

Wayne, W.P., I. Barnes, P. Biggs, J.P. Burrows, C.E. Canosa-Mas, J. Hjorth, G. Le Bras, G.K. Moortgat, D. Perner, G. Poulet, G. Restelli, H. Sidebottom, 1991, *Atmos. Environ.* **25A**, 1.

Chapter 3

Research on Clouds within EUROTRAC

Sandro Fuzzi[1] and Adolf Ebel[2]

[1]Institute of Atmospheric and Oceanic Sciences, ISAO - C.N.R..,
Via Gobetti 101, I-40129 Bologna, Italy
[2]Universität zu Köln, Institut für Meteorologie und Geophysik,
Aachener Straße 201-209, D-50931 Cologne, Germany

3.1 Introduction

a. Clouds in the troposphere

Although the first studies on cloud chemistry date back to the nineteen fifties, it was only at the beginning of the eighties that the key role played by clouds in tropospheric chemistry was clearly highlighted. Some fundamental papers then pointed out the overall effect on the cloud life cycle, of the incorporation of chemical components within cloud droplets and of the chemical reactions in the liquid phase. The study of clouds and their importance in the global atmosphere is now one of the main tasks of the atmospheric sciences research community. The following issues indicate the importance of clouds in the atmosphere:

* clouds redistribute trace compounds emitted at the Earth's surface in the vertical from the boundary layer to the free troposphere, and in some cases to the stratosphere;

* clouds interact with the incoming solar radiation and with the long wave radiation emitted by the Earth, thus affecting both photochemistry in the atmosphere and the radiation budget of the Earth;

* clouds are an efficient reaction medium for chemical transformations;

* clouds produce precipitation, which is a very efficient mechanism for removing trace components from the atmosphere.

The importance of these issues is emphasised by the fact that, at any given time, roughly 60 % of the earth's surface is covered by clouds.

The cloudy atmosphere constitutes a multiphase atmospheric system, in that gaseous species, atmospheric particulates and liquid droplets coexist at the same time. Interaction between the different phases must therefore be considered when discussing physical and chemical processes in clouds. Multiphase processes in the atmosphere act on a wide variety of spatial (and therefore temporal) scales varying from molecular processes acting at a spatial scale of a few tens of nanometres (corresponding to a microsecond temporal scale), to synoptic meteorology characterised by a spatial scale of thousands of km (a temporal scale of the order of days). This wide range of spatial and temporal scales involved in cloud processes (more than 15 orders of magnitude) is one of the main problems encountered in cloud modelling, since processes at the smallest scale cannot necessarily be neglected in describing larger scale processes.

The complexity of the cloud multiphase system can best be described with the graphical representation shown in Fig. 3.1. All the processes are interdependent and may proceed simultaneously. It should also be noted that the listed processes are essentially all of a physical nature; chemical transformations also occur, hence increasing the overall complexity of the multiphase system considerably.

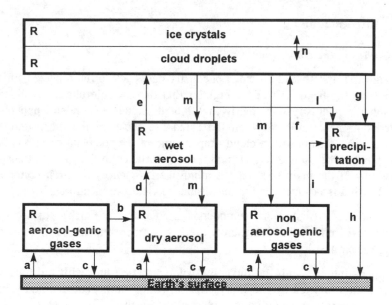

Fig. 3.1: Graphic representation of the multiphase atmospheric system. The arrows in the figure represent physical and chemical processes leading to mass exchange between the different phases. a) emission, b) gas-to-particle conversion, c) dry deposition, d) condensation, e) nucleation, f) dissolution, g) formation of precipitation, h) wet deposition, i) aerosol capture by falling droplets, m) evaporation, n) freezing/melting. Chemical reactions (R) occur in all phases of the system.

A further complexity in the study of clouds derives from their inhomogeneous nature, both at a microscale and macroscale level. It is now recognised that the simplistic view of cloud droplets being of homogeneous chemical composition is not realistic, and that the chemical composition of cloud droplets varies as a function of size. This is due both to inhomogeneities in the chemical composition of the cloud condensation nuclei (CCN) on which the droplets grow and to the dependence of in-cloud chemical processes on cloud droplet size distribution.

Macroscale cloud chemical inhomogeneities, on the other hand, depend on meteorology (*e.g.* temperature variation with height) and cloud dynamics (*e.g.* entrainment processes).

Last but not least, cloud research is an interdisciplinary field which requires joint contributions from a wide range of disciplines such as meteorology, photochemistry, aerosol and cloud physics, aerosol and cloud chemistry, as well as the integrated efforts of laboratory scientists, modellers and field experts.

b.   Cloud research within EUROTRAC

Within EUROTRAC, GCE (Ground-based Cloud Experiments) was a project specifically oriented toward the study of clouds, but other EUROTRAC subprojects also carried out field, laboratory and modelling research connected with cloud processes (Table 3.1).

Table 3.1:   EUROTRAC subprojects connected with the study of cloud processes and main research tasks.

GCE (Ground-based Cloud Experiments)	Aerosol scavenging by clouds
	Scavenging of gases by clouds
	Cloud chemistry and its dependence on microphysics
	Modification of aerosol particles passing through clouds
	Cloud droplet deposition
	Instrument development for cloud research
HALIPP (Heterogeneous and Liquid-Phase Processes)	Aqueous phase oxidation of S(IV)
	Transition metals in atmospheric aqueous phase
	Free radicals in the aqueous phase
	Gas/liquid interactions
ALPTRAC (High Alpine Aerosol and Snow Chemistry Study)	Alpine precipitation chemistry
	Gas and aerosol scavenging by super cooled cloud droplets
	Instrument development for cloud research
EUMAC (European Modelling of Atmospheric Constituents)	Cloud chemistry modelling
	Modelling of wet deposition processes
	Development of parameterisations for cloud modelling
BIATEX (Biosphere-Atmosphere Exchange of Pollutants)	Cloud water deposition
GLOMAC (Global Modelling of Atmospheric Chemistry)	Cloud modelling at the global scale

A real interdisciplinary approach to the study of clouds was attempted only towards the end of EUROTRAC with the appointment of an inter-project panel: the Cloud Group (see below).

An attempt will be made in the following description to integrate the results achieved by the different EUROTRAC subprojects. This description will, in fact, be provided with reference to the main processes of concern in the study of clouds, avoiding as much as possible a mere repetition of the results achieved within each subproject, which can be found in the specific reports.

3.2   Main results of cloud research within EUROTRAC

a.   Aerosol and clouds

(i)   Aerosol as cloud condensation nuclei

Cloud droplets are formed in the atmosphere by the condensation of water vapour onto aerosol particles when the relative humidity (r.h.) exceeds the saturation level. The term nucleation scavenging refers to the process through which some atmospheric particles, the cloud condensation nuclei (CCN), grow into cloud droplets in a supersaturated air parcel (heterogeneous nucleation). Particles which do not act as CCN remain suspended in the air between the droplets are called interstitial particles. The mathematical formulation of how water droplets grow in high relative humidity conditions and supersaturated environments (Köhler equation) has long been known and is used in all models (Pruppacher and Klett, 1978). While nucleation scavenging is the main mechanism for the incorporation of particulate species into cloud droplets, other aerosol particle scavenging mechanisms occur in cloud (attachment of particles to cloud droplets by Brownian motion, phoretic effects, collisional capture of particles by the droplets), but their contribution in terms of extra mass scavenged within the droplets is generally low (Flossmann et al., 1985).

The cloud droplet nucleating properties of an aerosol population depend on the distribution of sizes and hygroscopic properties among the particles: an increase in hygroscopicity, i.e. particle soluble fraction, causes an increase in the number of cloud droplets for a given particle size distribution and number concentration. However, in view of the rather complex chemical composition of atmospheric particles, their hygroscopic behaviour cannot analytically be described by the Köhler equation.

By using a coupled Tandem Differential Mobility Analyser (TDMA), with accurately controlled relative humidity in between the two DMAs, it was possible to determine within the subproject GCE (Fuzzi et al., 1997) the growth in size of particles exposed to increasing r.h. This allowed the growth in particle size to be measured with a very high precision and accuracy. The results of TDMA measurements performed at different locations in Europe have always shown a bimodal hygroscopic distribution in continental aerosol: with particles growing either ca. 5 % or ca. 45 % in size at 85 % r.h., compared to the size at 20 % r.h. The data available today cover so many areas in the world and such long time periods that there is no doubt that continental aerosol, perturbed by anthropogenic sources, consists of two major groups of particles characterised by different hygroscopicity.

Unfortunately, not much is known about the chemistry of the two hygroscopic types of particles besides the inorganic composition and the amount of soot. This means that approximately 50 % of the mass of the aerosol in all size ranges is not speciated and it is not known whether the sources are anthropogenic or natural. In this respect it should be mentioned that the present knowledge of the amount and composition of the organic fraction of the aerosol is very poor. All these factors introduce a large degree of uncertainty in describing the aerosol life cycle in the atmosphere and thus in predicting the actual influence of anthropogenic emissions on the physical and chemical properties of aerosols and clouds.

One additional aspect which was extensively investigated during the GCE experiments (Fuzzi *et al.*, 1997) is that the presence of clouds produces a chemical fractionation of aerosol particles, with chemical species associated with more soluble particles being incorporated in the liquid phase of cloud, and species associated with less soluble particles remaining interstitial: for example the scavenging efficiency of elemental carbon is much smaller than that of sulfate.

(ii)   Aerosol modification after cloud processing

It is well known that most clouds do not lead to precipitation but evaporate. Upon cloud evaporation, gas and particles are released back to the atmosphere. The aerosol resulting from the evaporation of cloud droplets is likely to be quite different (in physical and chemical properties) from that which entered the cloud, due to in-cloud processes (Pruppacher, 1986).

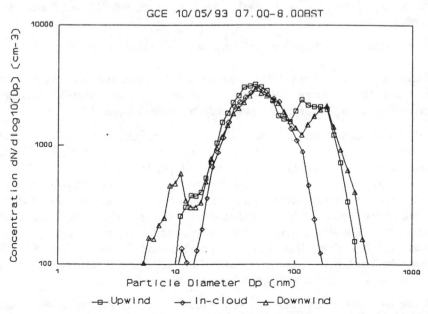

Fig. 3.2:   Aerosol size spectra measured by three DMPSs operated simultaneously upwind of, within and downwind of a hill cloud.

Both modelling and experimental studies (Fuzzi *et al.*, 1997) have shown that sulfate aerosol loading before and after passage through a cloud differs significantly, with much larger concentrations in the outflow of the cloud system. (Fig. 3.2) shows the data obtained from the three Differential Mobility Particle Sizing (DMPS) instruments operating upwind of, within and downwind of a cap cloud. At the downwind site the cloud droplets evaporate, returning the CCN to the aerosol phase with some growth due to sulfate production. At the smallest sizes, it is sometimes observed that new particles are produced in the air stream leaving the cloud.

b.   Trace gases and clouds

Major advances in the understanding of the atmospheric multiphase system and its model representation have been achieved within EUROTRAC by considering the chemical equilibria existing between the different phases of the system. A basis for assuming chemical equilibrium between gas and liquid (droplet) phases was that characteristic times for mass transfer across the gas-liquid interface, ionisation, diffusion into the aqueous phase and chemical equilibration are usually much shorter than the lifetime of cloud droplets. Henry's law equilibrium is therefore routinely used to describe the partitioning of species between gas and aqueous phases for both wet aerosols and cloud droplets (Warneck, 1986; Schwartz, 1986; Lelieveld and Crutzen, 1991).

Different sophisticated techniques were developed within the HALIPP subproject for laboratory studies of gas/liquid water interaction (Warneck *et al.*, 1996): droplet train; liquid jet; wetted-wall flow tube; wind tunnel; and the aerosol chamber.

These experimental set-ups are applicable to a wide range of conditions allowing the measurement of reliable uptake rates. The techniques have been employed for different gases, leading to the determination of mass accommodation coefficients larger than 10^{-2}, meaning that in most cases interfacial mass transfer does not represent a limiting factor for the gas uptake by cloud droplets.

Henry's Law equilibrium for NH_3, H_2O_2, SO_2, HCOOH and CH_3COOH and formaldehyde was also studied in real clouds within the subproject GCE by simultaneous measurements in both liquid and gas phases (Fuzzi *et al.*, 1997). Large deviations from Henry's law equilibrium, up to two orders of magnitude in some cases, were encountered in all the studies. The pH dependency of these deviations was clearly highlighted showing, for example, that formic and acetic acids were generally supersaturated in the liquid phase at low pH and subsaturated at high pH (Fig. 3.3), as opposed to the behaviour of NH_3. Similar non equilibrium conditions were evidenced in mixed-phase clouds studied within the ALPTRAC subproject (Wagenbach *et al.*, 1997).

Because the reported deviations from equilibrium cannot be fully explained either by bulk sampling and time integration artefacts (which can only account for deviations up to a factor of 3), or by the formation of additional compounds, other hypotheses have been advanced such as a shift in equilibrium due to the presence of chemical substances not taken into account in the Henry's Law calculation, or a kinetic inhibition due to mass transfer limitation by an organic film coating cloud droplets.

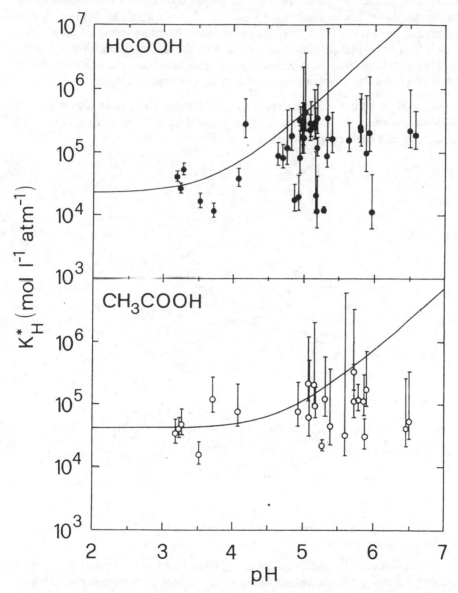

Fig. 3.3:   Large deviations of experimental data (dots with error bars) from theoretical pseudo-Henry's Law coefficient(solid line) for HCOOH and CH3COOH in fog droplets.

Future research should certainly address this issue and aim at integrating laboratory and field results in order to obtain better parameterisations of the processes involved for modelling studies.

c. Chemical composition and transformations in clouds

(i) In-cloud chemical reactions

Laboratory studies within the HALIPP subproject on chemical transformations in the atmospheric liquid phase have mainly dealt with S(IV) to S(VI) oxidation (Warneck *et al.*, 1996). In fact, S(IV) oxidation reactions occur in clouds at a much faster rate than in the clear air: model calculations (Langner and Rodhe, 1991) have shown that, on a global scale, tropospheric in-cloud SO_2 oxidation is from two to five times more important than out-of-cloud oxidation.

Appreciable progress has been made in identifying the reactions responsible for S(IV) oxidation in the atmospheric liquid phase (Fig. 3.4) and in determining the associated rate coefficients (Warneck *et al.*, 1996).

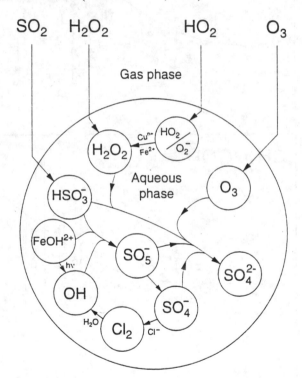

Fig. 3.4:   Schematic summary of the different pathways for S(IV) to S(VI) oxidation in the aqueous phase of clouds.

It was shown that the direct oxidation of HSO_3^- and SO_3^{2-} by H_2O_2 is a non-radical process, and mechanism and rate coefficient as a function of pH have been established for this reaction.

The reaction of S(IV) with O_3 has been better defined, but the important question of whether the reaction involves radicals or not has not been answered.

Transition metal ions play an important role in the chemistry of atmospheric liquid phase. S(IV) oxidation catalysed by Fe has been shown to proceed by a chain reaction. The rate coefficient for the initiation reaction between Fe(III) and S(IV) has been defined, and other important reactions with Fe(II) and Fe(III) have been identified. Effects of other catalytically active transition metals such as Mn, Co and Cu have also been studied and interpreted in terms of a mechanism similar to that of Fe. Further studies are however necessary for the extrapolation of the laboratory data to atmospheric conditions, especially in terms of concentration ranges and pH.

The mechanism of the OH radical induced chain oxidation of S(IV) has been fully established, and rate coefficients were obtained for the chain initiating, chain propagating and chain terminating reaction steps.

The chemistry of cloud droplets is also dependent on radicals produced in the aqueous phase by photochemical reactions and/or scavenged from the gas phase. The entrainment of HO_2 radicals followed by dissociation to form O_2^- and reaction with O_3, which was thought to be the major source of OH in the atmospheric liquid phase, cannot probably compete with the scavenging of HO_2/ O_2^- by transition metals. Cu ions appear to be the most efficient scavengers, converting HO_2/O_2^- either to H_2O_2 or O_2 depending on the oxidation state of Cu. Another source of OH radicals in solution is provided by photodecomposition reaction of dissolved species as NO_3^-, HNO_2/NO_2^-, H_2O_2, $FeOH^{2+}$ and $FeSO_4^+$. The Fe(III)-hydroxo complex is most efficient in terms of quantum yield and abundance in cloud water.

Chemical transformations within the aqueous phase of clouds were also studied through large integrated ground-based field experiments carried out within the subproject GCE which were among the larger experimental efforts in cloud research of the last decade (Fuzzi *et al.*, 1992; Wobrock *et al.*, 1994; Choularton *et al.*, 1997).

Experiments were carried out in both oxidant-limited and oxidant-rich conditions. In this latter situation changes in atmospheric acidity were simply due to advection of acidic air masses to the sampling sites, rather than to S(VI) or HNO_3 production processes. On the contrary, in experiments where high levels of oxidants were encountered an efficient transformation of both sulfur and nitrogen species was measured.

In oxidant-limited conditions, it was shown that the evolution of fog and clouds can be well described using the concept of atmospheric acidity, defined as the base neutralising capacity of a unit volume of an atmosphere including gas, interstitial aerosol and liquid phase. In fact, cloud and fog systems exhibited active exchanges of acidic and basic components among the different phases: advection of HNO_3-rich air resulted in an acidification of fog and cloud systems after exhaustion of their neutralising capacity (Fig. 3.5).

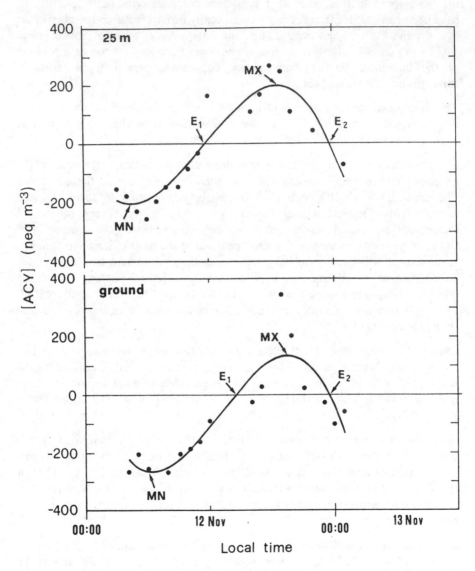

Fig. 3.5:   Temporal evolution of atmospheric acidity ([ACY]) at two different heights during a fog episode. ([ACY] is the sum of gas, interstitial aerosol and fog droplet acidities). The experimental data (dots) are fitted with a cubic regression function (solid line). The figure indicates a net addition (ascending branch of the function) of acidic species to the system, which was originally alkaline (negative [ACY]), followed by a net addition of basic species (descending branch of the function). It was shown that the acidic species added to the system was actually gaseous HNO_3 entrained from outside the system. When the acid input stopped, NH_3 from local sources drove back the system to an alkaline character.

In the oxidant-rich situation, sulfate aerosol loading before and after passage through a cloud showed significant differences with much larger concentrations in the outflow of the cloud system. Although part of the increase in aerosol sulfate mass results clearly from S(IV) to S(VI) oxidation by H_2O_2 in the liquid phase, difficulties in quantifying the mechanisms leading to sulfate production arose from the fact that dynamical mixing in the cloud was also found to be a key process in supplying oxidants (H_2O_2) and possibly particulate sulfate to the system. In addition, liquid phase formation of HCOOH, possibly through formaldehyde hydrolysis, may have interfered in the odd hydrogen cycle leading to liquid phase production of H_2O_2.

Detailed studies were also carried out on the transformations of oxidised nitrogen species in cloud (Fuzzi et al., 1997). During one of the cloud events, significant conversion of NO_x to HONO and HNO_3 in cloud was monitored, followed by degassing of HNO_3 as the cloud dissipated. In fact, degassing of HNO_3 from dissipating clouds was often monitored during other cloud events, but no conclusions could be drawn on the mechanisms by which HNO_3 was formed or absorbed into cloud droplets.

The important outcome of these field studies is that of in-cloud chemical conversion requires that both dynamical, microphysical and chemical aspects of the cloud system be taken into account in evaluating in-cloud chemical transformations.

Model investigations on aqueous phase chemistry in clouds were also performed within the subproject GLOMAC (Rodhe et al., 1997) in order to study the effect of clouds on global tropospheric chemistry (Lelieveld and Crutzen, 1991). The results of this study have shown that atmospheric gas phase chemistry cannot properly be assessed without a comprehensive knowledge of multiphase processes. In fact, clouds exert a major influence on the photochemistry of the troposphere, affecting the concentration of atmospheric radicals and ozone. This in turn affects the so-called oxidation capacity of the atmosphere which determines the chemical lifetimes of atmospheric trace substances.

(ii)   Inhomogeneities in cloud chemical composition

The investigation of the size dependence of cloud droplet chemical composition and solute concentrations was an important part of the research within the subproject GCE from the outset (Fuzzi et al., 1997), following the suggestion of Ogren and Charlson (1992) who stressed the importance of extending cloud water chemical analysis from the volume-weighted bulk level to a size-resolved one.

The differences in concentration of various chemical species in different cloud droplet size ranges were found to vary both in amount and sign among the various types of cloud, providing evidence of the strong influence of microphysical and chemical processes on the size dependence of cloud droplet chemical composition.

The results collected within the different GCE campaigns led to the formulation of the following generalised picture:

*   increasing solute concentrations with increasing size during the initial stage of a cloud, *e.g.* near the cloud base where the droplets have just formed;

*   decreasing solute concentrations with increasing diameters in aged cloud parcels which can be observed, for example, high above the cloud base in cumuliform clouds or in stratiform clouds advected to the observation point.

Further research is however needed on this issue which is of great importance for process parameterisations to be used in cloud models.

d. Precipitation and deposition

(i) High-elevation studies

The determination of the chemical deposition flux due to precipitation at high elevation sites was one of the main goals of the subproject ALPTRAC (Wagenbach *et al.*, 1997). The importance of the riming process in determining the chemical composition of precipitation was particularly studied. It was found that the degree of riming of a cloud is the predominant mechanism in determining the chemical composition of precipitating snow. In fact, high correlation coefficients (between 0.64 and 0.92) were found between the degree of riming and cloud water NH_4^+ NO_3^- and SO_4^{2-} concentration and pH. A positive linear correlation was also found between degree of riming and cloud liquid water content, LWC.

Wet deposition fluxes of NH_4^+, NO_3^- and SO_4^{2-} were also determined at high elevation sites within ALPTRAC. Although the ionic loading of the precipitation was up to a factor of five lower at high elevation with respect to lower level sites, the annual deposition flux of chemical substances tends to be higher in the alpine environment (Wagenbach *et al.*, 1997). The ionic loading of precipitation in the alpine region was found to be considerably lower during winter than in summer, as a result of precipitation originating from cleaner air masses in winter.

Snow pit studies were also performed on selected glaciers of the Alps to investigate the spatial and temporal variability of total deposition of chemical species. A west-to-east increasing deposition pattern was clearly seen throughout the period of the EUROTRAC project. On the other hand, the SO_4^{2-}/NO_3^- ratio decreases from west to east, and this is not in agreement with the current emission scenario for the precursors. No clear spatial deposition pattern was evidenced in the north-south direction. On average, calcium was one of the four major ions in the alpine snow pack, and exhibited an extremely high temporal variability. The maxima in calcium concentration were clearly associated to long-range transport of Saharan dust. In one of these cases, a complete neutralisation of the winter snow acidity inventory occurred due to the alkaline input from Saharan dust (Wagenbach *et al.*, 1997).

Long-term temporal trends of deposition in the alpine region were also derived from ice-core drillings (Fig. 3.6) The onset of a systematic increasing trend for NH_4^+ NO_3^-, SO_4^{2-} and heavy metals was determined around the year 1860. However, a dramatic increasing trend for all the above species took place in the period 1959-1969, producing high concentration peaks in the early seventies. A strong downward trend in deposition started in the middle of the seventies for SO_4^{2-} and lead, while NH_4^+ NO_3^- deposition trend was still increasing in the last decade (Wagenbach et al., 1997).

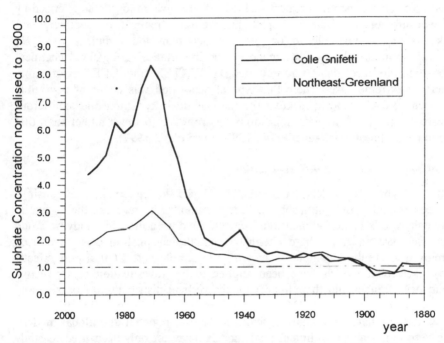

Fig. 3.6:   The twentieth century change of mean sulfate concentrations in alpine and Greenland snow. Both records are smoothed by a decadal running mean and normalised to the level at the turn of the century.

A useful directory of relevant ALPTRAC data was produced (Kromp-Kolb et al., 1993) where the information reported above can easily be retrieved.

(ii)   Cloud droplet direct deposition

The importance of cloud droplet deposition by interception of coniferous trees in the hydrological cycle at high elevations has been known for several years (Linke, 1916; Grunow, 1955). However, although it was regarded as a possible cause of forest decline in mountain areas, the input to the forests of trace substances dissolved in cloud droplets was not well quantified, due to the variability of droplet solute concentrations,. A causal link between direct cloud droplet deposition and forest decline was deduced from observations that forest decline in North America (Johnson, 1987) increased with altitude where the most likely contributory factor was interception of fog and orographic clouds containing high levels of pollutants.

The issue of direct deposition of cloud droplets was addressed within both subprojects BIATEX (Gallagher *et al.*, 1997) and GCE (Fuzzi *et al.*, 1997). As expected, the relative importance of cloud water chemical deposition with respect to precipitation deposition is very site specific, depending on fog/cloud occurrence, fog/cloud chemical composition and land use. The turbulent component dominates the sedimentation component at high elevation sites where wind speeds are generally greater. At lowland sites, where fogs, characterised by a lower turbulence and larger droplet size, are more frequent, the sedimentation flux may be comparable to or even greater than the turbulent flux. No generalisation can therefore be made on the importance of fog/cloud chemical deposition; in fact, data were reported by BIATEX and GCE investigators reporting extreme cases where this type of deposition was either of negligible importance, even in forested areas, or the chemical deposition onto coniferous forests due to cloud droplets impaction was from 3 to 6 times higher than that due to precipitation (Gallagher *et al.*, 1997; Fuzzi *et al.*, 1997).

e.   Cloud modelling at different scales

Different types of models have been developed to investigate relationships between clouds and atmospheric composition. As regards their spatial dimensions, box or zero-dimensional models have been used for studying cloud processes like aqueous phase chemistry; one-dimensional models have been employed to combine complex chemistry and simple vertical transport effects; two-dimensional models have been applied to the investigation of convective cloud development and their impact on the vertical distribution of passive and chemically reactive tracers. Finally, three-dimensional large scale phenomena have been simulated and analysed with the help of regional and global models. The zero-, one- and two-dimensional models have not only been used to study isolated or a small number of interrelated processes in clouds. They have also been applied for testing and improving parameterisations of cloud processes as implemented in computationally demanding three-dimensional chemical transport and meteorological models. In this way episodic and budget studies of ozone in the troposphere focusing on the sensitivity to clouds has become possible.

(i)   Process modelling

An advanced explicit aqueous-phase mechanism has been developed (Flossmann *et al.*, 1995; Liu *et al.*, 1997) to replace simpler approaches in three-dimensional chemistry transport models, *e.g.* in EURAD (European Air Pollution Dispersion model system; Ebel *et al.*, 1997). The mechanism includes detailed sulfur, nitrogen and transition metal chemistry. Gas-phase and aqueous-phase reactions are coupled to mass transfer processes between the gas and liquid phase.

Using the column version of the EURAD Chemical Transport Model (1-D EURAD-CTM), Mölders *et al.* (1994) and Liu *et al.* (1997) demonstrated that vertical mixing caused by clouds plays a crucial role for the modification of vertical and temporal changes of concentrations of trace constituents. While single events may show small effects, repeated processing of contaminated air by clouds, which often happens, may lead to strong changes, *e.g.* enhanced decreases of ozone, in such models.

Clouds strongly influence the solar radiation flux and thus the photolysis rates. To understand and simulate this phenomenon more thoroughly, the one-dimensional model STAR (System for Transfer of Atmospheric Radiation; Ruggaber *et al.* 1994, 1997) has been developed. It allows one to study the influence of various atmospheric parameters, including aerosols and their characteristics, on actinic fluxes. Chemical reactions within droplets can be simulated in addition to chemical transformation processes in the interstitial medium. Applications of the model to fog revealed increases of sulfate production, up to a factor of about 2, when compared with situations without fog. Another important process which can be studied to greater depth in such models is the wet and moist deposition of gases and particles.

The need of better understanding of the role of mixing by convective clouds on the interaction of the atmospheric boundary layer and the free troposphere led to the application of two-dimensional convective cloud models with an emphasis on microphysical processes such as condensation, evaporation, freezing, collection *etc.* (Laube *et al.*, 1991). In addition, a simplified gas and aqueous-phase chemical module was included by Chaumerliac and Cautenet (1993). The effect of accumulation of polluted air from the planetary boundary layer near the tropopause has also been addressed. There are indications that appreciable amounts of NO_x-rich air from below can reach the upper troposphere in areas with strong convection.

(ii)   Cloud type and distribution

Whereas global chemical transport models (CTM) in EUROTRAC preferably incorporated empirical models of cloud type and distribution, episodic regional CTMs usually generate the clouds internally and self consistently employing one of the many available cloud parameterisations. The clouds can be of convective and stratiform nature. However the treatment of cirrus clouds and their impact on atmospheric chemistry through modifications of photolysis rates on the one hand and heterogeneous reactions in the tropopause region attracted little attention in EUROTRAC. This aspect is of particular importance for treatment of aircraft emissions and may play a larger role in future.

(iii)   Sensitivity studies and model results

Numerical calculations have been carried out

*   to test the plausibility of models, modules and parameterisations for the treatment of cloud effects,

*   to explore the sensitivity of the contaminated atmosphere to cloud processes, and

*   to perform episodic simulations of realistic episodes or long-term runs of climatological and/or anthropogenic trend scenarios.

These studies have mainly been conducted employing three-dimensional models though box and column models also represent convenient tools for sensitivity experiments. There have been some simulations using column and three-dimensional models in combination, an approach chosen to overcome the limitations of available computational resources.

Such an approach was chosen by Liu et al. (1997) controlling the environment of a column (1D EURAD-CTM) with meteorological and chemical output from the original three-dimensional version of the EURAD system. In this way the effects of clouds due to mixing, actinic flux modifications, aqueous-phase chemistry and wet deposition could be studied separately. The results point towards radiation and mixing effects as the most important processes causing changes of chemical composition and transformation in clouds and their neighbourhood.

Another combined application of a 1-D model designed for simulations enabled a first approach to three-dimensional modelling of fog chemistry effects (Tippke, 1997) showing that increased sulfate production takes place on a larger scale. This is in agreement with conclusions by Forkel et al. (1997) using an elaborated version of a 1-D fog chemistry model.

The modification of ozone concentrations through clouds has been studied employing regional as well as global CTMs in EUROTRAC. Lelieveld and Crutzen (1991) report strong liquid water effects, i.e. ozone reductions, after implementing cloud chemistry in the MOGUNTIA model (Zimmermann 1988). Using the LOTOS model Builtjes and Matthijsen (1997) find for European conditions isolated areas with reductions up to 20 % due to cloud cover. Yet on the average over a larger domain the reduction amounts only up to about 5 %.

Mölders et al. (1994) have pointed out the need for reliable simulation of air mass mixing due to clouds. Including the effect of cloud mixing the episodic net result may even be an increase of ozone when averages over a larger area are taken (Memmesheimer et al. 1997). This is demonstrated in Fig. 3.7 for the main part of Europe during a period of high ozone concentrations. Obviously, the complex case of cloud effects on ozone deserves further investigations on regional and global scales and for episodic and climatic effects.

An apparent problem of standard chemical mechanisms is the over prediction of SO_2 paralleled by an under prediction of sulfate. Dennis et al. (1993) demonstrated that this shortcoming is sensitive to the parameterisation of clouds in a chemical transport model. Hass et al. (1993) partly confirmed this finding using the EURAD system, yet they also pointed to the problem that clouds do not always and not completely cure the deviation of observed and modelled sulfur components. This seems to indicate problems of emission data and emission modelling as confirmed by Langmann and Graf (1997).

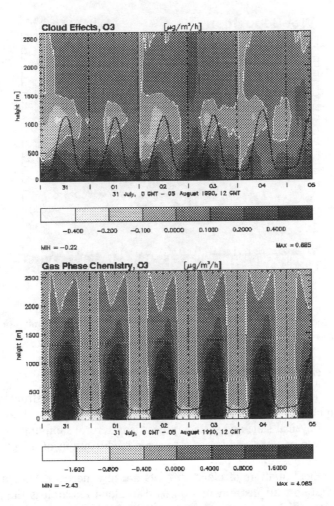

Fig. 3.7:   Contribution of cloud processes (upper panel) to the ozone budget of the lower troposphere over Europe in comparison with the contribution from gas phase chemistry (lower panel). Domain averages in μg m^{-3}h^{-1} are shown as a function of height (in m, vertical axis) and time (horizontal axis) for the Central European photosmog episode from 31st July 1990, 00 GMT, to 5th August 1990, 12 GMT. The continuous line represents the estimate of the mean height of the atmospheric boundary layer. The simulation was performed with the EURAD system.

The sensitivity of nitrogen deposition to the presence of precipitating clouds is demonstrated in Fig. 3.8 exhibiting results from a simulation of a wet episode in 1986. Wet deposition exceeds dry deposition to the North Sea by a factor of about 4 in this case.

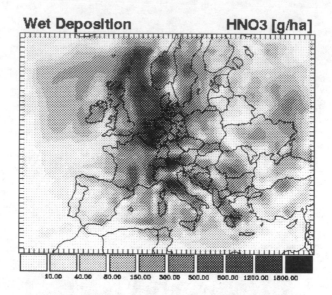

Fig. 3.8:   Wet deposition of HNO$_3$ in units of g(HNO$_3$)/ha accumulated in Europe during the episode from 23rd April, 00 GMT, to 7th May 1986, 24 GMT. Simulation with the three-dimensional air quality model system EURAD.

f.   Instrument development for cloud studies

In order to improve the knowledge of cloud properties, instrumentation is required which is able to provide size-dependent aerosol chemical composition, morphology, physical state and surface properties. Aerosol sampling for chemical analysis needs to be improved in order to avoid modifications of the physical and chemical properties of the particles during sampling. Instruments for size-resolved chemical analysis of cloud droplets are also needed. Also, a common drawback of almost all instrumentation used in cloud research is the poor time resolution at which measurements in the real atmosphere are presently performed, as compared to the characteristic times of cloud processes.

The EUROTRAC subprojects GCE and ALPTRAC were particularly active in the effort of developing and testing new instrumentation and techniques for cloud studies (Fuzzi *et al.*, 1997; Wagenbach *et al.*, 1997). These activities are briefly summarised below.

* Development and testing of collectors for both bulk and size-segregated cloud droplet sampling;

* Intercomparison of liquid water content measuring techniques;

* Development and testing of the Droplet Aerosol Analyser (DAA, Fig. 3.9) which provides the relation between the size of a cloud droplet and that of its residue after evaporation (Martinsson, 1996);

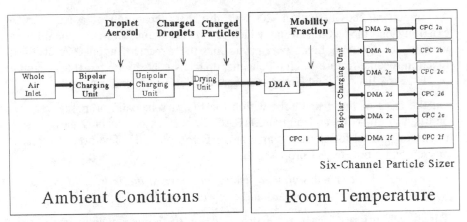

Fig.3.9:   Block diagram of the Droplet Aerosol Analyser (DAA). The droplets collected from the atmosphere are introduced in a charging unit, where each droplet is electrically charged according to its size. The water is then evaporated by diffusion drying, and a residue is obtained from each droplet which carries a charge determined by the size of the original droplet. This original size is then determined by electrostatic spectrometry. The residues are then re-charged by a bipolar charging unit and classified a second time according to their actual size. Using these two measurements, the size of the residue and of the associated droplet can be determined.

* Development of the epiphaniometer, a sensitive low consumption instrument for aerosol concentration measurement (Gäggler *et al.*, 1989);

* Development and testing of collectors for supercooled cloud water droplets

* Development of a measuring station for air chemistry and nivological studies specifically designed to work unattended in harsh meteorological conditions.

3.3   The Cloud Group, the integration of competencies required for cloud research and the assessment of future research needs

The Cloud Group (CG) was established two years before the end of EUROTRAC and consisted of representatives of the all subprojects engaged in cloud-related studies: GCE, HALIPP, ALPTRAC, EUMAC, GLOMAC. The aim of the CG was

* to exchange experiences and ideas among scientists involved in laboratory, field and modelling studies,

* to integrate the different competencies in terms of the work already performed within different subprojects, and

* to start planning for future activities in cloud research, taking up the challenge of interdisciplinary collaboration involving laboratory and field studies, instrument development and modelling.

An important activity within CG was also the preparation of a summary of existing cloud models in Europe, which reports the main features of the different models, the required input parameters and the expected outputs. A chemical reaction scheme for use in cloud chemical models is also reported (Flossmann *et al.*, 1995). A list of Cloud Group members is given in Appendix A, table 7.

On the basis of the discussions within the CG, a schematic proposal was then outlined towards the end of EUROTRAC for the discussion of a new possible integrated project for cloud research at the European scale. The overall objective of such future project was proposed to

> *"... provide parametrisations of cloud processes at the individual cloud scale for application in mesoscale and global models ..."*

Three high priority issues were proposed for the project which are of great novelty and importance and where the lack of knowledge is particularly high:

(a) *The organic component of CCN and the organic chemistry in cloud droplets.* This issue addresses the scientific questions on how the organic component of atmospheric aerosol affects the ability of atmospheric particles to act as CCN and on how the organic components of CCN affect the chemistry of cloud droplets.

(b) *The ice chemistry in clouds and its connection to precipitation.* This issue addresses chemical processes in super-cooled and iced clouds. Since this type of cloud is closely related to the formation of precipitation over Europe, the initialisation of precipitation should be studied in relation to the chemical characteristics and processes in icing clouds.

(c) *The photochemistry within clouds.* This issue addresses the influence of liquid water clouds on photochemical processes taking place in the troposphere. The budget of photo-oxidants, their chemical reactions and multiphase products, and their effect on major biogeochemical cycles should be studied.

It was recommended that the project should include (Fig. 3.10):

* field experiments, both ground-based and airborne organised under the guidance of careful modelling work and in order to test model results;

* models using the field-derived data as input parameters;

* laboratory work on heterogeneous atmospheric processes to provide modellers with carefully tested reaction schemes

* model-derived results enabling laboratory scientists to find existing gaps in knowledge which need to be investigated;

* focus on processes and chemical species which are expected to be important on the basis of field observations;

* development of tools for laboratory and field investigations on cloud processes.

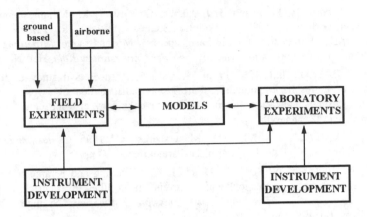

Fig. 3.10:   The organisational structure of a proposed future project for interdisciplinary cloud research.

In spite of the wealth of results and new understanding obtained within the different EUROTRAC subprojects, it is unfortunately true that an even more complicated picture has emerged compared with our knowledge at the beginning. It is clear that much remains to be done in the field of cloud/aerosol interaction and cloud chemistry before we can be satisfied that clouds and aerosols are adequately understood and that their processes can be properly accounted for in the chemical transport models.

3.4   References

Builtjes P., J. Matthijsen; 1997, Regional-scale modelling of cloud effects on atmospheric constituents, in: A. Ebel, R. Friedrich, H. Rodhe (eds), *Tropospheric Modelling and Emission Estimation,* Springer Verlag, Heidelberg, pp. 147–155.

Chaumerliac N., S. Cautenet; 1993, The redistribution of ozone precursors by convective clouds, *EUROTRAC Annual Rep. 1993,* Part 5, EUROTRAC ISS, Garmisch-Partenkirchen, pp. 19–25.

Choularton T.W., R.N. Colvile, K.N. Bower, M.W. Gallagher, M. Wells, K.M. Beswick, B.G. Arends, J.J. Möls, G.P.A. Kos, S. Fuzzi, J.A. Lind, G. Orsi, M.C. Facchini, P. Laj, R. Gieray, P. Wieser, T. Engelhardt, A. Berner, C. Kruisz, D. Möller, K. Acker, W. Wieprecht, J. Lüttke, K. Levsen, M. Bizjak, H-C. Hansson, S-I. Cederfelt, G. Frank, B. Mentes, B. Martinsson, D. Orsini, B. Svenningsson, E. Swietlicki, A. Wiedensohler, K.J. Noone, S. Pahl, P. Winkler, E. Seyffer, G. Helas, W. Jaeschke, H.W. Georgii, W. Wobrock, M. Preiss, R. Maser, D. Schell, G. Dollard, B. Jones, T. Davies, D.L. Sedlak, M.M. David, M. Wendisch, J.N. Cape, K.J. Hargreaves, M.A. Sutton, R.L. Storeton-West, D. Fowler, A. Hallberg, R.M. Harrison, J.D. Peak; 1997, The Great Dun Fell Cloud Experiment 1993: an overview. *Atmos. Environ.* **31**, 2393–2405.

Dennis R.L., J.N. McHenry, W.R. Barchet, F.S. Binkowski, D.W. Bryun; 1993, Correcting RADM's sulfate underprediction: discovery and correction of model errors and testing the corrections through comparisons against field data, *Atmos. Environ.* **27A**, 975–997.

Ebel A., H. Elbern, H. Feldmann, H.J. Jakobs, C. Kessler, M. Memmesheimer, A. Oberreuter, G. Piekorz; 1997, Air Pollution Studies with the EURAD Model System (3): EURAD - European Air Pollution Dispersion Model System, *Mitteilungen aus dem Institut für Geophysik und Meteorologie der Universität zu Köln*, Heft Nr. 120.

Flossmann A.I., W.D. Hall, H.R. Pruppacher; 1985, A theoretical study of the wet removal of atmospheric pollutants. Part I: the redistribution of aerosol particles captured through nucleation and impaction scavenging by growing cloud droplets. *J. Atmos. Sci.* **42**, 583–606.

Flossmann, A.I., Cvitaš, T., Möller, D., Mauersberger, G.; 1995, *Clouds: models and mechanisms*, EUROTRAC ISS, Garmisch-Partenkirchen, 93 pp.

Forkel R., A. Ruggaber, W. Seidl, R. Dlugi; 1997, Modelling and parameterisation of chemical reactions in connections with fog events, in: A. Ebel, R. Friedrich, H. Rodhe (eds), *Tropospheric Modelling and Emission Estimation*, Springer Verlag, Heidelberg, pp. 139–145.

Fuzzi, S., Facchini, M. C., Orsi, G., Lind, J. A., Wobrock, W., Kessel, M., Maser, R., Jaeschke, W., Enderle, K. H., Arends, B. G., Berner, A., Solly, I., Kruisz, C., Reischl, G., Pahl, S., Kaminski, U., Winkler, P., Ogren, J. A., Noone, K. J., Hallberg, A., Fierlinger-Oberlinninger, H., Puxbaum, H., Marzorati, A., Hansson, H. C., Wiedensohler, A., Svenningsson, I. B., Martinsson, B. G., Schell, D., Georgii, H. W.; 1992, The Po Valley Fog Experiment 1989: an overview. *Tellus* **44B**, 448–468.

Fuzzi, S., Hansson, H-C., Choularton, T.W., Jaeschke, W., Wobrock, W., Schell, D.; 1997, European contribution to advances in the study of cloud multiphase processes. in: Fuzzi, S., Wagenbach, D. (eds), *Cloud Multiphase Processes and High Alpine Air and Snow Chemistry*, Springer Verlag, Heidelberg, pp. 3–59.

Gäggler, H.W., Baltensperger, U., Emmenegger, M., Jost, D.T., Schmidt-Ott, A., Haller, P., Hoffmann, M.; 1989, The epiphaniometer, a new device for continuous aerosol monitoring. *J. Aerosol Sci.* **20**, 557–564.

Gallagher, M., Fontan, J., Wyers, P., Ruijgrok, W., Duyzer, J., Hummelshøj, P., Pilegaard, K., Fowler, D.; 1997, Atmospheric particles and their interactions with natural surfaces. in: Slanina, S. (ed) *Biosphere-Atmosphere Exchange of Pollutants and Trace Substances*, Springer Verlag, Heidelberg, pp. 45–92.

Grunow J.; 1955, Der Niederschlag im Bergwald-Niederschlagszurückhaltung und Nebelniederschlag, *Forstwiss. Cbl.* **74**, 21–36.

Hass H., A. Ebel, H. Feldmann, H.J. Jakobs, M. Memmesheimer; 1993, Evaluation studies with a regional chemical transport model (EURAD) using air quality data from the EMEP monitoring network, *Atmos. Environ.* **27A**, 867–887.

Johnson, A.H.; 1987, Deterioration of red spruce in the northern Appalachian Mountains. in: Utchinson, T.C., Meema, K.M. (eds) *Effects of Atmospheric Pollutants on Forests, Wetlands and Agricultural Ecosystems, NATO ASI series*, Vol. 916, Springer Verlag, Heidelberg, pp. 83–99.

Kromp-Kolb, H., Schöner, W., Seibert, P.; 1993, *ALPTRAC data catalogue*, EUROTRAC ISS, Garmisch-Partenkirchen, pp. 137.

Langmann B., H.F. Graf; Another meteorological driver (HIRHAM) for the EURAD Chemistry Transport Model (CTM): validation and sensitivity studies with the coupled system, in: A. Ebel, R. Friedrich, H. Rodhe (eds), *Tropospheric Modelling and Emission Estimation*, 1997, Springer, pp. 46–51.

Langner, J., Rodhe, H.; 1991, A global three-dimensional model of the tropospheric sulfur cycle. *J. Atmos. Chem.* **13**, 255–263.

Laube M., A. Ebel, E. Kälicke, P. Scheidgen; 1991, Vertical transport and scavenging of pollutants by convective clouds, in: H. van Dop, D.G. Stein (eds), *Air Pollution Modelling and Its Application,* Plenum Press, New York, pp. 515–520.

Lelieveld J., P.J. Crutzen; 1991, The role of clouds in tropospheric photochemistry. *J. Atmos. Chem.* **12**, 229–267.

Linke F.; 1916, Niederschlagsmessungen unter Bäumen. *Meteorol. Z.* **33**, 140–141.

Liu X., G. Mauersberger, D. Möller; 1997, The effects of cloud processes on the tropospheric photochemistry: an improvement of the EURAD model with a coupled gaseous and aqueous chemical mechanism, *Atmos. Environ.* **31**, 3119–3135.

Martinsson, B.; 1996, Physical basis for a droplet aerosol analysing method. *J. Aerosol Sci.* **27**, 997–1013.

Mölders N., H. Hass, H.J. Jakobs, M. Laube, A. Ebel; 1994, Some effects of different cloud parameterizations in a mesoscale model and a chemistry transport model, *J. Appl. Met.* **33S**, 845.

Ogren J.A., R.J. Charlson; 1992, Implications for models and measurements of chemical inhomogeneities among cloud droplets. *Tellus* **44B**, 208–225.

Pruppacher H.R., J.D. Klett; 1978, *Microphysics of clouds and precipitation.* Reidel, Dordrecht.

Rodhe, H., Crutzen, P., Kanakidou, M, Kelder, H., Lelieveld, J., Raes, F., Roeckner, E.; 1997, An overview of global atmospheric chemistry modelling. in: Ebel, A., Friedrich, R., Rodhe, H. (eds) *Tropospheric Modelling and Emission Estimation,* Springer Verlag, Heidelberg, pp. 361–372.

Ruggaber A., R. Dlugi, T. Nakajima; 1994, Modelling of radiation quantities and photolysis frequencies in the troposphere, *J. Atmos. Chem.* **18**, 170–210.

Ruggaber A., R. Dlugi, A. Bott, R. Forkel, H. Herrmann, H.-W. Jacobi; 1997, Modelling of radiation quantities and photolysis frequencies in the aqueous phase in the troposphere, *Atmos. Environ.* **31**, 3137–315.

Schwartz S.E.; 1986, Mass-transport consideration pertinent to aqueous phase reactions of gases in liquid water clouds. in: W. Jaeschke (ed) *Chemistry of the Atmospheric Multiphase Systems,* Springer Verlag, Heildelberg, pp. 451–471.

Tippke J.; 1997, Die Behandlung von Nebelereignissen in einem mesoskaligen Chemietransportmodell, *Mitteilungen aus dem Institut für Geophysik und Meteorologie der Universität zu Köln,* A. Ebel, M. Kerschgens, F.M. Neubauer, P. Speth (eds) Heft Nr. 117.

Wagenbach, D., Gäggler, H., Kromp-Kolb, H., Kuhn, M., Puxbaum, H.; 1997, High alpine air and snow chemistry. in: Fuzzi, S., Wagenbach, D. (eds), *Cloud Multiphase Processes and High Alpine Air and Snow Chemistry,* Springer Verlag, Heidelberg, pp. 165–199.

Warneck P.; 1986, The equilibrium distribution of atmospheric gases between the two phases of liquid water clouds. in: W. Jaeschke (ed), *Chemistry of Multiphase Atmospheric Systems,* Springer Verlag, Heidelberg.

Warneck, P., Mirabel, P., Salmon, G.A., van Eldik, R., Vinckier, C., Wannowius, K.J., Zetzsch, C. (1996) Review of the activities and achievements of the EUROTRAC subproject HALIPP. in: Warneck, P. (ed), *Heterogeneous and Liquid-Phase Processes*, Springer Verlag, Heidelberg, pp. 7–74.

Wobrock, W., Schell, D., Maser, R., Jaeschke, W., Georgii, H. W., Wieprecht, W., Arends, B. G., Möls, J. J., Kos, G. P. A., Fuzzi, S., Facchini, M. C., Orsi, G., Berner, A., Solly, I., Kruisz, C., Svenningsson, I. B., Wiedensohler, A., Hansson, H. C., Ogren, J. A., Noone, K. J., Hallberg, A., Pahl, S., Schneider, T., Winkler, P., Winiwarter, W., Colvile, R. N., Choularton, T.W., Flossmann, A. I., Borrmann S.; 1994, The Kleiner Feldberg Cloud Experiment 1990. An overview. *J. Atmos. Chem.* **19**, 3–35.

Zimmermann P.H.; 1988, MOGUNTIA: a handy global tracer model, in: H. van Dop (ed), *Air Pollution Modelling and its Applications*, VI NATO/CCMWS, Plenum, New York.

Chapter 4

Surface Exchange

Niels Otto Jensen[1] and Casimiro A. Pio[2]

[1]Risö National Laboratory, Postbox 49, DK-4000 Roskilde, Denmark
[2]Universidade de Aveiro, Departamento de Ambiente e Ordenamento,
P-3810 Aveiro, Portugal

4.1 Introduction

With surface exchange we mainly think of the dry flux of various chemical compounds between the terrestrial or aquatic surfaces of the earth and the overlying atmospheric turbulent boundary layer.

The fluxes can be upwards, *i.e.* from the surface to the air if the surface is the source of the compound. Examples are volatile organic compounds (VOCs) such as α-pinene emitted from plants or dimethylsulfide (DMS) deriving from algae living in the surface waters. The fluxes can also include particulate matter (aerosols) from either soil dust or sea spray.

The dry flux can also be downwards *i.e.* from the air to the surface. In this case the exchange is normally called deposition. Examples are air pollutants emitted from anthropogenic sources and then depositied from the air on the earth's surface. It could of course also be trace substances emitted from natural sources, or indeed compounds that are not emitted but created by air chemistry, such as ozone.

In EUROTRAC experimental data on surface exchange were obtained mainly in the sub-projects ASE and BIATEX plus some in TRACT (see volumes 4 and 9 in this series; Slanina *et al.*, 1997; Larsen *et al.*, 2000).

The primary effect of deposition is a cleansing of the air. The effect on the surface (*e.g.* plant communities or surface waters) can be harmful as, for example, in the case of ozone which is toxic to plants. In the case of deposition of nitrogen compounds or other nutrients the effect can be both beneficial and harmful. They can provide extra plant nutrients but, for example in estuarine areas, the extra nutrients can cause eutrophication, large growth of algae populations, and subsequent oxygen deficiency in the bottom waters when the dead algal material decomposes.

a.   Wet deposition

When discussing deposition it is necessary, in addition to the turbulent dry flux to the surface, to include the amount that is deposited by precipitation. The significance of wet deposition depends on the climatic situation, but at Northern European temperate latitudes it provides about half of the atmospheric input of nitrogen to natural ecosystems.

(i)   Wet-deposition methods to determine dry-deposition fluxes

Traditionally the "wet input" has been determined extensively, mainly because the collection of rainwater is quite simple. Indeed, because more complicated techniques are required for dry deposition measurements, wet deposition techniques have also been used for estimating dry deposition to forest canopies by putting rain water collectors on the forest floor. The excess concentration of chemical species in these collectors, compared with the concentration in collectors placed outside or above the forest is then taken to be the result of wash-off of material that has been dry deposited between rain events. The mehtod assumes that everything, which is dry deposited, is also washed off which is unfortunately not true for compounds absorbed in the leaf stomata. It also assumes that everything which is washed off, derives from dry deposition, which is not always the case because material on the outside of the leaves' cuticula can derive from the plant's interior.

With the advance of technology, we would recommend that this very approximate technique of determining dry deposition fluxes is abandoned all together. The errors are so large that they even may lead to wrong conclusions, apart from the fact that it only works for this (small) part of the surface that consists of permanent tall vegetation.

(ii)   Determination of POPs by wet-deposition methods

This is not to diminish the role of wet deposition measurements as such. The wet input is large (at temperate latitudes it is about 50 % as mentioned above) and the sensitive analyses performed on collected rain water reveals deposition of "strange" compounds such as PCBs (polychlorinated biphenyls), chlorinated pesticides like Lindane (γ-isomer of hexachlorocyclohexane, or γ-HCH) and other persistent organic compounds (POPs).

The depositions are very low and cannot be measured with dry-flux techniques. However, the concentration of Lindane in rainwater has been determined at a number of Danish sites (remote from local sources) to be 27–122 ng/L during Spring and 7 ng/L in the remaining part of the year (Cleemann et al., 1995). The variation from site to site was not significant. Hence this input must be assumed to originate from long-range transported material. The annual average including all sites was 15 ng/L corresponding to a deposition flux of 12 μg/m^2 per year. On a European scale these are not particularly high numbers. Thus Galassi (1987) reports concentrations up to 400 ng/L in rain water collected near Milan, Italy.

Nevertheless, 12 μg/m^2 extrapolated to the area of Denmark, including the coastal and inner Danish waters (1.25×10^5 km^2) gives 1.5 ton per year compared to annual sales of Lindane of between 8 and 15 tons in the corresponding years. To this figure we should probably add an additional portion resulting from dry deposition. Thus the input of Lindane to natural ecosystems is significant compared to the amounts used in the agricultural sector.

So this technique is of utmost importance in determining inputs to the ecosystems of less usual compounds. It also possesses a unique possibility of giving a longer term perspective on deposition and thereby an estimate of the then prevalent air concentrations of various atmospheric compounds. Here we think of the historical records preserved in for example the alpine glaciers. This aspect is dealt with in volume 5 in this series (Fuzzi *et al.*, 1997), and particularly in Gäggeler *et al.* (1997)

b.   Plant physiology and dry deposition

If we now revert to dry surface exchange and concentrate on vegetated terrestrial surfaces, then it appears to be essential to consider the physiological state of the plant cover, at least for those compounds that are emitted or absorbed through the stomata of the plants. Thus the regulation of the emission/uptake is determined by the photosynthetic state of the plants. In the case where the emission/uptake is from the external surfaces of the leaves (cuticula) this plant physiological control may not be relevant. In the case of ozone it is both; while about half of the total dry flux depends on stomatal activity, the other half deposits on the outer surfaces or reacts with NO emitted from the soil (Pilegaard *et al.*, 1997), see section 4.2 below.

When the flux is dependent on stomatal activity, it is essential to have an indication of this in order to be able to parameterise the dry flux. Two significant indicators are the CO_2 flux and the water vapour flux which are both tracers of the plant activity. Large CO_2 downward fluxes and large evaporative losses generally go together to indicate that the stomata are open and thus favour deposition of certain compounds. However, a one to one correlation is not possible since CO_2 results from microbial activity and water vapour exchange also occur with the soil itself.

For a full parametrisation of the influence of plant physiology on the deposition process it is necessary to include a sub-model describing the plant's activity (CO_2 and water vapour exchange) as a function of external parameters such as solar radiation, temperature, air humidity, soil water content, season, *etc.*

4.2   Ozone deposition; the effect of NO emission from soil bacteria

Bacteria in the soil produce a number of intermediate compounds in the nitrification and denitrification processes. One is N_2O, although not very reactive and therefore not of interest in tropospheric chemistry, is a greenhouse gas and therefore of interest for other reasons. However, on the way to this compound some N remains as NO and is emitted as such to the atmosphere. NO is an active compound reacting with O_3 to make NO_2; chapter 2, section 2.2.a. In urban areas this is hardly of importance because of most of the NO is emitted from combustion or by traffic, but in rural areas, especially agricultural areas with application of large amounts of fertiliser, the ozone budget may be altered appreciably (Pilegaard *et al.*, 1997). The size of the NO input to the atmosphere is estimated to be 5–10 % of the anthropogenic emissions (Fowler *et al.*, 1997, p. 144) but the estimate is quite uncertain. However it is evident that, in air pollution studies on the regional scale, this effect cannot be ignored and that, in air-chemistry transport models, it is necessary to include land-use maps and data about the type of utilisation of the individual areas.

4.3   Emission of volatile organic compounds from plants

Vegetation synthesises a variety of hydrocarbon compounds that are not used in the plants own metabolism. The reason for this expenditure is not yet fully understood (Harborn, 1988). A fraction of the more volatile organic compounds (VOCs) produced is lost by evaporation and the ensuing emissions intervene in photochemical processes in the atmosphere.

Most of the VOC emission by plants passes through the stomatal openings of the leaves. Usually two main groups of VOCs are considered in vegetation emissions: isoprene and monoterpenes. Isoprene is mainly synthesised and emitted by broad-leaved-vegetation species while monoterpenes are the most frequent emission product from conifers.

a.   Isoprene

Generally it is considered that isoprene is emitted exclusively through the plant stomata and that emission happens immediately, or in a short period, after isoprene synthesis within plant cells. Since there is no significant pool of isoprene inside the plant, emission rates are closely associated with the isoprene-biosynthesis rate and by variability in the resistance to diffusion to the ambient atmosphere associated with the degree of stomatal opening. As a consequence light intensity reaching the plant and leaf temperature are the two main environmental variables affecting and controlling isoprene emission fluxes on a short-term basis (hours–weeks). Isoprene emission fluxes vary rapidly with the daily solar cycle and decrease to insignificant values during night after the rapid exhaustion of reserves inside the plant. However sometimes the emission of isoprene has a delay of more than an hour depending on the environmental conditions. This type of behaviour suggests that, in some species such as *Quercus frainetto*, the biosynthesis of isoprene is not switched on until a primary pool of precursors is filled (Steinbrecher *et al.*, 1997a).

Algorithms have been developed and used with success to describe isoprene emission fluxes from vegetation. Isoprene emission is correlated with temperature and photosynthetically active radiation (PAR), with saturation at high temperatures and radiation intensities, following a pattern described by the Guenther model (Guenther *et al.*, 1993).

b.   Monoterpenes

A large number of monoterpene compounds have been identified in emissions from vegetation, mainly conifers. The more common ones are α-pinene, β-pinene, limonene, myrcene, Δ^3-carene, camphene, 1,8-cineole and β-phelandrene (Guenther *et al.*, 1994). Monoterpene emissions are usually considered to be dependent mainly on temperature, and independent of light on the short-time basis. A large pool of monoterpene compounds is always present inside the trunk, branches and leaves, in quantities that depend only on the slow variation in the balance between loss from the plant and biosynthesis resulting from photosynthetic activity. Furthermore for these species the effect of light on monoterpene emission rates through the control of stomatal opening is also minor because the main resistance to monoterpene diffusion is in transport from terpene pools to the stomatal cavities, (Lerdau, 1991). The transfer rate is dependent on the monoterpene concentration in the pool and monoterpene characteristics such as diffusivity, solubility and vapour pressure. The emission flux of monoterpenes is often expressed as

$$F = F_0 \exp[c(T - T_0)]$$ (1)

referred to as the Guenther expression. This is equivalent to

$$\ln F = F_0 + c(T - T_0)$$ (2)

where F is the flux, T is the absolute leaf temperature, subscript zero refers to a reference state and c is an empirical dimensional constant. In other cases time expressions of the type

$$\ln F = \alpha + \frac{b}{T}$$ (3)

are given where a and b are empirical constants. These expressions are all variations on the Clausius-Clapeyron equation

$$\frac{e}{e_0} = \exp\left[\frac{L}{R}(\frac{1}{T_0} - \frac{1}{T})\right]$$ (4)

which describes the variation with temperature of the saturated vapour pressure e of a volatile compound (which is proportional to the concentration of the compound in the liquid phase inside the plant via Henry's law). L is the compound's heat of evaporation, R is the universal gas constant and subscript zero again indicates some reference state. Assuming saturated conditions inside the leaf cavities and a partial pressure in the ambient air of near zero, then the flux is

$$F = \frac{1}{r}(e - 0) = \frac{e_0}{r} \exp\left[\alpha \frac{T - T_0}{TT_0}\right]$$ (5)

where r is the resistance to transfer and $\alpha = L/R$. The quantity e_0/r can be identified as F_0.

In general this equation can be written:

$$\ln F = (\ln F_0 + \frac{\alpha}{T_0}) - \frac{\alpha}{T} \qquad (6)$$

from where the parameters a and b in equation (3) can be identified. If T is not too different from T_0, the parameter c in eqs (1) and (2) can be identified as $c = \alpha/T^2_0$. If furthermore $(T - T_0 \ll T^2_0)$, a linear expression can be obtained by expanding the exponential in eq. (5) (exp [ε] ≈ 1 + ε), viz.:

$$F = F_0[1 + \beta(T - T_0)] \qquad (7)$$

In contrast to isoprene, monoterpene emissions also occur at night, although the rate isusually lower than during the day because of the lower nocturnal ambient temperature. As a result of the different emission behaviour of isoprene and monoterpenes, the concentrations of isoprene in the ambient air above forested areas decrease significantly at night while the relative concentration of monoterpene is frequently higher during night-time periods (Torres *et al.*, 1997).

c.   *Variations from the general pattern*

As previously mentioned, the general pattern of biogenic VOC emissions from vegetation do not represent the specific behaviour of every vegetation species in Europe and important deviations have been observed. Using laboratory screening to examine a large number of vegetation species, characteristic for northern as well as southern Europe, several studies have revealed that while most of the deciduous trees investigated always emit *traces* of monoterpene, a small percentage of deciduous tree species emit isoprene in *large* amounts (Steinbrecher *et al.*, 1997a; Pio *et al.*, 1997; Street *et al.*, 1997). The screening tests also showed that some conifers, such as Sylvester pine, Sitka spruce and Norway spruce, emit isoprene in addition to monoterpenes. In another atypical example some evergreen oak species, such as the Holm oak and Cork oak, which are common or predominant in Mediterranean regions, were found to be important emitters of monoterpene compounds, but not to emit isoprene.

Detailed studies performed, either in the laboratory using young specimen in controlled chambers, or in the field with adult specimen in natural ambient conditions, showed repeated clear differences in VOC composition. They also showed differences in emission rates between young and adult specimens, with young plants usually emitting at much higher rates than adult specimens (Street *et al.*, 1997; Pio *et al.*, 1997). These results raise questions about the accuracy of algorithms based on laboratory experiments with young trees only.

Both the absolute and relative amounts of monoterpenes emissions (magnitude and composition) change with leaf development. For example in Norway spruce monoterpenes emissions are low before bud burst and consist mainly of limonene and 1,8-cineole. When needles are fully developed α-pinene is the main compound emitted at a much greater emission rate (Steinbrecher *et al.*, 1997b).

In several conifer species not only temperature but also solar radiation influences monoterpene emission rates, again in contrast to the general picture given above. Norway spruce twigs located in the shadow crown may emit only at one tenth of the rate of similar twigs located in the sun at the same leaf temperatures. Experiments with ^{13}C tracer did not show any clear direct link between the monoterpene pool and monoterpenes emissions, indicating that, for these species, emitted monoterpenes originate primarily from direct photosynthetic products (Steinbrecher et al., 1997b). Emission of monoterpenes by some evergreen broad leaf species, such as the holm oak, also shows a dependency on light and temperature with emission rates obeying the Guenther model (Kesselmeier et al., 1996).

For conifer species emitting both monoterpenes and isoprene, the isoprene emission rate follows a daily emission pattern similar to that observed for emissions of isoprene from broad leaf plant species, with a strong dependence of emissions on solar radiation intensity and a smaller dependence on temperature. A similar behaviour was observed for eucalyptus trees, which predominantly emit isoprene at a strong rate during the day but also emit monoterpene compounds at a lower rate, day and night, with a predominant dependency on temperature (Street et al., 1997; Pio et al., 1997).

Although being predominant, leaves are not the exclusive source of biogenic VOC from vegetation. For example the cobs of Norway spruce emit eight different monoterpene compounds with a rate dependent on temperature. The contribution of cobs is only minor and constitutes about 2 % of the total plant emission. In contrast the trunks of Norway spruce are strong emitters of monoterpenes with rates that are quite variable from tree to tree. The trunks of undisturbed trees may contribute with up to 64 % to the total amount of α-pinene emitted by the spruce canopy (Steinbrecher et al., 1997b). Injury to bark resulting in resin loss dramatically increases emission rates and changes the relative amounts of different monoterpenes emitted by coniferous species. The practice of resin tapping in coniferous forests increases monoterpene emissions by more than one order of magnitude (Pio et al., 1997). Monoterpenes emissions from trunks are well correlated with temperature of bark and resin.

d.   Other compounds

Some vegetation species also release other volatile organic compounds such as sesquiterpenes and oxygenated organic species (alcohols, organic acids, aldheydes and ketones). Much less is known about the emission flux characteristics of these compounds from vegetation than of isoprene and monoterpenes. Maize, which is a low emitter of isoprene and monoterpenes, emits important quantities of formaldehyde, acetaldehyde, acetone and acrolein from leaves and cobs (Street et al., 1997). Spruce twigs release acetic and formic acid at rates which follow a typical daily pattern with maximum emissions during early afternoon (Kesselmeier et al., 1997). Experiments with spruce, beech and ash showed the existence of seasonal effects with maximum organic acid emissions during fall when leaves start discolouring.

e.   Conclusions

There have been large advances recently in our knowledge concerning the emission of VOCs by vegetation. However there is not yet a complete and precise view of the emission levels of the various compounds from the various plant species as a function of all the environmental variables controlling the emission processes. Factors such as inter-specimen variability within each species and seasonal variation, although recognised as important have not yet been fully quantified. Spatial and seasonal distribution of vegetation species and biomass density in the various European regions is still poorly known. As a consequence our present knowledge of biogenic VOC emissions to the atmosphere is still highly uncertain. Errors of as much as 500 % in the determination of emission fluxes have been estimated in US studies where biogenic emission inventories are very advanced (Simpson *et al.*, 1995; Lamb *et al.*, 1997).

4.4   Input of nitrogen and eutrophication of coastal waters

The atmospheric input of a pollutants and trace substances to the marine environment is of importance in policy development. Even in coastal areas the atmospheric input (wet and dry) can be of the same order as the riverine input, typically 30–40 % of the total load, and generally of the same order of importance regardless of the particular compound. Studies of nutrient input from the atmosphere to regional seas has become of special interest because of recent algae blooms in these waters (see for example Asman *et al.*, 1995). The knowledge of this and the processes behind has greatly improved as a result of the ASE subproject (Larsen *et al.*, 2000).

The interest in nutrient deposition derives from the environmental problems it can cause in terms of severe oxygen depletion induced by the decomposing algae. This happens in cases where the nutrient transfer is beyond what is beneficial for the aquatic ecosystem. The possibly beneficial thing about the atmospheric input is that it is in forms (like nitrate) that are readily available for biological uptake unlike organically bound nitrogen.

Other essential compounds of interest for the biological activity in the seas are completely dominated by the atmospheric input. Thus it is claimed that iron, the source of which is wind-blown dust from continental areas, is a limiting factor for the primary production in the Antarctic Ocean. It has even been hypothesised that this is a major factor in the oceanic carbon sequestration and thus of importance for the earth's climate (Martin *et al.*, 1994).

4.5   Emission from the coastal waters by spray formation

Breaking waves form spray. To a small extent some spray is emitted directly by wind action, but the most important mechanism is the entrainment of air into water. Bubbles are formed that rise to the surface and burst, ejecting droplets into the air. The larger droplets quickly gravitate out of the air again. However, the smaller droplets may stay airborne long enough to dry out and we are then left with an aerosol consisting of the material that originally was in solution in the water. For seawater this is mainly salt but the aerosol will also contain the pollutants and other trace materials present in the water.

When an air bubble rises through seawater it collects micro-organisms and other surface active materials on its surface and, when it breaks at the surface, the bubble film is thrown into the air as small droplets. Thus the composition of these droplets are not the same as in bulk seawater and enrichment factors of 500 are quite normal (Monahan and van Petten, 1989). Further enrichment of the particular substances can derive from fractionation by the living organisms before they are scavenged and thrown into the air. Recreation near the seashore may thus not be particularly healthy!

The spray production is an important process in the emission to the atmosphere of compounds dissolved in the ocean. Some emissions are of importance for the chemistry of the atmosphere: not only indirectly as a source of aerosols and condensation nuclei but also as a source of chemical compounds such as sulfates derived from biogases (*e.g.* dimethyl-sulfide, DMS). The subproject ASE has contributed significantly to the advancement of our knowledge in this field (Larsen *et al.*, 2000).

4.6   New developments in experimental techniques for flux measurements and future needs

The various methods for measuring surface exchange can roughly be divided into two categories: cuvette methods and micro-meteorological methods.

a.   Cuvette methods

With the cuvettes a particular, relatively small, part of the surface is enclosed and then the concentration in the cuvette is monitored. If the cuvette is of the closed type, then the rate of change of the concentration is a measure of the flux. If the cuvette is of the flow-through type, the concentration difference between the inflow and outflow air is a measure of the flux.

One drawback of cuvette systems is that they interfere with what is to be measured since the exchange conditions inside the cuvette are artificial. Another problem can be absorption on the chamber walls. A final uncertainty is that the measurements require up-scaling to obtain area estimates. For example, with a small cuvette mounted on a twig or branch it is necessary to assume that the compound of interest only comes from twigs or branches and not from other parts of the canopy and that the area index of the active parts can be estimated. Part of this is also a sampling problem, namely to decide how many sampling spots are necessary to get a representative typical value.

Cuvette techniques are mainly interesting for estimating fluxes out of the soil or other specific compartments of the surface, for example for estimating the relative importance for the total exchange of the various parts of the plants.

b.   Micro-meteorological methods

Micro-meteorological methods on the other hand do not in principle interfere with the emission/deposition conditions, and because of the turbulent diffusion between the upstream surface and the level of measurements, the measured flux is an area average. The area from which the measured surface exchange has taken place is known as the "footprint". Its size depends on the height of the measurements and the level of turbulence, which in turn can be expressed as a function of the surface roughness and the atmospheric stability.

There are some basic conditions that must be fulfilled for all micro-meteorological methods. First the flow and chemical fields must be stationary; for instance if the mean concentration C changes with time, the estimated flux is not equal to the required surface flux. Secondly the underlying terrain must be flat, with homogeneous plant cover in the footprint area; otherwise there will be horizontal advection, and again the estimated flux will deviate from the required surface flux. If the an emission flux is being measured, a special problem can occur in night-time stable conditions on a sloping terrain. Here catabatic flows can carry the emissions away under the micro-meteorological flux level. It is a known phenomenon for night-time CO_2-flux measurements over forests.

c.   The profile method

There are two classical methods for flux determination in micro-meteorology: the profile method and the eddy-correlation method.

The profile method in its simplest form uses the mean concentration difference between two levels. To convert this into a flux requires a model for the exchange coefficient (eddy diffusivity). It can for example be obtained from surface-layer-similarity theory using empirical functions of the Monin-Obukhov length. Alternatively it can be obtained by measuring the concentration difference of another scalar for which it is possible to determine the flux independently. Thus

$$F_1 = F_2 \frac{\Delta C_1}{\Delta C_2} \tag{8}$$

where F_1 is the flux to be determined and F_2 is the known flux, and ΔC_1, ΔC_2 are the mean concentration differences. A slightly rough but very robust estimate can be obtained by using momentum for the "known scalar". A disadvantage of the method is that it requires very long homogeneous fetches because the height of observation must be relatively large for similarity theory to be valid (one to two orders of magnitude larger than the aerodynamic surface roughness - a problem for observations over forests). Another consequence is that the footprints of the two measurement levels are different.

d.   Eddy correlation methods

The eddy-correlation method on the other hand does not suffer from any of these limitations. It requires no assumptions or empirical input but gives the flux directly. It is the preferred method. The only constraint is that the level of measurements must be sufficiently away from local roughness elements such that the mean flow streamlines are reasonably straight. However, if the scalar spectrum contains very low frequency fluctuations, the flux calculation procedure becomes very sensitive to the external alignment of the sonic (rotation of the co-ordinate system in order to make the mean vertical velocity $\overline{w}$ equal to zero). This also depends on the choice of the length of the averaging time.

This potential error can become very large if this procedure is neglected. The fluxes can become lost altogether if a levelling instrument is placed near the sonic while still neglecting the alignment procedure: the only effect of the "inclinometer" is to distort the streamlines through the anemometer more than they otherwise would be. In Fontan *et al.* (1997) it is claimed that the low frequency variations in the scalar spectrum should be removed before computing the eddy correlation in order to avoid spurious fluxes. We doubt that this conclusion can be made if a proper alignment procedure is followed.

The disadvantage of the eddy-correlation method is that it requires rapid response sensors. While this is no problem for the determination of the vertical wind speed fluctuations using a sonic anemometer, a suitable analyser for the scalar concentration fluctuations does not exist for many compounds.

e.   Analysers for use in flux measurments

Analysers using chemiluminescent techniques with response times well under one second exist for O_3, NO_2 and NO. For CO_2 and water vapour, which has high concentrations in the atmosphere compared to trace elements, it is possible to utilise infrared absorption spectroscopy. A new detector utilises solid state laser diodes (see Bösenberg *et al.*, 1997) tuned to an absorption band characteristic for the molecule of interest. However, to keep the tuned frequency stable the emitter and detector must be kept at a very precise cryostatic temperature. Thus in practice the apparatus is quite complex, not very rugged, expensive, and each apparatus is able to detect only one particular compound. However improvements can soon be expected.

FTIR (Fourier transform infra red) spectroscopy is another optical method that has been applied in flux measurements of atmospheric gases. In this technique a atmospheric broad-band infrared absorption spectrum is analysed and a large number of interesting compounds like N_2O, CO_2, CO, CH_4, H_2O, NH_3 *etc.* can be measured simultaneously with good specificity. By using long optical path lengths (100–1000 m) and sophisticated multi-regression evaluation algorithms, detection limits of the order of a few ppb (10^{-9} volume per volume) can be obtained. Long optical path lengths can be obtained either by applying closed multi-reflection cells, or open long-path systems. In the latter application the additional advantages of integration along the path length, and non-intrusive measurements are achieved. Because the time resolution in the measurements is limited, the technique has mainly been applied in profile or cuvette measurements. An advantage with the spectroscopic techniques is that accurate difference measurements can be made by taking the ratio of the spectra to be compared. Thus all instrument factors and common concentrations errors cancel leaving only a spectrum of the concentration differences. With this technique differences less than 1/1000 of the ambient concentrations can be detected and it has been applied in profile measurements and Relaxed Eddy Accumulation measurements (see below) of greenhouse gas fluxes.

Another successful application of FTIR spectroscopy is flux measurements using tracer gas release techniques. A known flux of a synthetic trace gas is released from the area under study, and by measuring simultaneously the trace gas and the gas under study the flux of the latter may be determined. This technique has been applied to studies of the emission of ammonia from liquid manure spreading and emissions of methane from waste depositories.

f.   Relaxed eddy accumulation

A relatively new technique which may have a lot of potential, is the so-called Relaxed Eddy Accumulation (REA) technique. In its simplest form it consists of a pump, a fast two-way valve, and two collector reservoirs (for example teflon bags), together with a sonic anemometer to control the valve. When the air motion is upwards, air is led to the "up" reservoir and *vice versa*. After a suitable period of time (say 30 min.), the concentrations in the two reservoirs of all components of interest can be determined by using conventional slow response analysers. The flux is then determined from

$$F = \beta \sigma_w (C_+ - C_-) \tag{9}$$

where C_+ and C_- are the concentrations in the "up" and "down" reservoirs respectively, σ_w is the standard deviation of the fluctuations of the vertical wind velocity, w, and β is an empirical coefficient of order one. It is not a constant though, but depends on the statistical properties of the turbulence and thus indirectly upon the stability).

There are many variations to the actual design. On-line analysers can replace the bags and analysers based on a differential principle are ideal. Introducing an interval for w (a so-called dead band) in which air is not sampled in either of the reservoirs has the advantage of reducing the activity of the valve (better life time) and increasing the difference between C_+ and C_-. If the dead band is determined as $\pm 1/2\ \sigma_w$, β turns out to be 0.42 independent of atmospheric stability (Jensen *et al.*, 1998).

If the compounds of interest are very reactive, one may *not* want them to pass through the pump and valve before detection. In this case a design could be used as shown in Jensen *et al.* (1998), built to measure VOC fluxes or fluxes of NH_3.

In their requirements for chemical analysis REA systems are similar to those required when using mean profile or gradient techniques. The advantages of REA is that, just like the eddy-correlation method, no assumptions are made concerning similarity profiles and that it has a well defined footprint.

4.7   Conclusions

During the lifetime of EUROTRAC there has been much development in our experimental capability to measure the exchange of chemical compounds between the atmosphere and the natural surfaces, terrestrial or aquatic and this has led to a significant increase in our knowledge base of the magnitude of these fluxes. Prior to BIATEX and ASE not even the sign of many of these fluxes was known. In turn the data have been used to validate and improve parametrisation schemes or construct new ones and, in some cases, have given an impulse to the development of completely fresh theoretical ideas about how to treat the physical/chemical processes involved. Despite of the large steps made in this direction more work, both experimental and theoretical, is unfortunately still needed, before we can be satisfied that our descriptions of biogenic emissions and deposition to the biosphere are good enough to be include with confidence in the models used for policy development.

4.8   References

Asman, A.H., O. Hertel, R. Berkowicz, J. Christensen, E.H. Runge, L.-L. Sørensen, K. Granby, H. Nielsen, B. Jensen, S.E. Gryning, A.M. Semprevivia, S.E. Larsen, P. Hummelshøj, N.O. Jensen, P. Allerup, J. Jørgensen, H. Madsen, S. Overgaard, F. Vejen; 1995. Atmospheric nitrogen input to the Kattegat. *Ophelia* **42**, 5–28.

Bösenberg, J., D. Brassington, P. Simon (eds), 1997; *Instrument Development for Atmospheric Research and Monitoring*, Springer Verlag, Heidelberg; Vol. 8 of the EUROTRAC final report.

Cleemann, M., Poulsen, M.E., Hilbert, G.; 1995. Deposition of Lindane in Denmark. *Chemosphere* **30**, 2039–2049.

Fontan, J., Lopez, A., Lamaud, E., Druilhet. A.; 1997. Vertical flux measurements of the submicronic aerosol particles and parametrisation of the dry deposition velocity, in J. Slanina (ed), *Biosphere-Atmosphere Exchange of Pollutants and Trace Substances*, Springer Verlag, Heidelberg, pp. 381–390.

Fowler, D., Meixner, F.X., Duyzer, J.H., Kramm, G., Granat, L.; 1997. Atmosphere-Surface Exchange of Nitrogen Oxides and Ozone, in J. Slanina (ed), *Biosphere-Atmosphere Exchange of Pollutants and Trace Substances*, Springer Verlag, Heidelberg, pp. 135–166.

Fuzzi, S., D. Wagenbach (eds), 1996, *Cloud Multi-phase Processes and High Alpine Air and Snow Chemistry*, Springer Verlag, Heidelberg; Vol. 5 of the EUROTRAC final report.

Galassi, S., Camusso, M., De Paolis, A., Tartari, G.; 1987. Trace organic pollutants in wet deposition of Brugherio (Milan). (In Italian). *Inquinamento* **29**, 116–120.

Guenther A. B., Zimmerman P. R., Harley P. C., Monson R. K., Fall R.; 1993. Isoprene and monoterpenes emissions rate variability: model evaluations and sensivity analysis. *J. Geophys. Res.* **98**, 12609–12617.

Guenther A. B., Zimmerman P. Wildermuth M. (1994) Natural volatile organic compound emission rate estimates for U. S. woodland landscapes. *Atmos.Environ.*, **28**, 1197–1210.

Gäggeler, H.W., Schwikowski, M., Baltensperger, U., Jost, D.T.; 1997. Transport, Scavenging and Deposition Studies of Air Pollutants at HighAlpine Sites, in S. Fuzzi, D. Wagenbach (eds), *Cloud Multi-phase Processes and High Alpine Air and Snow Chemistry*, Springer Verlag, Heidelberg.

Harborn J.B.; 1988. *Introduction to Ecological Biochemistry*, Academic Press, London, UK.

Jensen, N.O., Courtney, M.S., Hummelshøj, P.; 1998. A REA System for Reactive Gases. *J. Appl. Meteor.* submitted.

Kesselmeier J., Schaffer L., Ciccioli P., Brancaleoni E., Cecinato A., Frattoni M., Foster P., Jacob V., Denis J., Fugit J. L., Dutaur L., Torres L.; 1996, Emission of monoterpenes and isoprene from Mediterranean oak species *Quercus ilex L.* measured within the BEMA (Biogenic Emission in the Mediterranean Area) Project. *Atmos. Environ.* **30**, 1841–1850.

Kesselmeier J., Ammann C., Beck J., Bode K., Gabriel R., Hofmann U., Helas G., Kuhn U., Meixner F. X., Rausch Th., Schafer L., Weller D., Andrea M. O.; 1997. Exchange of short chained organic acids between the biosphere and the atmosphere, in: J. Slanina (ed) *Biosphere-Atmosphere Exchange of Pollutants and Trace Substances,* Springer Verlag, Heidelberg, pp. 327–334.

Lamb B., Hopkins B., Westberg H., Zimmerman P.; 1997. Evaluation of biogenic emission estimates using ambient VOC concentrations in Western Washington. *Workshop on Biogenic Hydrocarbons in the Atmospheric Boundary Layer,* August 24–27, University of Virginia, USA. American Meteorological Society, Boston MA, pp. 53–56.

Larsen, S., F. Fiedler, P. Borrell (eds), 2000; *Exchange and Transport of Air Pollutants over Complex Terrain and the Sea* Springer Verlag, Heidelberg (in preparation); Vol. 9 of the EUROTRAC final report.

Lerdau M.; 1991. Plant function and biogenic terpene emissions. in: Shrakey T. D., Holland E. A., Mooney H. A. (eds), *Trace gas emissions by plants*, Academic Press, pp. 121–134.

Martin, J.H., K.H. Coale, K.S. Johnson, S.E. Fitzwater, R.M. Gordon, S.J. Tanner, C.N. Hunter, V.A. Elrod, J.L. Nowicki, T.L. Coley, R.T. Barber, S. Lindley, A.J. Watson, K. Van Scoy, C.S. Law, M.I. Liddicoat, R. Ling, T. Stanton, J. Stockel, C. Collins, A. Anderson, R. Bidigare, M. Ondrusek, M. Latasa, F.J. Millero, K. Lee, W. Yao, J.Z. Zhang, G. Friederich, C. Sakamoto, F. Chavez, K. Buck, Z. Kolber, R. Greene, P. Falkowski, S.W. Chisholm, F. Hoge, R. Swift, J. Yungel, S. Turner, P. Nightingale, A. Hatton, P. Liss, N.W. Tindale; 1994. Testing the iron hyhothesis in ecosystems of the Equatorial Pacific Ocean. *Nature* **373**, 123–129.

Monahan, E.C., Margaret A. Van Patten; 1989. Climate and Health Implications of Bubble-mediated Sea-Air exchange Connecticut Sea Grant College Program, Marine Sciences Institute. University of Connecticut at Avery Point, Groton, CT 06340, USA.

Pilegaard, K., Jensen, N.O.. Hummelshøj, P.; 1997. Ozone fluxes over forested ecosystems. In: The oxidising capacity of the troposphere, in: Larsen. B, Versino, B. and Angeletti, G. (eds), *Proc. 7th European symp. on physico-chemical behaviour of atmospheric pollutants*, Report EUR 17482 EN, pp. 401–405.

Pio, C. A., Nunes, T., Valente, A.; 1997, Forest emissions of hydrocarbons. in: J. Slanina (ed), *Biosphere-Atmosphere Exchange of Pollutants and Trace Substances,* Springer Verlag, Heidelberg, pp. 335–341.

Simpson D., Guenther A., Hewitt C.N., Steinbrecher R.; 1995. Biognic emissions in Europe. I- Estimates and uncertainties. *J. Geophys. Res.* **100**, 22875–22890.

Slanina, J., (ed); 1997, *Biosphere-Atmosphere Exchange of Pollutants and Trace Substances,* Springer Verlag, Heidelberg; volume 4 of the EUROTRAC final report.

Steinbrecher R., Hahn J., Stahl K., Eichstadter G., Lederle K., Rabong R., Schereiner A., Slemr J.; 1997a. Investigations on emissions of low molecular weight compounds (C_2–C_{10}) from vegetation, in: J. Slanina (ed), *Biosphere-Atmosphere Exchange of Pollutants and Trace Substances,* Springer Verlag, Heidelberg, pp. 342–351.

Steinbrecher R., Ziegler H., Eichstadter G., Fehsenfeld U., Gabriel R., Kolb Ch., Rabong R., Schonwitz R., Schurmann; 1997b. Monoterpenes and isoprene emission in Norway spruce forests, in: J. Slanina (ed) *Biosphere-Atmosphere Exchange of Pollutants and Trace Substances,* Springer Verlag, Heidelberg, pp. 352–365.

Street R. A., Duckham S. C., Boissard C., Hewitt C.N.; 1997. Emissions of VOCs from stressed and unstressed vegetation, in: J. Slanina (ed), *Biosphere-Atmosphere Exchange of Pollutants and Trace Substances,* Springer Verlag, Heidelberg, pp. 366–371.

Torres L., Clement B., Fugit J. L., Haziza M., Simon V., Riba M.L.; 1997. Isoprenoic compounds: emissions and atmospheric concentration measurements, in: J. Slanina (ed), *Biosphere-Atmosphere Exchange of Pollutants and Trace Substances,* Springer Verlag, Heidelberg, pp. 372–380.

Chapter 5

Chemical Transfer and Transport Modelling

Adolf Ebel

Universität zu Köln, Institut für Meteorologie und Geophysik,
Aachener Straße 201-209, D-50931 Cologne, Germany

5.1   Introduction

Chemical transfer and transport modelling of atmospheric constituents in EUROTRAC has been applied to numerous processes and phenomena covering a large range of scales from local to global. The EUROTRAC subprojects GLOMAC (Global Modelling of Atmospheric Chemistry) and EUMAC (European Modelling of Atmospheric Constituent) were exclusively devoted to the task of global and mesoscale model development and model application to scientific as well as environmental policy problems. Other subprojects needing support from chemical transport modelling co-operated with the modelling groups particularly in EUMAC or initiated their own modelling activities as, for instance, TOR and TRACT. Yet despite the spread of modelling activities, co-ordination between the subprojects managed to avoid the duplication of numerical work in general and generated efficient collaboration. Considerable synergism was achieved in this way in the area of air quality modelling within EUROTRAC.

A subproject, GENEMIS, dealing with the generation of emission of data for mesoscale models was established some time after the start of EUROTRAC. It played a crucial role in the provision of reliable emission estimates required for advanced air pollution models with respect to temporal and spatial resolution as well as the completeness of relevant anthropogenic and natural emissions. Due to the delay of formation of GENEMIS, the anthropogenic emission work was a part of the mesoscale modelling subproject at the beginning of EUROTRAC and very close links have continued to exist between the former EUMAC and GENEMIS groups. This is especially true for the estimation of biogenic emissions. In this context it should be mentioned that it was an aim of BIATEX (Slanina, 1997) and EUMAC to collaborate on the development new parameterisations for biogenic emissions and for the surface resistance of deposition fluxes. Yet, due to the complexity of the problems and approaches, this goal could only partly be achieved and the implementation of improved deposition modules in chemical transport models (CTMs) used in EUROTRAC had to be delayed. The establishment of global emission inventories remained a task of GLOMAC itself. Actually, long-term inventories of anthropogenic sulfur emissions for climate modelling, improved treatment of natural emissions of the earth's crust due to volcanic and other geological processes and a novel DMS emission parameterisation are genuine products of the global modellers in EUROTRAC.

The overall scientific objectives of the modelling activities were the development of reliable advanced models for the simulation of chemistry and transport of trace constituents in the troposphere and lower stratosphere and the application of the models to the exploration of formation, distribution and impact of photo-oxidants on the atmospheric system. Whereas GLOMAC mainly addressed the long-term and planetary scale aspect of this issue, EUMAC focused on episodic treatment of regional conditions.

The distinctive goals regarding the scales of the air pollution problem lead to different strategies and priorities of model development. For instance, global models need accurate treatment of NO_x and ozone chemistry in clouds (Lelieveld and Crutzen, 1990; 1991; Dentener, 1993) for planetary scale budget estimates of O_3, whereas regional episodic simulations of the ozone budget are more sensitive to vertical redistribution of O_3 and its precursors (Mölders et al., 1994) and thus have to put stronger emphasis on the parameterisation of vertical mixing in clouds (Walcek and Taylor, 1986). Of course, both types of models require as realistic actinic flux estimates as possible. Table 5.1 summarises the principal time and space scales covered by the hierarchy of chemistry transport models in EUROTRAC and, for illustration, some typical scale-dependent meteorological and tracer dispersion phenomena in the atmosphere.

The dependence of tracer distributions and budgets on meteorological processes of all scales, including the microscale, enforced a considerable investment in the dynamical part of the chemical transport issue. It led to a strategy of joint development and improvement of the meteorological and chemistry transport modules of on-line models or off-line model systems in both modelling subprojects and several individual contributions. Major emphasis was put on off-line calculations, i.e. sequential estimation of meteorological parameters and simulation of chemical tracer distributions with no feedback of atmospheric chemistry on atmospheric dynamics. As regards global modelling the process of increasing investment in meteorological calculations is characterised by a most intensive use of the MOGUNTIA, which is dynamically forced with predefined (empirical) meteorological fields, at the beginning of EUROTRAC and a gradual progression to the implementation and use of ECHAM, a highly advanced global circulation model (GCM), as the dynamic basis for chemistry transport simulations.

The usual but not exclusive approach to regional (mesoscale) air quality modelling was the application of intimately coupled chemical transport models and meteorological models like the EURAD system, KAMM/DRAIS or MEMO/MARS to off-line meteorological and air pollution simulations.

Table 5.1: Typical space and time for motions and selected related meteorological and chemical phenomena. Examples of CTMs (Chemistry Transport Models) used in EUROTRAC which can resolve the respective scales.

scale	approximate space	approximate time	meteorological phenomena	chemistry-transport phenomena	CT models*
planetary	global to 5000 km	> 12 h	long waves, tides, cyclones anticyclones	hemispheric transport, ozone hole	MOGUNTIA ECHAM TMK
synoptic	5000 – 2000 km	10 – 1 h	cyclones, anticyclones	long-range transport	LOTOS EURAD, MCT
meso α	2000 – 200 km	200 – 12 h	fronts, hurricanes	summer and winter smog	LOTOS EURAD, MCT
meso β	200 – 20 km	24 – 1.h	cloud cluster, coastal circulation	geographic effects, city plume	KAMM/DRAIS MEMO/MARS EURAD, CIT
meso γ	20 – 2 km	10 – 0.1 h	thunderstorms, urban circulation	road pollution, extended plumes from point sources	KAMM/DRAIS MEMO/MARS

* For short model descriptions refer to section 5.2

Though the main objective of EUMAC was the development and application of models for chemical transformation and three-dimensional transport of reactive species, the need for high quality meteorological input for transport calculations lead to considerable feedback between chemical and meteorological model development. Improved treatment of local circulation controlled by geographical structures, like land-sea breezes, or cities generating heat islands provide a good example (Moussiopoulos, 1994). Another result of challenging meteorological models with transport problems, namely those of stratospheric ozone fluxes into the troposphere, was the first successful mesoscale simulation of a tropopause fold and cut-off low and related cross-tropopause ozone transport (Ebel *et al.*, 1991).

The characteristics of the models and model systems mainly used in GLOMAC and EUMAC are briefly described in the next section. Then follows a discussion of model input data and problems they may cause for tracer transport simulations in section 5.3. section 5.4 contains an overview of scientific and applied results proceeding from smaller to planetary scales. Efforts to evaluate the models and their output are addressed in section 5.5.

5.2   Models

For completeness a short description of the models and model system which were mainly developed and applied in EUMAC and GLOMAC is given in this section. The reader is referred to volume 7 of this series (Ebel *et al.*, 1997b) and references therein for more detailed information.

a.   MOGUNTIA

The model of global universal tracer transport in the atmosphere (MOGUNTIA; Zimmermann, 1988; Crutzen and Zimmermann, 1991) simulates the distribution and chemistry of trace constituents between the earth's surface and 100 hPa (about 16 km altitude). It is an Eulerian model with horizontal resolution of 10 degrees lat. × 10 degrees long. with 10 layers in the vertical. The large-scale transport in the model is derived from monthly mean temperature and wind fields obtained from observations over a 10-year period (Oort, 1983) or from ECMWF analyses. It contains parameterised vertical diffusion and convective clouds (Feichter and Crutzen, 1990). The tropospheric chemistry mechanism accounts for CH_4–CO–NO_x–O_3–OH photochemistry, pertaining to the background global troposphere. Heterogeneous reactions have also been included to simulate the effects of aerosols and clouds (Dentener and Crutzen, 1993). In particular the long-lived trace species are well represented by the model. Trace gas emissions have recently been updated, and the model has been coupled to a radiative transfer scheme to simulate anthropogenically-induced global change (Lelieveld *et al.*, 1998). The chemical mechanism was adjusted to include VOCs in the free troposphere (Kanakidou *et al.*, 1997).

b. TM3, TMK

The Tracer Model version 3 (TM3) is the successor of the atmospheric transport model TM2 (Heimann, 1995). It computes, in an off-line mode, the transport of a tracer based on the fields from three-dimensional meteorological analyses or from the output of an atmospheric general circulation model. A horizontal resolution of about 5 degree long. by 4 degree lat. and a hybrid (sigma-pressure) vertical coordinate with 19 layers from the surface to the top (100 hPa) is used. The meteorology is usually taken from ECHAM with a temporal resolution of 6 hours. Thus diurnal cycles are only marginally resolved in the present version. The TM employed at KNMI (TMK, Kelder *et al.*, 1997) has been coupled with a tropospheric chemistry mechanism developed at the Max-Planck-Institute Mainz.

c. ECHAM

The dynamics and part of the model physics of the Hamburg spectral climate model ECHAM have been adopted from the European Centre for Medium-Range Weather Forecasts (ECMWF) model (Roeckner *et al.*, 1996a). Prognostic variables are vorticity, divergence, temperature, the logarithm of the surface pressure, and the mass mixing ratios of water vapour, and total cloud water (liquid and ice together). The model equations are solved on 19 vertical levels in a hybrid (sigma-pressure) system by using the spectral transform method with triangular truncation at wavenumber 40 (T42). Advection of water vapour, cloud water and chemical species is treated with a semi-Lagrangian scheme. Cumulus clouds are represented by a bulk model including the effects of entrainment and detrainment on the updraft and downdraft convective mass fluxes. The turbulent transfer of momentum, heat, water vapour, and total cloud water is calculated on the basis of a higher-order closure scheme. The radiation code is based on a two-stream solution of the radiative transfer equation. Gaseous absorption due to water vapour, CO_2, O_3, CH_4, N_2O, and CFCs is included, as well as scattering and absorption due to prescribed aerosols and model-generated clouds. ECHAM has been applied to study anthropogenic impacts on climate. It includes two atmospheric chemistry codes, one focussing on the tropospheric background chemistry (Roelofs and Lelieveld, 1995) and another on the chemistry of the lower stratosphere (Steil, 1997; Steil *et al.*, 1998).

d. LOTOS

The LOTOS (long term ozone simulation) model is a three-layer Eulerian grid model for the simulation of the formation, transport and deposition of oxidants within the lower part of the troposphere up to about 3 km. The model covers Europe in grids of 0.5×1.0 degrees (optionally 0.25×0.5). LOTOS uses the concept of dynamical layers which expand or contract locally in response to changes of meteorological conditions. The lowest layer represents the well-mixed boundary layer, the middle layer represents the capping inversion, and the top layer represents the lower free troposphere reservoir layer. A purely diagnostic surface layer is used to handle the dry deposition (resistance analogy).

The chemical mechanism employed is CBM-IV with 24 reactive species and 52 reactions. Boundary and initial conditions are provided by a global two-dimensional model. The meteorological winds, temperature and specific humidity are provided as 3-hourly layer averages. Other input data includes gridded fields of mixing heights, cloud cover and precipitation rates. These data are provided by the output of the Norwegian NWP model or by the output of the model from the Free University Berlin. The main focus of LOTOS is the calculation of ozone and related species over extended periods, such as a growing season or a year (Builtjes, 1992). At the moment, model results are available for the complete years of 1990 and 1994, and for the summer months of 1997.

e.   EURAD model system

The European Air Pollution Dispersion (EURAD) model system (Ebel et al., 1997d) consists of three main parts, namely the EURAD-CTM (Hass, 1991) originating from the Regional Acid Deposition Model (RADM, Chang et al., 1987), the Pennsylvania State University/NCAR Mesoscale Model, version 5 (MM5; Grell et al., 1993; Anthes and Warner, 1978) and the EURAD Emission Model (EEM, Memmesheimer et al., 1991). Its standard version covers a height range up to 16 km (100 hPa) but was recently extended to about 30 km (10 hPa). Horizontal domains covering the whole of Europe or just parts of it can be chosen. Using nesting techniques (Jakobs et al., 1995) the horizontal resolution can be varied between about 2 and 80 km. For the vertical, the terrain following σ-coordinate is used.

The EURAD system is designed for episodic calculations. Meteorological initial and boundary conditions are obtained from ECMWF analyses. Anthropogenic emissions are calculated from EMEP inventories or taken from scenarios provided by GENEMIS (Lenhart et al., 1997). The RADM2 chemical mechanism (Stockwell et al., 1990) containing 140 reactions with 63 species is mainly used. The vertical resolution is high in the boundary layer and decreases with increasing height except for a special version developed for the simulation of the tropopause region. Usually 16 model levels have been chosen for model application to the troposphere.

A system similar to EURAD was applied to the simulation of two episodes by Langmann and Graf (1994, 1997) who used the high resolution model HIRHAM as meteorological driver for the EURAD-CTM.

f.   MCT

To produce the meteorological data necessary as input to the MCT (Mesoscale Chemistry Transport Model) a NWP model based on the limited area model NORLAM from the Norwegian Meteorological Institute, described in Gronas et al. (1997) is used. The horizontal domain, grid resolution, position and number of vertical layers and model top can easily be changed. The grid resolutions normally used are between 25 and 150 km at 60° N with 10–30 unequally spaced vertical layers extending up to levels between 100 and 30 hPa. σ-coordinates are employed. When the MCT model is run operationally during measurement campaigns, analysis and forecast fields arriving every 12 hours from the ECMWF are used as initial and boundary conditions for the NWP prognoses. Since a chemical analysis based on observations is not available, each new forecast with the MCT model must be initialised with data from the previous forecast. For normal runs ECMWF analysis with 6h resolution are used both as boundary and initial conditions for 12 hour runs with the NWP model.

The transport part of the MCT model is described by Flatøy *et al.* (1995), and the chemistry part by Flatøy and Hov (1997). The chemistry scheme is primarily for photochemical oxidants. Aerosol and liquid phase chemistry is parameterised. The chemistry is quite similar to the EMEP (European Monitoring and Evaluation Program) Lagrangian photo-oxidant model, documented in Simpson (1992a; 1992b) and to the chemistry used in the global 2-D model by Strand and Hov (1994) to study global ozone.

Surface emissions are specified for NO_x, SO_2, CO and VOC. The most recent EMEP emissions inventory for 1995 from Berge (1997) is utilised for Europe and Russia west of the Urals. The NO_x and SO_2 outside the EMEP domain are taken from Benkovitz *et al.* (1997). Further details about annual and diurnal variations are given by Flatøy and Hov (1996). Natural VOC emissions represented by isoprene are derived by an emission-temperature relationship from Lübkert and Schöpp (1989) using the forest cover and the variable surface temperature from the result of the NWP model calculations. Aircraft NO_x emissions are taken from ANCAT/ECAC (Abatement of Nuisances Caused by Air Transport/European Civil Aviation Conference). The MCT model is self-nesting.

g.   KAMM/DRAIS

The model system KAMM/DRAIS consists of the meteorological model KAMM (Karlsruhe Atmospheric Mesoscale Model, Adrian and Fiedler, 1991) and the chemistry transport model DRAIS (three dimensional regional atmospheric dispersion model, Nester and Fielder, 1992; Nester *et al.*, 1995). The meteorological model KAMM is forced by the large scale meteorological conditions (basic state) which can be derived from observations or from larger scale model results. The non-hydrostatic Eulerian model KAMM applies a terrain following coordinate system which allows a better resolution close to the ground compared with the upper levels. The interaction between the soil and the atmosphere is described by a soil-vegetation model (Schädler *et al.*, 1990; Lenz, 1996). The chemistry transport model DRAIS solves the diffusion equation on a Eulerian grid using the same terrain following coordinate system as the KAMM model. Dry deposition of the different species is parameterised by their dry deposition velocity (Baer and Nester, 1992). The gas phase chemical reaction mechanism of the RADM2 model is implemented in DRAIS. In the stable and neutral boundary layer, a local first order approach for the diffusion coefficients is chosen. In the convective boundary layer a non-local similarity approach is used. A new module in the model system allows the calculation of the mass budgets in predefined volumes for all species considered (Panitz *et al.*, 1997). A particular mode of application is the use of EURAD results for the derivation of the meteorological and chemical initial and boundary conditions for the KAMM/DRAIS model system (Nester *et al.*, 1995). The size of the model domain and the horizontal resolution can be varied from a few hundred km with grid lengths of a few km in both horizontal directions to much smaller areas and grid sizes.

h.   MEMO/MARS (EZM)

The system MEMO/MARS or EUMAC Zooming Model (after EUROTRAC renamed European Zooming Model, EZM; Moussiopoulos, 1994; Moussiopoulos *et al.*, 1997) is a comprehensive model system for simulations of wind flow and pollutant transport and transformation (Moussiopoulos, 1995). The EZM may be used either in conjunction with a regional scale model or as a stand-alone model system driven directly with measured data. Core models of the EZM are the non-hydrostatic prognostic mesoscale model MEMO and the photochemical dispersion model MARS.

MEMO allows describing the air motion and the dispersion of inert pollutants over complex terrain (Kunz and Moussiopoulos, 1995). The code allows multiple nesting (Kunz and Moussiopoulos, 1997). Within MEMO, the conservation equations for mass, momentum, and scalar quantities as potential temperature, turbulent kinetic energy and specific humidity are solved. The governing equations are solved in terrain-influenced coordinates. The discrete pressure equations are solved with a fast elliptic solver in conjunction with a generalised conjugate gradient method (Moussiopoulos and Flassak, 1989). Turbulent diffusion can be described with either a zero-, one- or two-equation turbulence model.

MARS is a 3-D Eulerian dispersion model for reactive species in the local-to-regional scale (Moussiopoulos, 1989). Processes of emission, dispersion, transformation and deposition of pollutants are calculated on a staggered grid in terrain-influenced coordinates. Thanks to the modular structure of MARS, chemical transformations can be treated using any suitable chemical reaction mechanism. The dry deposition process is described with the resistance model concept. MARS offers the possibility of selecting the level of complexity with regard to the numerical algorithm used and allows performing `telescopic' simulations.

i.   CIT

The CALTECH/Carnegie Mellon Institute of Technology (CIT) Eulerian photochemical model (Harley *et al.*, 1993) is driven by meteorological, air quality and emission data. Depending on the application the horizontal resolution can vary between 1 km and 20 km. The height range of 6 vertical levels reaches from the surface up to about 2.5 km. The vertical resolution is highest at the bottom with a height of 36 m for the lowest layer. Usually episodes of a few days are simulated. The chemistry module of the model is based on the LCC mechanism

(Lurman *et al.*, 1987). The original gas phase mechanism was developed to represent the photo-oxidation of non-methane organic compounds and nitrogen oxides under urban conditions. It has been modified to include biogenic emissions which represent an important contribution to the atmospheric hydrocarbon content. Altogether, the concentrations of 52 chemical species are calculated. 36 of them are prognostic and 9 (fast reacting radicals) in steady state. The hydrocarbons are partly lumped and partly used as surrogates in 106 gas phase chemical reactions. The model has been further extended during the time of the EUMAC project (Giovannoni *et al.*, 1997). A module for aqueous chemistry can be included optionally. It includes 9 additional prognostic species and 12 reactions (Müller *et al.*, 1996). The calculations of photolysis rates are carried out following Ruggaber *et al.* (1994, 1997). One-way-nesting can be applied in order to provide boundary conditions for a domain with high resolution (Clappier *et al.*, 1997, Krüger *et al.*, 1998)

Special model versions have also been developed for application to specific problems. Isaksen *et al.* (1997) used a two-dimensional channel model covering the latitudinal belt between 30° and 60° N to study the large-scale effect of inhomogeneous release of ozone precursors in this belt. A few versions of vertical column models (one-dimensional models) have been developed and used for studies of the atmospheric boundary layer and lower free troposphere.

1-D models which have mainly been employed are STAR (System for Transfer of Atmospheric Radiation, Ruggaber *et al.*, 1994, 1997) and various forms of EURAD-1D which is based on the EURAD-CTM. Processes and phenomena investigated with the column models include modifications of actinic fluxes through changes of the atmospheric background, chemical reactions in fog (Forkel *et al.*, 1990), formation of aerosol and its chemical impact (Ackermann *et al.*, 1995; 1996), cloud mixing and scavenging, and the dependence of the photochemical ozone creation potential (POCP, Derwent and Jenkins, 1991) on atmospheric composition in a vertically structured atmosphere (Ebel *et al.*, 1997a; 1997d).

5.3   Model input, initialisation and boundary conditions

Atmospheric modelling, *i.e.* the solution of a larger set of non-linear equations for the conservation of momentum, energy and atmospheric constituents, is a complex and demanding initial value problem. Successful realistic time dependent simulation of the atmosphere requires as accurate as possible initialisation. This had already been shown by Ertel (1944) and others in the early days of mathematical weather forecast and is still a great challenge today, in particular, for numerical modelling of physical and chemical transformations and transport of atmospheric constituents. Apart from this, a considerable effort is necessary - and has been made in EUROTRAC - for the derivation and formulation of various categories of input data for CTMs. In principle, all the parameters chosen for the formulation of chemical mechanisms and other process modules like those of boundary layer transport and deposition are among these categories. They are treated elsewhere in this volume, so that we can restrict ourselves to the class of two- and three-dimensional fields of input data in the following. This section also includes an overview of emission estimation in EUROTRAC aiming at the generation of emission data sets suitable as input for CTMs. These data together with meteorological and photolytic input data itself usually require the application of modelling since observational data fulfilling the requirements of global and mesoscale models are generally, with some exceptions, not available.

a. Meteorological initial and boundary conditions

Different formulations of initial and boundary conditions have been used throughout the models which have been employed in EUROTRAC. Global simulations have been performed using climatological data based on observations of wind, temperature, humidity and clouds thus avoiding the problem of meteorological initial and boundary value definition (MOGUNTIA). Similarly, GCM fields have been used to drive chemistry and transport processes in global calculations (ECHAM). First approaches towards episodic planetary scale simulations of realistic situations have been made employing analysed global meteorological fields as provided by the European Centre of Medium-Range Weather Forecast (ECMWF, Kelder *et al.*, 1997).

Lateral boundary and initial conditions for regional modelling have also been based on meteorological analyses provided by the ECMWF (as a standard procedure of the EURAD system) or the Norwegian Meteorological Institute (LOTOS, MCT). Smaller scale models also made use of meteorological observations. Yet a method favoured towards the end of EUROTRAC was the nesting of these models into larger scale ones. Refined methods for large and small mesoscale models of different structure have been developed and tested by Gantner and Egger (1997). In a larger number of cases the meteorological output of the EURAD system, *i.e.* MM5, has been used for coupled simulations. Another way successfully exploited not only for the meteorological component of the model systems but also for their chemistry and transport part is the method of internal nesting, *i.e.* the application of the same model in sub-domains with higher resolution (Jakobs *et al.*, 1995). One- and two-way nesting procedures have been employed where the latter one takes into account the feedback of the solutions obtained for the sub-domain on those for the larger model domain.

b. Photolysis rates, minor constituents

Sensitivity studies in EUMAC and GLOMAC revealed that four-dimensional treatment of photo-dissociation could considerably improve the reliability of chemistry and transport simulations. Yet they have also shown that such calculations require immense computational resources and only allow example studies with fully interactive dynamic, chemical and radiative schemes (*e.g.* Hass and Ruggaber, 1995). A convenient method to avoid this problem is the tabulation of photolysis rates for a given sample of chemical species as a function of time, latitude and height for clear sky conditions and average atmospheric composition. Actinic flux calculations of the EURAD system are based on Madronich (1987). Modification of clear sky fluxes by clouds has been realised through parameterisations depending on the nature of the model.

As mentioned in the previous section chemical initial and boundary conditions for smaller scale models may be derived from larger scale ones. The situation for regional models covering larger areas like Europe is still rather problematic. In principle global or hemispheric two- or three-dimensional models might be used to generate initial and boundary conditions as it has been done, *e.g.*, for LOTOS. Yet for complex modelling as with the EURAD system input data for a large spectrum of reactive species and meteorological situations are needed which presently cannot yet be derived from existing global scale CTMs. Also observations of tracer concentrations and their spatial and temporal variations are still insufficient for direct formulation of consistent chemical boundary and initial conditions on continental scales. Therefore, *ad-hoc* solutions have to be chosen. A solution that provides sufficiently reasonable results under present conditions is to choose tentative tracer distributions (often characterised by a single vertical profile for the whole model domain) in overall accordance with the existing observations, to run the model for a short period allowing the chemical fields to achieve a certain degree of consistency, and then to use the resulting distributions as initial conditions (Chang *et al.*, 1987). Evidently, this is a crucial problem area of air pollution modelling. Improvements are needed in future not only from the experimental and monitoring side but also through the development of improved model based initialisation methods. Some work in this direction was started during the first phase of EUROTRAC (see section 5.3.e).

c. The surface

The availability of high resolution topographic data can still be problematic in various regions, particularly when the data is classified for military reasons. This can seriously hamper the flexibility of regional models as used in EUMAC. As shown by Gantner and Egger (1997) it is necessary to take into account ever more details of orography when nesting procedures are applied.

Gaps also exist in the available land-use data sets which may cause serious deficiencies in air quality simulations. For parameterisations of important processes like radiative fluxes, natural emissions, deposition of minor constituents and boundary layer formation are crucially dependent on the quality of such surface data. An improvement of the situation regarding the European land cover data base was achieved by GENEMIS (Köble and Smiatek, 1997) which also started a study about the applicability of the leaf area index (LAI) to air quality simulations. There are indications that the ratio vegetation index which can be derived from satellite measurements (LANDSAT) may be used to estimate the LAI (Fig. 5.1).

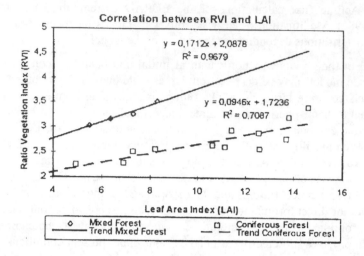

Fig. 5.1:   Variation of the ratio vegetation index (RVI) derived from satellite data (LANDSAT TM5) as a function of observed leaf area index (from Köble and Smiatek, 1997).

d. Emissions

Much of the success of chemical transport simulations depends on the quality of emission data used as input to the models. The situation regarding emission inventories was poor at the beginning of EUROTRAC and so considerable effort had to be devoted towards the improvement of the situation for global as well as regional modelling. Emission research and inventory development was carried out in close connection with other ongoing activities, e.g. with those embedded in IGAC (International Global Change Program) on global scale or EMEP, LOTOS and CORINAIR for Europe. Also detailed inventories from regional environmental agencies or other sources have been used (e.g. Grosch, 1996).

In the framework of planetary scale modelling, improvements regarding the sulfur scheme have been achieved by introducing the following features in scenarios of anthropogenic emissions: a seasonal cycle, height dependence, estimate of decadal changes since 1860. Furthermore, the inventory of natural sulfur emissions has been completed by treating eruptive and non-eruptive lithospheric emissions separately and applying an improved parameterisation of DMS emission. Studies were carried out to improve global methane and NO_x emission estimates and the spatial structure of the sources of these species. Only rough estimates of the global NMHC emissions and their source distributions are available at present. A model of black carbon aerosol emissions was developed and tested in MOGUNTIA based on an annual emission rate from industrial sources and biomass burning of 5.5. and 7.9 Tg/yr, respectively.

The need for emission inventories with high spatial and temporal resolution as a prerequisite for regional air quality modelling, in particular for complex episodic simulations, was motivation for a large and efficient group of projects within

GENEMIS to develop a unique set of European emission data as input for regional chemical models (Friedrich, 1997). Such models require regionally consistent gridded data with at least hourly time spacing. Therefore, one of the major tasks became the harmonisation and combination of existing emission inventories which usually contain annual estimates, and their transformation to discrete values in space and time, down to horizontal resolutions of, say, 1 km or less. For this purpose models have been developed to disaggregate annual emissions from different sources and regions and also for inventory updates. To achieve this goal technical, economic and meteorological data that can be used as parameters for annual emission inventory disaggregation had to be collected. The product is a growing set of European emission data for episodes in different years, seasons and regions. Evaluation studies have contributed to continuous improvement of the quality of this data basis.

First studies of ozone formation in the boundary layer over Europe revealed a significant deficit of NO_x emissions among other problems in the existing inventories (Memmesheimer et al., 1991; 1997d). Therefore, filling the gaps existing in central and eastern Europe became a major goal of GENEMIS (Heymann, 1997). Since dramatic economic changes during the period of investigation (1990–1992) took place in this part of Europe the completed inventory forms an excellent base for European air quality trend studies employing regional models.

Chemical mechanisms as employed in the EUROTRAC model hierarchy need rather detailed non-methane VOC splits. Since standard inventories usually supply total VOC emissions, methods had to be developed to derive VOC profiles as required for the application of more complex chemical mechanisms. The GENEMIS data base provides 34 classes of non-methane VOC. It is obvious that the structure of this VOC profile (like others) depends on region, country, season, time of the day and changes of technology and economy making the update of such complex emission inventories a difficult and expensive task. What is needed in future is an emission model with forecast skills since otherwise difficulties will arise for the prediction of photo-smog and long-term trends of boundary layer ozone concentration.

Of particular value for regional episodic simulations of the nitrogen budget is information on ammonia emissions (Asman et al., 1997). It is also of special relevance to secondary aerosol treatment in regional models (Ziegenbein et al., 1994a; 1994b; Ackermann et al., 1995). Aerosols are becoming of increasing concern with respect to air quality assessment. Obviously, the weakness of existing emission inventories with regard to this component of air pollution needs to be overcome as soon as possible.

Various attempts have been made to evaluate GENEMIS emission estimates. Staehelin und Schäpfer (1997) conducted a tunnel experiment aiming at the assessment of emission factors for road traffic. In comparison with other experiments carried out in eastern Europe they find considerable differences of measured factors indicating the importance of fleet composition and vehicle maintenance standards. Another strategy is the use of chemical transport models to assess the quality of emission input from the resulting concentration fields. This has been attempted on global and regional scales in the EUROTRAC modelling subprojects with some success. A problem, which deserves more attention with regard to the use of emission data in chemical transport models, is the process of subgrid mixing of a primary pollutant after the moment of release. Usually, instantaneous mixing is assumed even in the case of relatively rough spatial resolution, say 50×50 km^2. If horizontal mixing and chemical lifetimes are of similar magnitude, one can expect initial subgrid chemical transformations which may be of significance for chemical transformation estimates on the scales resolved by a model. It appears that a part of the well-known problem of sulfur over prediction and sulfate under prediction in regional air quality models can originate from the assumption of instantaneous mixing (Hass *et al.*, 1993).

e.   Chemical data assimilation

In contrast to meteorological modelling the development of algorithms for atmospheric chemical data assimilation gained little attention before the end of EUROTRAC in the field of chemistry and transport modelling. The main reason is that suitable chemical data sets are rare. Furthermore, it is often argued that pollutant fields are strongly controlled by emissions and assimilation procedures may not be applicable. Nevertheless, there exist problems which can advantageously be treated with assimilation methods, and the development of suitable methods was among the goals of EUMAC since it was established. It is expected that advanced satellite experiments will provide measurements which can be used for four-dimensional data assimilation. First applications have been published by Fisher and Lary (1995) for the stratosphere.

The method of optimum interpolation (Daley, 1991) has been used by Petry (1993) to study the dependence of mesoscale air pollution simulations with the EURAD system on the initial distribution of reactive trace gases including a realistic scenario as given by the existing European networks for air pollution monitoring. Focusing on ozone this method has been modified by combining it with the principal component analysis (Preisendorfer, 1988) for the characterisation of different ozone regimes in the atmospheric boundary layer. The development of four-dimensional variational data assimilation algorithms for regional chemical data analysis and modelling with EURAD was initiated during the final phase of EUROTRAC I (Elbern *et al.*, 1996).

5.4   Application of models and results

The models employed in EUROTRAC have been applied to a wide range of scientific as well as practical problems. Regarding the aspect of policy applications the reader is referred to volume 10 of this series (Borrell *et al.*, 1997). According to the character of questions addressed with the help of numerical methods the whole hierarchy of EUROTRAC models has extensively been used in a great variety of studies. Only a limited choice of scientific applications carried out during the eight years of the first phase of EUROTRAC, can be presented in this section. Examples are selected according to the role they played for the development of modelling activities within EUROTRAC as well as their significance for the understanding of atmospheric chemistry and transport at their time or the identification of problems for future research.

The motivation to use sub-models or stand-alone versions of specific modules alongside complex three-dimensional models was twofold. On the one hand there was the methodological aspect of exploring the reliability, sensitivity, the efficiency and range of applicability of modules, and algorithms or mathematical techniques for a large parameter space, before implementing them in Eulerian models. On the other hand, a large number of scientific problems are encountered which can more conveniently be treated with a box or a one-dimensional sub-model than with the full complexity of three-dimensional schemes.

After discussing process oriented modelling, results of limited area simulations will be presented beginning with smaller scale models and proceeding to models applied to larger regions. Examples from global modelling will conclude this section.

a.   Process modelling

(i) Chemical gas-phase mechanisms

A comparison of various gas-phase mechanisms which are used in advanced chemical transport models has been carried out (Kuhn et al., 1998). In addition, the performance of the RADM2 mechanism used in the EURAD system was assessed using observations of the real atmosphere. Field data for nitrogen oxides, a whole suite of hydrocarbons, photo-oxidants like ozone, H_2O_2 and alkyl peroxides, photolysis frequencies and meteorological parameters were measured during a summer smog event. Agreement between modelled and measured ozone concentrations could only be achieved when the substantial impact of biogenic hydrocarbons was taken into account (Poppe et al., 1997). The development and tests of improved solvers of the ordinary differential equation system of atmospheric chemistry mechanisms (e.g. Kessler, 1995, Sandu et al., 1997) helped to increase the efficiency and reliability of chemical calculations.

There is still a considerable lack of understanding chemical processes in the perturbed atmosphere, especially in the turbulent range of motions. Usually it is assumed that complete and instantaneous mixing occurs when box calculations are carried out. Therefore, perturbed chemistry simulations employing large-eddy modelling were initiated in the framework of EUMAC to check the uncertainty of simulated results originating from this assumption. Stockwell (1995) demonstrated that non-uniform mixing of gas-phase trace species may limit the accuracy of the prediction of mechanisms assuming instantaneous uniform mixing. Unfortunately, research on this topic did not get to the point where definite conclusions could be drawn from the numerical experiments, particularly those using the method of large eddy simulation, except that surprises regarding our understanding of chemical transformation of polluted air in the presence of turbulence cannot be excluded.

(ii) Clouds, fog and photolysis rates

Clouds are not only responsible for redistribution of water, trace substances and heat in the atmosphere and soil but also act as chemical factories. Cloud processes, which are effective on the relatively small scale of an individual cloud or a cloud system also have a substantial impact on the regional and even the global scale. These problem areas have been attacked by the EUMAC group on 'cloud processes' and the EUROTRAC working group on clouds (Flossmann et al., 1995). Beyond indicating weak points in the current understanding of clouds and the associated multiphase processes they pointed out directions for an improvement, and first steps into this direction were taken (e.g. Liu et al., 1997; Mauersberger, 1997; Matthijsen et al., 1995; 1997; Stockwell and Schönemeyer, 1997). From the work on aqueous phase chemistry a significant improvement of our knowledge about in-cloud reactions can be expected for the future. Cumulus convection parameterisations have also been a major issue because their improvement is required for better understanding and simulation of wet deposition properties of clouds. Here, the investigation of the ice phase and its implications for wet phase chemistry needs special attention as very little is known regarding this topic.

Orographic clouds (Chaumerliac et al., 1990), either precipitating or non-precipitating, have a poly-disperse feature that can influence tropospheric photochemistry and lead to deviations from Henry's law for the most soluble species. These deviations vary along the mountain slopes due to different partitioning processes among the gas and aqueous phases. In order to take into account the drop-size dependency of the wet chemical processes, spectral representation of drop populations should be accounted for by means of semi-spectral schemes that give reasonable accuracy rather than using fully spectral schemes since they are easier to include in CTMs. A large number of uncertainties remain in describing aqueous phase photochemistry since only a few dominant reactions are considered in existing models and because the numerical methods used for solving gas and aqueous phase systems separately are not necessarily adequate for solving coupled gas/aqueous phase systems. Model sensitivity studies should be compared with in-situ measurements carried out at mountain summits where cap clouds can provide the natural chemical laboratory conditions needed to improve our knowledge of cloud chemistry.

In addition to the generation of input data for three-dimensional CTMs with stand-alone photolysis models, as mentioned in section 5.3.e, sensitivity studies have been carried out to investigate the interrelation of cloud or fog and radiation effects quantitatively. One-dimensional simulations reveal that strong enhancement of photolysis rates can occur, not only above optically thick clouds and fog, but also above layers with reduced optical thickness (Forkel et al., 1995, 1997). Employing these results Tippke (1997) has shown that the inclusion of fog processes in three-dimensional models can strongly modify the simulated tracer distributions through a combined effect of radiation, wet chemistry and wet deposition in areas with fog formation.

(iii) Other process studies

The efficiency of chemical mechanisms with respect to photo-oxidant formation under realistic conditions has been studied using the concept of photochemical ozone creation potential (POCP). Three- and one-dimensional sensitivity studies were performed employing the LOTOS model (Builtjes, 1992) and 1-D EURAD (the one-dimensional version of the EURAD-CTM; Heupel 1995). The CTM-IV (LOTOS) and RADM2 mechanism (EURAD) exhibit POCPs similar to the Lagrangian experiments by Derwent and Jenkin (1991) or Andersson-Sköld *et al.* (1991). Yet a clear dependence on emission and meteorological conditions as well as on the age of a polluted air mass is found from the 1-D calculations within a wide but realistic range of parameters. Obviously, this may cause problems for finding the optimum of VOC emission mixtures to reduce photo-oxidant creation from anthropogenic activities. For example, terminal alkenes show a larger POCP in urban than in rural air and are more effective under normal than under high pollution conditions by a factor of about 1.5, while their POCP decreases with the increasing age of air. Other VOCs like toluene may show an inverse dependence of ozone production efficiency on the degree of air pollution.

For those substances whose deposition velocities are large enough to generate noticeable effects during its residence time in the model domain, deposition essentially controls the budgets of many trace substances not only on the global scale but also on regional scales. Uncertainties in the parameterisation of deposition appear to be large. Comparing global and regional simulations it seems that significant differences exist, for instance, for nitrogen deposition as an important factor of atmosphere-biosphere interaction. As regards mesoscale models, in EUMAC the sensitivity of air pollution calculations to new values of surface resistance and other parameters for dry deposition estimation (*e.g.* Erisman *et al.*, 1994; Walmsley and Weseley, 1995) have been tested employing the EURAD system. Significant modifications arose from improved parameterisations for the dry deposition of PAN and HNO_3.

Dry deposition is strongly controlled by the state of the atmospheric boundary layer, *i.e.* through the intensity of mechanically and convectively generated turbulence. Strong differences are found between the types and classes of models in EUROTRAC regarding the parameterisation of the atmospheric boundary layer (ABL; Stull, 1998) and the transport in it. Comparative studies are needed in future to determine the reliability and applicability of available parameterisations not only to dynamic and thermodynamic processes in the ABL but also to near surface tracer transport.

b.   Simulation studies for limited areas

(i) Smaller scale simulations

Eastern Germany. The large variability of conditions controlling air quality in different parts of Europe represents a strong challenge for the high resolution modelling of chemistry and transport of air pollutants. The various smaller scale models developed and applied within the subproject EUMAC have been employed for limited areas in central Europe with flat and complex terrain, including the Alpine regions, as well as for coastal areas of the Mediterranean Sea, the Atlantic and the North Sea. All these areas exhibit considerable differences in land use, climate, local wind systems and, most of all, in emission characteristics. For instance, numerical simulations were carried out for East Germany with decreasing emissions of sulfur and increasing emissions of NO_x and VOC in order to study the impact of changing economic conditions in a part of Europe with neighbouring areas having high emission levels. As expected and observed concentration levels remain high if abatement measures are not extended to larger areas.

The Heilbronn experiment. In contrast to this unintended and uncontrolled experiment, a closely controlled pollution-reduction study was carried out in Heilbronn, Germany, providing a good opportunity to evaluate smaller scale models and use them to interpret the field observations (Moussiopoulos *et al.*, 1997b). The EUMAC Zooming Model (EZM) showed good agreement with measurements during the Heilbronn experiment, revealing low NO_x concentrations in the area with emission control but little response of ozone to locally reduced nitrogen emissions. Predictions of such an unspectacular result by modellers did not impress the proponents of the experiment.

Athens.   Several studies have been conducted to explore the role of local circulation for the development of photo-smog, *e.g.* in the Upper Rhine Valley (Schneider *et al.*, 1995), the Athens basin (Moussiopoulos *et al.*, 1997a) or the greater Lisbon area (Borrego *et al.*, 1997). A dominant feature of pollution dispersion in coastal areas like Athens (Giovannoni and Russel, 1995; Giovannoni *et al.*, 1995), Thessaloniki, Lisbon (Borrego *et al.*, 1995) and other pollution hot spots at southern latitudes in Europe are land/sea breezes. Correct simulation of this phenomenon is required for accurate estimates of ozone and other pollutant concentrations. To check the performance of smaller scale models in such areas an international study, Athens Photo-Smog Intercomparison Study (APSIS), was organised (Moussiopoulos, 1993; Kunz and Moussiopoulos, 1995) revealing remarkable differences in the ability of the various smaller mesoscale meteorological models when applied to a complex photo-smog situation.

Nesting procedures. As pointed out in the previous section, the success and accuracy of a simulation is heavily dependent on the quality of the boundary and initial conditions. This is especially true for models covering smaller domains where transport through lateral boundaries is the controlling factor for chemical processes with time constants comparable to the time an air parcel needs to traverse the domain. A way to overcome this difficulty is to nest high resolution models within larger scale models. Another reason to apply nesting techniques is the difficulty of coarse resolution simulations to generate realistic features of tracer distributions if smaller scale processes are important. Thus the reproduction of the night-time ozone minimum which is difficult to obtain with coarse resolution is considerably improved when finer grids are used (Nester *et al.*, 1997).

Various approaches to optimise nesting procedures have been explored, and used for both the meteorological and chemical parts of the model system in EUMAC. The simplest way is the use of output from a different larger scale model as initial and boundary conditions. This has been done with good success for a series of simulations with KAMM/DRAIS using meteorological and chemical fields generated by the EURAD system. A more consistent approach is to apply the same model in a coarse and refined version and nest the latter in the coarse grid version. This can be done by consecutive application (one-way nesting) or with feedback from the nested model to the coarse grid calculations (two-way nesting). Schemes that allow the incorporation of orography of higher resolution in a meteorological mesoscale model (MM5) have been developed and successfully tested. Tests within the APSIS initiative revealed a better performance of the EZM with regard to meteorology in the greater Athens area when one-way nesting was applied.

The technique of one- and two-way nesting has been used to extend the range of application of various models to more explicit treatment of smaller scale phenomena. Kunz and Moussiopoulos (1995) applied one-way nesting to the calculation of wind flow fine structures. The method of two- and one-way nesting has been widely exploited for dynamical and chemical simulations using the EURAD-system (Jakobs *et al.*, 1995). Plumes of polluted air originating from industrial centres of limited extent, which could not be resolved by coarser grids, could realistically be reproduced as first shown in a study of air pollution in East Germany.

Iberian Peninsula. Zooming of smaller areas with specific orographic, land use or emission characteristics is possible. For instance, in an application to the Iberian Peninsular it was demonstrated that the hypothesised net transport of ozone from the boundary layer to the free troposphere through large-scale convective motions under photo-smog conditions does not take place, or is only marginally efficient in this region (Ebel *et al.*, 1997c). This is because to subsidence of air from the free troposphere prevails under such conditions which leads to downward transport of ozone to the lower tropospheric layers and an overcompensation of the vertical ozone flux out of the boundary layer. Only gases which exhibit lower concentrations in the free troposphere and have their sources mainly in the planetary boundary layer (*e.g.* SO_2) can generate noticeable contributions to free tropospheric pollution through upward fluxes under anticyclonic, *i.e.* subsidence, conditions.

(ii) Larger scale simulations

The EUROTRAC models designed for application to larger scale phenomena have mainly been used for the simulation of episodes of air pollution usually extending over a few days up to a few weeks. The choice of episodes was governed by a broad range of reasons: scientific, logistical, applications or methodological. One of the first simulations on the continental scale in the framework of EUMAC was stimulated by the Chernobyl nuclear reactor accident. To coordinate model development, interactive application of larger and smaller mesoscale models so-called joint cases of air quality simulation, *i.e.* a 'wet' and a 'dry' case in spring and summer 1986, respectively, were chosen. They served as a convenient basis to explore the performance of different models with respect to their scientific and numerical formulation. Several episodes were selected according to the needs of experimental projects within EUROTRAC. The TRACT campaign in September 1992 was not only supported by smaller scale models but also by European scale simulations to provide information about the time dependent background composition of the troposphere. Furthermore, special tracer transport situations were investigated within the context of High Alpine and Snow Chemistry Studies (ALPTRAC, Kromb-Kolb *et al.*, 1997). In collaboration with TOR photo-smog episodes with a dense and high quality data basis generated by this subproject were chosen as the so-called EUMAC/TOR cases (ETC, July and August 1990; Memmesheimer *et al.*, 1997c) for joint intensive studies of photo-oxidant formation and other processes controlling ozone generation and transport of pollutants. In particular, budget studies were carried out for ozone and its precursors over Europe (Memmesheimer *et al.*, 1997b; Roemer *et al.*, 1997). The main emphasis of larger scale episodic modelling in EUROTRAC used to be the simulation of the spatial distribution of photo-oxidants and the spatial and temporal change of their concentrations. Yet other phenomena being of scientific and practical interest were included as well if they could contribute to the development of modelling tools or if sensible contributions to the solutions to environmental problems beyond that of photo-smog could be made with the tools already existing in EUROTRAC. Examples are the deposition of acidifying and nutrifying airborne tracers and the spread of the radioactive cloud after the Chernobyl accident over Europe mentioned above.

The Chernobyl accident. The event provided an opportunity to investigate, from the viewpoint of model performance and understanding of processes, several curios effects which are of interest not only for the treatment of dispersion of radioactive debris but also for the simulation of other pollutants in general. There was the strong westward transport of radioactive material from a remote East European location to densely populated areas of central and western Europe, which came surprised many used to thinking in climatological terms of average eastward winds. The simulation of this event was used to study the interplay of a big variable point source and wind fields which depended on height (Hass *et al.*, 1990). The problem of treating a point source in a model with relatively coarse resolution (80 km × 80 km in the horizontal) became evident from the unreliable results for Caesium deposition near the source (Hass *et al.*, 1995a). Nevertheless, calculations of the contamination of areas further away from the sources lead to quite reasonable qualitative results and contributed to the assessment of transport effects during the event. Yet they also showed that reliable simulation of precipitation, especially its convective component, is a prerequisite for improving the accuracy of such simulations. A later evaluation project (ATMES, Klug *et al.*, 1991) carried out for several meteorological transport models, both Eulerian and Lagrangian, yielded another important result: that comparison of models and observations should only be done with quality controlled data (as, for instance, from the recent ETEX project, the TRACT campaign, or some of the SANA field experiments).

Transport of ozone in the troposphere. Europe appears to be a considerable source of background ozone in the free troposphere of the northern hemisphere and attempts have been made to quantify the photo-oxidant production of the continent using mesoscale models. A study with LOTOS revealed that Europe is contributing about one third to the net photochemical generation of ozone in the lowest 2 km of the northern mid-latitudes (Roemer *et al.*, 1997). Simulations with EURAD illuminate the complexity of transfer of polluted air from its source regions to remote regions and to the free troposphere. Significant geographical differences exist, partly controlled by climatological, topographic and land type (especially land-sea distribution) features. When a smog episode over central or western Europe is terminated by a low pressure system approaching the continent from the Atlantic, which replaces the continental anticyclone causing elevated photo-oxidant levels, there is frequently a pre-frontal streak of air with high concentrations of ozone and other pollutants. Under such conditions the streak tends to extend north east, transporting polluted air from the western and central European sources to the Baltic Sea and Scandinavia (Memmesheimer *et al.*, 1994). An example is shown in Fig. 5.2 with simulated PAN as a tracer. The calculations also hint at an extension of the central and west European cloud of pollutants to the Mediterranean and its merging with polluted air from northern Italy.

Emission reduction scenarios were studied using a relatively simple conventional approach, namely defining emission changes proportional to the presently estimated source strengths of NO_x and VOC. As expected the reduction of the anthropogenic emission of both groups of primary pollutants leads to an overall decrease of photo-oxidants, yet locally the response can complicated by the VOC/NO_x ratio and availability of the two ozone precursors. Since both factors are not only controlled by anthropogenic and natural sources but also by atmospheric transport and mixing, the details of the reduction pattern are often affected by the meteorological situation. Work was done to prepare the models for more sophisticated emission control studies; for example:. using different mixtures of VOCs resulting from road traffic and solvent use, or regionally varying emission reduction or enhancement factors.

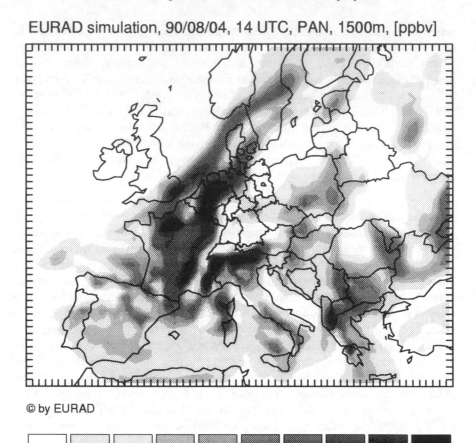

© by EURAD

0.10   0.20   0.40   0.60   0.80   1.00   1.50   2.00   5.00

Fig. 5.2: Distribution of PAN over Europe on 4th August 1990, 14:00 UTC. Mixing ratios are given in ppb. Results from a simulation with the EURAD system are shown for a height of approximately 1500 m, *i.e.* the upper part of the atmospheric boundary layer.

Effect of aerosols. The long-range transport and deposition of nitrogen compounds and related impacts on ecosystems, for instance eutrophication, may be modified by the formation of secondary aerosols, particularly ammonium sulfate, nitrate and chloride. Runs with the EURAD system indicated that the modelled wet deposition of nitrate is improving particularly in remote areas when secondary aerosol chemistry is taken into account (Ziegenbein *et al.*, 1994; Ackermann *et al.*, 1995). Such simulations are first approaches to a more comprehensive treatment of aerosols in mesoscale chemical transport models. They have to be extended to primary aerosols and to the organic fraction of particulate matter in the troposphere, taking into account aerosol dynamics. Unfortunately, EUROTRAC lacked the stronger stimulation of this kind of research. Its relevance for a better understanding of atmospheric chemistry and radiative forcing of dynamical processes became gradually to be more appreciated in the course of the project, at least on larger, *i.e.* hemispheric and global, scales. The need for an intensification of mesoscale aerosol studies with numerical models is obvious.

Cloud effects. The integral effect of clouds on ozone over polluted regions like Europe is only poorly known. During photo-smog episodes numerical simulations often lead to a net increase of the ozone content in a larger atmospheric volume due to the appearance of clouds which indicate the dominance of mixing processes in and through clouds (Memmesheimer *et al.*, 1997b). A detailed analysis of different effects by Builtjes and Matthijsen (1997) employing a modified version of the LOTOS model for a summer month (August 1990) revealed that reduction of ozone formation within clouds, due to aqueous phase chemistry, radiative changes and scavenging, may become rather large locally. Yet the overall net reduction of ozone in the planetary boundary layer over a larger domain showed to be the order of about 5 %. This estimate deviates considerably from global estimates of cloud effects on the free tropospheric ozone budget (Lelieveld and Crutzen, 1991). It should be noted that a scenario of permanently present and chemically active clouds in the global troposphere is different from that of sporadic cloud appearance over a limited area. Perhaps, this can explain the differing model results. Certainly more studies of mesoscale scenarios with inclusion of cloud processes are needed to arrive at definite conclusions about the magnitude and variability of net cloud effects.

Convective clouds. Convective redistribution by clouds is not only a significant process for ozone but also for other trace constituents (*e.g.* Mölders *et al.*, 1994; Flatøy *et al.*, 1995) and can affect the whole troposphere up to the tropopause (Ehhalt *et al.*, 1992; Brunner, 1998). Yet the significance of the effect clearly depends on the meteorological situation in a contaminated region, the distribution of emission sources of primary pollutants, the background field of the transported species together with other factors.

Budgets of trace constituents. Cloud effects are one process among others contributing to the total budget of a trace constituent over limited areas as treated with mesoscale models. Several model studies have been conducted to provide a deeper insight into the interdependence of different budget terms for a larger variety of situations, *i.e.* episodes (Ebel *et al.*, 1997c; Hov and Flatøy, 1997; Memmesheimer *et al.*, 1997a; 1997b). Results of two such studies employing different models and addressing different situations are compiled in Table 5.2. Tendencies of ozone and NO_x in the ABL are shown. To make the estimates of both studies comparable, the tendencies are normalised with the value for chemical production of the gas phase. MER (Memmesheimer *et al.*, 1997b) simulated a photo-smog episode in central Europe, whereas Hov and Flatøy (1997) treated a full summer month. The results nicely show how different transport processes (horizontal and vertical advection, convection and vertical turbulent diffusion) affect the budgets of a) a tracer like NO_x with a relatively short chemical lifetime, strong emission sources in the ABL and low background concentrations in the free troposphere, and b) a secondary species like ozone with longer chemical lifetime and higher background concentrations. NO_x

emissions are mainly consumed through photochemical reactions in the general vicinity of their sources or by being partly exported horizontally with an efficiency depending on the strength of chemical losses. Dry deposition, intimately related to vertical turbulent diffusion, proves to play a major role for the ABL ozone budget but not for that of NO_x in both cases. Yet it should be noted that different results would be found for different meteorological and land-use conditions, as discussed by Hongisto and Joffre (1997) for Finland and adjacent parts of the Baltic Sea. Studies of nitrogen deposition to a maritime environment including dry and wet deposition of HNO_3 were made with the EURAD system in the context of a study on the contribution of atmospheric transport of pollutants to the eutrophication of the North Sea (Memmesheimer and Bock, 1995).

Table 5.2:   Budgets (tendencies) of ozone and NO_x in the atmospheric boundary layer over Europe as simulated by Memmesheimer et al. (1997b, columns MER) and Hov and Flatøy (1997, columns H&F) for a photo-smog episode (31st July–5th August 1990) and a summer month (July 1991), respectively. Values normalised with the net chemical gas phase production (ozone) or loss (NO_x) estimates are given to enable easier comparison of both simulations. X: No estimate available. [ ]: Horizontal advection term estimated from the total balance of tendencies. ( ): Estimated from graphical representations. *Note that dry deposition is also included in the turbulent diffusion term in MER. Negative (positive) values indicate losses (gains) of the domain under consideration.

	ozone		NO_x	
process	MER	H&F	MER	H&F
gas phase chemistry	1.00	1.00	−1.00	−1.00
horizontal advection	−1.10	[−0.30]	−0.07	[−0.33]
vertical advection	0.00	−0.12	0.00	−0.08
dry deposition	−0.42*	−0.33	−0.07*	small
vertical turbulent diffusion	−0.43	X	−0.07	X
emission	---	---	1.14	1.36
accumulation	0.16	(0.04)	−0.02	(−0.07)

If in Table 5.2 only the transport and deposition part of the NO_x budget is considered one finds that in the MER case dry deposition with vertical eddy diffusion and horizontal advection cause equally large losses whereas in the H&F case it is mainly NO_x flux divergence of the horizontal flow. The terms for vertical advection and convection are a measure of the exchange of pollutants between the ABL and the free troposphere (Beck et al., 1997) and therefore of special interest. Results obtained in the framework of the GLOMAC project revealed that ozone of the free troposphere has considerably been increasing due to anthropogenic activities, pointing to an overall transport of photo-oxidants from their precursors from the ABL to the troposphere above on the global scale. Yet the finding that the ABL/free troposphere exchange may work in the opposite direction on the regional scale and become a major term for the ABL ozone budget under photo-smog conditions, as shown with the EURAD system, belongs to the more appealing results of EUROTRAC modelling work. In the MER case of Table 5.2, the contribution to the ozone budget through import to the ABL from the free troposphere amounts to about 70 % of that of chemical gas phase production. Situations may be encountered where the convergence of vertical ozone flux exceeds chemical production in limited areas.

Vertical fluxes. Obviously, strong convergence of the vertical ozone flux is largely compensated by divergence of the horizontal flux. This is a consequence of the meteorological situation frequently encountered during photo-smog episodes. Anticyclonic flow conditions prevail. They are characterised by subsidence or Ekman pumping (Stull, 1988) and horizontal outflow of air, *i.e.* divergent air mass flux, out of the high pressure system. The monthly episode simulated by H&F shows that, over longer periods, net vertical and horizontal outflow can also coexist. Furthermore, there may exist significant geographical differences during the same episode over the European continent (Ebel *et al.*, 1997c). Evidently, free troposphere concentrations of NO_x are too low to allow convergent downward fluxes to the ABL even under pronounced subsidence conditions as in the case discussed by MER. The downward ozone flux and its convergence are significantly controlled by the magnitude of the background concentration and the form of the vertical ozone profile, as explored in mesoscale sensitivity studies. Furthermore, the ozone intrusion into the ABL modifies the efficiency of ozone formation in the gas phase. This efficiency and the increase through vertical advection are anti-correlated (Ebel *et al.*, 1998).

Stratosphere-troposphere exchange. The subprojects TOR and EUMAC co-operated in exploring stratosphere-troposphere exchange processes at middle latitudes by combining experimental and modelling activities (Beekmann *et al.*, 1997a; 1997b). The EURAD model system, particularly its meteorological part MM5, has intensively been used to support climatological as well as diagnostic studies of the phenomenon on the mesoscale. A global climatology of tropopause folding and cut-off lows has been established using ECMWF meteorological data (Ebel *et al.*, 1996; Elbern *et al.*, 1998). [7]Be and other observations on the Zugspitze and Wank   (47° N, 11° E, 2964 m and 1780 m a.s.l.) have been analysed with the help of model case studies with respect to intrusions of stratospheric air into the free troposphere and ABL (Elbern *et al.*, 1997). Case studies of tropopause folding and cut-off lows were initialised and performed by means of mesoscale modelling in the framework of EUROTRAC (Ebel *et al.*, 1991, 1993). Later modelling work by other authors yielded downward cross-tropopause fluxes of ozone in tropopause folds ranging between about 1.5 and 10 $\times 10^{32}$ molecules per day (Lamarque and Hess, 1994; Spaete *et al.*, 1994; Meyer, 1997). It was shown that such values represent net fluxes, consisting of an upward and a downward component in the region affected by folding, cut-off lows and streamers (Appenzeller and Davies, 1992; Kowol-Santen, 1998). The use of mesoscale CTMs for the exploration of the effect of stratospheric intrusions on the chemistry of the free troposphere was still at its beginning when the first phase of EUROTRAC ended. Preliminary results have become available only recently (Ebel *et al.*, 1998; Kowol-Santen, 1998).

c.   Global chemical transport studies

Whereas regional episodic model studies treat longer lived species like methane and CO as ozone precursors of secondary importance for episodic events, these gases are a focus of research in global long-term simulation studies together with the other precursors NO_x and non-methane hydrocarbons (NMHC). Furthermore, the role of sulfur in global models is viewed from the perspective of its role in aerosol formation on a global scale and thus for the radiative forcing of the climate whereas, in regional studies, the main emphasis lies on the problem of transboundary transport, acidification and secondary aerosol formation. As we shall see in the following section, SO_2 is a convenient trace gas for regional model comparison and evaluation studies. Aims which global and regional models had in common in EUROTRAC were the search for answers to the questions about how anthropogenic activities contribute to the change of the composition of the background troposphere, and which role the exchange of air masses across the tropopause plays for the tropospheric ozone budget.

Another overlap of interests and research areas developed in the course of EUROTRAC was the study of the impact of air traffic on the chemistry of the tropopause region. The sensitivity of the ozone budget to aircraft NO_x exhaust was investigated with the global MOGUNTIA model (Zimmermann, 1994; Velders *et al.*, 1994) aiming at an assessment of the world wide change of the ozone distribution around the tropopause and possible climatic consequences. On the other side, regional studies were carried out to improve the understanding of the role of synoptic scale processes and the fine structure of aircraft emissions for the change of the ozone budget at cruise altitudes focusing on the North Atlantic flight corridor (Petry *et al.*, 1994; Flatøy and Hov, 1996). The simulations have yielded increases of ozone concentrations between 1 and 10 % of the background values depending factors like background concentration, height and season.

Dynamics and transport in the tropopause region have been investigated with different versions of global models. The results of transport calculations were compared with measurements of radioactive tracers in the troposphere. Numerical experiments seem to indicate that larger scale transport effects may well be captured with a horizontal resolution of $4° \times 5°$ (van Velthoven and Kelder, 1996). This differs from results obtained with mesoscale models indicating that for the representation of smaller scale structures of cross-tropopause fluxes a horizontal resolution of 50 km $\times$ 50 km should be chosen (Meyer, 1997).

Aircraft emissions may contribute to the greenhouse effect and have therefore been studied in connection with the change of radiative forcing due to the general increase of greenhouse gases, particularly ozone, in the context of EUROTRAC research goals (Kelder *et al.*, 1997). It has been found that radiative forcing due to aircraft emissions may range between –0.5 and +0.3 W m^{-2}. Assessing the radiative forcing effect by the overall increase of tropospheric ozone due to anthropogenic emissions values around 0.5 W m^{-2} have been derived, showing that air traffic may contribute significantly to this kind of anthropogenic perturbations of climate.

As regards the climate forcing, the role of anthropogenic aerosols and their dependence on man-made sulfur emissions was a major focus of research. Large increases of the sulfate aerosol burden resulting from anthropogenic SO_2 emissions have been estimated for the northern hemisphere in particular (Langner *et al.*, 1992; Feichter *et al.*, 1996; Rhode *et al.*, 1997). They range between a factor of about 2 and 100. The larger values were found for highly polluted European areas in winter. Calculating the contribution of anthropogenic aerosols to radiative forcing taking into account possible changes of optical cloud thickness on the global scale one arrives at reductions of the radiative flux of around 1 W/m^2 counteracting the increase, *i.e.* the global warming effect, due to anthropogenic greenhouse gas emissions (*e.g.* Charlson *et al.*, 1991; Roeckner *et al.*, 1996b; Rhode and Crutzen, 1995). Recognising the importance of reliable aerosol treatments in global models, work was done to improve the sulfate aerosol modules used so far and to include black-carbon aerosols in order to estimate possible impacts of this product of industrial activities and biomass burning on climate (Raes *et al.*, 1997).

A better understanding of nitrogen chemistry and ozone formation in the troposphere is also required to obtain realistic estimates of the aerosol background. Surface reactions of NO_3 and N_2O_5 are essential for NO_x transformations and thus ozone and OH concentrations (Crutzen *et al.*, 1997). Global simulations showed that the NO_x budget is significantly overestimated by chemical mechanisms not including heterogeneous reactions of N_2O_5 and NO_3.

Studies of possible changes of tropospheric chemistry through emissions of HCFCs, *i.e.* halogenated hydrocarbons used as a replacement of CFCs, were conducted on the basis of published emission scenarios (Kanakidou *et al.*, 1997). It was possible to reduce existing uncertainties in global emission scenarios of methane employing inverse techniques of chemistry transport modelling in the treatment of the CH_4 cycle.

The progress in understanding aerosol and cloud effects, achieved in GLOMAC and other projects dealing with global air quality issues, considerably influenced the assessment of the global and hemispheric budgets of photo-oxidants, particularly ozone. Improvement of the quantification and geographical structure of the emission sources of the main ozone precursors, namely NO_x, CO, CH_4 and NMHC lead to increasing reliability of global distribution and temporal change estimates for ozone. Taking into account the complexity of the three-dimensional tracer distributions and their non-linear interaction, it was found that a reduction of previous two-dimensional model estimates of the efficiency of photochemical O_3 production in the troposphere, and the magnitude of estimated O_3 increases due to growing anthropogenic activity since pre-industrial times (Crutzen *et al.*, 1997). Estimates of surface ozone trends based on observations (Volz and Kley, 1988; Kley *et al.*, 1988; Staehelin *et al.*, 1994) and global simulations generally agree in the magnitude of the man-made effect, namely an increase of 2-3 times (depending on season and region) of the concentrations over the last 120 years (Crutzen, 1995). These scenario calculations did not take into account episodic anthropogenic photo-smog events as discussed in Section 4.b. It remains a task of future research to combine global and regional chemical transport simulations in order to explore the interdependence of regional and global processes explicitly. In the case of the global tropospheric ozone trend induced by anthropogenic emissions, it can be expected that the estimates of the magnitude of the effect will increase (Rodhe, 1997). It is again emphasised that the ozone budget of the ABL depends on the background concentration of ozone during photo-smog episodes as discussed in the previous section.

5.5   Model comparison and evaluation

Several episodes, characterised by increased air pollution or the availability of high quality measurements, have been exploited for mesoscale model comparison. The EUMAC joint cases of spring and summer 1996 have already been mentioned. The latter case was used by Nester *et al.* (1995) to compare the KAMM/DRAIS and EURAD systems. The so-called EUMAC/TOR cases of summer 1990 also proved to provide a convenient basis for comparative studies (Langmann, 1995; Hass and Berge, 1996; Hass *et al.*, 1996, 1997). Also field campaigns like those of TRACT and SANA provided an opportunity to compare models, particularly the larger and smaller mesoscale ones used in EUROTRAC (Ebel *et al.*, 1995). Problems of model performance and questions related to spatial resolution were treated, for instance the over prediction of night time minima of ozone during photo-smog periods when coarse resolution is applied (Nester *et al.*, 1997).

Process oriented model comparisons of this kind were supplemented by more systematic approaches within APSIS (Moussiopoulos, 1993) dealing with smaller mesoscale models and two special studies only including larger mesoscale models (Hass *et al.*, 1996; 1997; Hass and Berge, 1996). These comparisons also included diagnostic model evaluation using existing ground-based observational networks, particularly the one established by EMEP for European air quality measurements. Rich data sets were generated in all comparison studies which demonstrate the usefulness and applicability of air quality models in general. They could only partially be analysed because of simulation differences resulting from differing numerical, physical and chemical formulations, from different subgrid process parameterisations and, most importantly, from the differing treatment of the initial and boundary conditions.

The intercomparison study by Hass *et al.* (1996, 1997) was carried out with four larger mesoscale oxidant models: EURAD, LOTOS, EMEP (Simpson, 1992a; 1992b; 1993) and REM3 (Stern, 1994). The four models represent an appreciable range of possible CTM design. The authors confined themselves to the analysis of four core problems, namely, the role of differences in input data, *e.g.* emissions and meteorology, the sensitivity of the single models to particularities of parameterisations, optimal model improvement on the basis of intercomparisons and the sensitivity of emission-reduction assessments to model type. Identification of the causes of differing simulation results proved to be impossible in most cases due to the unsystematic character of the deviations. This was also true for a second study where the EURAD system and the EMEP Acid Rain model (Tuovinen *et al.*, 1994) were compared with respect to their performance of sulfur transport and chemistry simulations (Hass and Berge, 1996). Nevertheless, useful general recommendations could be derived from the three major intercomparison and additional diagnostic evaluation studies (*e.g.* Hass *et al.*, 1993; 1995b). They concern the harmonisation of European episodic emission data, the analysis and improvement of natural emission treatment, the definition of deposition velocities of chemical species, the parameterisation of chemical, radiative and transport processes in clouds and the simulation of the mixing height. The studies confirmed known, and unveiled, new problems in the available data sets for model evaluation and in the understanding of the atmospheric chemical system. The quality of measurements should be controlled more rigorously and, in the case of photo-oxidant models, episodic and routine measurements of more species, particularly PAN, H_2O_2, speciated VOCs and nitrogen compounds other than NO_x, would be helpful. Also, information on the vertical structure of air pollutant fields is largely missing for episodic evaluation studies.

Clearly, not only model uncertainties, but also problems stemming from the quality and reliability of input data and observations used for comparison with model output, make model evaluation and validation a difficult exercise. Model evaluation also requires evaluation of these data where the evaluation procedure of one component of the three-component system (model, input, comparative data) has to be carried out with the help of the other two. Consequently, differences between simulations and observations can result from shortcomings of any of the three components. This implies the concept of iterative evaluation of a model and the two different types of data sets employed in model validation studies (Ebel *et al.*, 1997a).

The studies described so far belong to the category of diagnostic evaluation, *i.e.* the intercomparison of observed and simulated data and their quantitative description by statistical measures like bias or standard deviation. A more difficult method is that of process-oriented evaluation since this needs complex observational information. Analyses of this kind are rare. Focusing on one of the most important minor constituents in the atmosphere, namely water vapour, Hantel *et al.* (1995) and Hantel and Wang (1997) applied a diagnostic model (DIAMOD) to study the processes of moisture, rain and total convective heat flux using meteorological analyses of the ECMWF. Applying the method to an episode with strong rain over parts of central Europe, it was found that the meteorological model used in EURAD tends to under predict the convective heat and rain flux in the selected case. How far this result can be generalised is difficult to say since extended statistics of such demanding analyses would be needed.

Evaluation of global models and global simulations follows a strategy similar to diagnostic evaluation of episodic mesoscale simulations. The details of the approach differ due to different types and characteristics of available trace gas observations and climatologies. Seasonal and geographical variations of the trace gases, in particular ozone, are exploited. Validation studies have been carried out for the three global models employed in GLOMAC, namely MOGUNTIA, TMK and ECHAM. Comparisons of the models have been performed using ozone as the representative tracer. The differences between the models are usually moderate while deviations between simulated and observed concentrations sometimes can become significant (Rodhe, 1997).

5.6   Conclusions

EUROTRAC was successful in stimulating the development and application of global and mesoscale models. Though a EUREKA, *i.e.* European, enterprise, it also thrived on co-operation beyond its continental borders and institutional limits. Strong links of many contributors existed, for instance, with projects of the International Geo-Biosphere Program (IGBP) like IGAC. Fruitful collaboration developed with the UN-ECE project EMEP, and stimulating interrelationships were established with groups contributing to North American environmental programs like NAPAP, in particular. The attempt to establish closer links of regional modelling in the first phase of EUROTRAC with EC programs remained remarkably unsuccessful though in the area of emission modelling the EC project CORINAIR preparing a new European emission inventory made substantial contributions.

Modelling in EUROTRAC became markedly more efficient with increasing access to the largest computers with ever growing in power and performance. Massive parallel processing helped to solve problems that previously prevented the models from progressing from the second to the third generation, where they will hopefully arrive during EUROTRAC-2. The increase of computer performance provided the basis for the improvement of the physical and chemical content of the models. This enabled environmental simulations of higher complexity which is unavoidable from the scientific perspective. This trend of model development was sometimes counteracted by application needs since greater complexity of process treatment increases the cost of model products but not necessarily the (often fictitious) clarity of simulation-based statements aimed at environmental assessment.

During the second half of EUROTRAC, after the establishment of GENEMIS, the subproject succeeded in establishing a new high quality emission inventory, serving the needs of mesoscale chemical transport modellers to an unprecedentedly high degree. Together with the EUMAC group it was possible to identify and quantify certain deficiencies in the emission inventories for eastern

Europe employing three-dimensional simulations for emission sensitivity studies. They resulted in improved anthropogenic emission estimates particularly for NO_x, but also for other pollutants including non-methane VOC and SO_x. Yet serious difficulties were encountered for data collection due to ongoing economic and organisational changes in Eastern Europe. The methods for temporal disaggregation of annual emission data have been refined to take into account the different temporal behaviour of different emission sources in different areas. This resulted in a significant change of estimated emission patterns in comparison to those employed in the past. The methods allow separate treatment of individual source categories. The GENEMIS inventories show that the ratio of VOC/NO_x emissions of anthropogenic nature is higher in urban than in rural regions, lower in winter than in summer and reduced in the night.

Clearly, more user-friendly methods of emission scenario preparation are needed. The evaluation of inventories remains a difficult task still to be completed, though progress was made with a tunnel study to improve the accuracy of road traffic emission factors for NO_x, total VOC, CO, SO_2 and a large number of individual VOCs. Problems and tasks for future research are the establishment of aerosol emission inventories for CTM use and the development of emission models which can be applied to numerical ozone forecasting.

The primary goal of the mesoscale modelling subproject was the study of smog episodes on a larger European scale. Yet strong activities finally developed aiming at the investigation of air quality on smaller regional and local scales. Zooming capabilities, like the EZM (EUMAC Zooming Model), were generated and applied, for instance, in the APSIS project of EUMAC. In parallel, nesting methods were developed or improved and applied in the larger mesoscale model system EURAD. This enabled consistent simulations with rough and fine horizontal resolution ranging from about 80×80 km^2, down to 2×2 km^2 in topographically unproblematic areas.

A particularly illuminating way of analysing air quality over polluted areas proved to be the method of budget calculations over wider areas. They can be performed for arbitrary layers in the vertical. The atmospheric boundary layer and the adjacent free troposphere were studied particularly. Separating spatially integrated tendencies due to gas phase chemistry, dry deposition, cloud effects, turbulent diffusion, horizontal and vertical advection it could be shown that during photo-smog episodes downward flux of ozone from the free troposphere to the ABL can significantly contribute to the ozone budget at lower levels. This is a consequence of subsidence under mainly anticyclonic conditions during photo-smog events. The efficiency of photo-oxidant production seems to be reduced through ozone intrusion from the free troposphere into the ABL.

First steps towards three-dimensional mesoscale modelling of the formation of aerosols and the role of fog for chemical transformation of pollutants were made. The method of four-dimensional variational data assimilation was applied to observations of pollutants in the lower troposphere in the first studies. The subproject EUMAC did not succeed in carrying out substantial work on the problem of chemical transformation in air with strong spatial inhomogeneities and temporal disturbances of reactive tracer distributions. This together with three-dimensional regional aerosol simulation and four-dimensional data assimilation are promising research topics of future mesoscale modelling projects.

Atmospheric ozone budget studies were also carried out for the globe and the northern and southern hemisphere. They have been connected with an analysis of the impact of man-made emissions on the chemical system, particularly photo-oxidant production, in the atmosphere. Comparing pre-industrial and industrial conditions, it was found that the production efficiency considerably increased with increasing man-made air pollution. The effect is most pronounced in the northern hemisphere. The relative contribution of the stratosphere to the global budget of tropospheric ozone through stratosphere-tropospheric air mass exchange decreased from about 75 % of the net chemical effect in pre-industrial times, to approximately 45 % at present (Crutzen, 1995).

Inclusion of cloud chemistry in global models led to new insights into the role of clouds for the balance of ozone and other chemical species in the global troposphere. Unexpected results were also obtained through the simulation of anthropogenic sulfate aerosols. It was found that they play an important role for climate forcing compensating the radiative effect of greenhouse gases to a certain degree.

Joint publications about mesoscale air pollution modelling in the first phase of EUROTRAC can be found in three special issues which appeared in *Meteorology and Atmospheric Physics* (Vol. **57**, No. 1–7, 1995), *Atmospheric Environment* (Vol. **31**, No. 19, 1997) and *Journal of Atmospheric Chemistry* (Vol. **28**, No. 1–3, 1997, subproject TOR, containing some articles about results of mesoscale simulations).

It is gratefully acknowledged that contributions to this overview about modelling in the first phase of EUROTRAC were made by P.J.H. Builtjes, J. Feichter, F. Flatøy, A. Floßmann, J. Lelieveld, M. Memmesheimer, N. Moussiopoulos, F. Müller, K. Nester, D. Poppe and P. Zimmermann.

5.7   References

Ackermann I.J., H. Hass, M. Memmesheimer, C. Ziegenbein, A. Ebel; 1995, The parametrization of the sulfate-nitrate-ammonia system in the long-range transport model EURAD, *Meteor. Atmos. Phys.* **57**, 101–114.

Ackermann I.J., M. Memmesheimer, F. Binkowski, U. Shankar, A. Ebel; 1996, Regional particulate dynamics modelling in EURAD, in: P.M. Borrell, P. Borrell, T. Cvitaš, K. Kelly, W. Seiler (eds.), *Proc. EUROTRAC Symp. '96*, Computational Mechanics Publ., Southampton, pp. 157–161.

Adrian G., F. Fiedler; 1991, Simulation of unstationary wind and temperature fields over complex terrain and comparison with observations, *Beitr. Phys. Atmos.* **64**, 27–48.

Andersson-Sköld Y., P. Grennfelt, K. Pleijel; 1991, Results from a study of different concepts for calculating photochemical ozone creating potentials, *Proc. EMEP Workshop on Photo-oxidant Modelling for Long-Range Transport in Relation to Abatement Strategies*, Berlin pp. 260–271.

Anthes R.A., T.T. Warner; 1978, Development of hydrodynamic models suitable for air pollution and other mesometeorological studies, *Mon. Wea. Rev.* **106**, 1045–1078.

Appenzeller C., H.C. Davies; 1992, Structure of stratospheric intrusions into the troposphere, *Nature* **358**, 570–572.

Asman W., H. Nielsen, H. Langberg, S. Sommer, S. Pedersen, H. Kjaer, L. Knudsen; 1997, Diurnal and seasonal variation in the NH_3 emission rate, in: A. Ebel, R. Friedrich, H. Rodhe (eds), *Tropospheric Modelling and Emission Estimation*, Springer Verlag, Heidelberg, pp. 238–242.

Baer M, K. Nester; 1992, Parametrization of trace gas dry deposition velocities for a regional mesoscale diffusion model. *Ann. Geophysicae* **10**, 912–923.

Beck J.P., N. Asimakopoulos, V. Bazhanov, H.J. Bock, G. Chronopoulos, D. De Muer, A. Ebel, F. Flatøy, H. Hass, P. van Haver, Ø. Hov, H.J. Jakobs, E.J.J. Kirchner, H. Kunz, M. Memmesheimer, W.A.J. van Pul, P. Speth, T. Trickl, C. Varotsos; 1997, Exchange of ozone between the atmospheric boundary layer and the free troposphere, in: Ø. Hov (ed), *Tropospheric Ozone Research*, Springer Verlag, Heidelberg, pp. 111–130.

Beekmann M., G. Ancellet, S. Blonsky, D. De Muer, A. Ebel, H. Elbern, J. Hendricks, J. Kowol, C. Mancier, R. Sladkovic, H.G.J. Smit, P. Speth, T. Trickl, Ph. van Haver; 1997a, Regional and global tropopause fold occurrence and related ozone flux across the tropopause, *J. Atmos. Chem.* **28**, 29–44.

Beekmann M., G. Ancellet, S. Blonsky, D. De Muer, A. Ebel, H. Elbern, J. Hendricks, J. Kowol, C. Mancier, R. Sladkovic, H.G.J. Smit, P. Speth, T. Trickl, Ph. Van Haver; 1997, Stratosphere-troposphere exchange - Regional and global tropopause folding occurence, in: Ø. Hov (ed), *Tropospheric Ozone Research*, Springer Verlag, Heidelberg, pp. 131–152.

Benkovitz C. M., M.T. Scholtz, J. Pacyna, L. Tarrason, J. Dignon, E.C. Voldner, P.A. Spiro, J.A. Logan, T.E. Graedel; 19996, Global gridded inventories of anthropogenic emissions of sulfur and nitrogen, *J. Geophys. Res.* **101**, 29239–29253.

Berge E (ed); 1997, Emissions, dispersion and trends of acidifying and eutrophying agents, EMEP MSC-W Report 1/97, Part 1, Norwegian Meteorol. Inst., Oslo, Norway.
Borrego C., N. Barros; 1997, Development and application of a mesoscale integrated system for photochemical pollution modelling, in: A. Ebel, R. Friedrich, H. Rhode (eds), *Tropospheric Modelling and Emission Estimation*, Springer Verlag, Heildelberg, pp. 105–110.

Borrego C., M. Coutinho, N. Barros; 1995, Intercomparison of two meso-meteorological models applied to the Lisbon region, *Meteor. Appl. Phys.* **5**, 21–29.

Borrell P., P.J.H. Builtjes, P. Greunfelt, Ø. Hov (eds); 1997, *Photo-oxidants, Acidification and Tools: Policy Applications of EUROTRAC Results*, Springer Verlag, Heidelberg, 216 pp.

Brunner D.; 1998, One-year climatology oxides and ozone in the tropopause region, PhD thesis , University of Zürich, ETH No. 12556.

Builtjes P.J.H.; 1992, The LOTOS - Long Term Ozone Simulation - project; summary report, TNO report R92/240, Delft, The Netherlands.

Builtjes P.J.H., J. Matthijsen; 1997, Regional-Scale Modelling of Cloud Effects on Atmospheric Constituents, in: A. Ebel, R. Friedrich and H. Rodhe (eds), *Tropospheric Modelling and Emission Estimation*, Springer Verlag, Heidelberg, pp. 147–156.

Chang J.S., R.A. Brost, I.S.A. Isaksen, S. Madronisch, P. Middleton, W.R. Stockwell, C.J. Walcek; 1987, A three-dimensional Eulerian acid deposition model: physical concepts and formation, *J. Geophys. Res.* **92**, 14681–14700.

Charlson R.J., J. Langner, H. Rhode, C.B. Leovy, S.G. Warren; 1991, Perturbation of the northern hemisphere radiative balance by backscattering from anthropogenic sulfate aerosols, *Tellus* **43AB**, 152–163.

Chaumerliac N., E. Richard, R. Rosset; 1990, Acidity production in orographic clouds and rain in a mesoscale model with semispectral microphysics, *Atmos. Environ.* **24A**, 1573–1584.

Clappier A., J. Kübler, F. Müller, V. Sahya, A. Martilli, B.C. Krüger, V. Simeonov, F. Jeanneret, B. Balpini, H. van den Berg; 1997, Modelisation du smog photochimique en territoire genevois et analyse de scenarios de reduction des emissions de pollutants. EPFL-LPAS-Report, Lausanne, Switzerland.

Crutzen P.J.; 1995, Ozone in the troposphere, in: H.B. Snig (ed), *Composition, Chemistry, and Climate of the Atmosphere*, van Nostrand Reinhold, ITP Incorp., New York, pp. 349–393.

Crutzen P.J., P.H. Zimmermann; 1991, The changing photochemistry of the troposphere, *Tellus*, **43AB**, 136–151.

Crutzen P., R. Brost, F. Dentener, H. Feichter, R. Hein, M. Kanakidou, J. Lelieveld, P. Zimmermann; 1997, Development of a time-dependent global tropospheric air chemistry model 'GLOMAC' based on the weather forecast model of the 'ECMWF', in: A. Ebel, R. Friedrich, H. Rodhe (eds), *Tropospheric Modelling and Emission Estimation*, Springer Verlag, Heidelberg, pp. 380–400.

Daley R.; 1991, *Atmospheric Data Analysis*, Cambridge University Press, Cambridge.

Dentener F.; 1993, Heterogeneous chemistry in the troposphere, Ph.D. Thesis, State University Utrecht, The Netherlands.

Dentener F.J., P.J. Crutzen; 1993, Reaction of N_2O_5 on tropospheric aerosols: impact on the global distributions of NO_x, O_3 and OH, *J. Geophys. Res.* **98**, 7149–7163.

Derwent R.G., M.E. Jenkin; 1991,Hydrocarbons and the long-range transport of ozone and PAN across Europe, *Atmos. Environ.* **25A**, 1661–1678.

Ebel A., H. Hass, H.J. Jakobs, M. Memmesheimer, M. Laube, A. Oberreuter; 1991, Simulation of ozone intrusion caused by a tropopause fold and cut-off low, *Atmos. Environ.* **25A**, 2131–2144.

Ebel A., H. Elbern, A. Oberreuter; 1993, Stratosphere-troposphere air mass exchange and cross-tropopause fluxes of ozone,in: E.V. Thrane *et al.* (eds), *Coupling processes in the Lower and Middle Atmosphere*, Kluwer Academic Publ., Dordrecht, pp. 87–94.

Ebel A., H. Elbern, H. Hass, H.J. Jakobs, M. Memmesheimer, H.J. Bock; 1995, Meteorological effects on air pollutant variability on regional scales, in: A. Ebel, N. Moussiopoulos (eds), *Air Pollution III*, Computational Mechanics Publ., Southampton, pp. 1–6.

Ebel A., H. Elbern, J. Hendricks, R. Meyer; 1996, Stratosphere-troposphere exchange and its impact on the structure of the lower stratosphere, *J. Geomag. Geoelectr.* **48**, 135–144.

Ebel A., H. Elbern, H. Hass, H.J. Jakobs, M. Memmesheimer, M. Laube, A. Oberreuter, G. Piekorz; 1997a, Simulation of chemical transformation and transport of air pollutants with the model system EURAD, in: A. Ebel, R. Friedrich, H. Rodhe (eds), *Tropospheric Modelling and Emission Estimation*, Springer Verlag, Heidelberg, pp. 27–45.

Ebel, A., R. Friedrich, H. Rodhe (eds), 1997b; *Tropospheric Modelling and Emission Estimation*, Springer Verlag, Heidelberg; Vol. 7 of the EUROTRAC final report, 440 pp.

Ebel A., M. Memmesheimer, H.J. Jakobs; 1997c, Regional modeling of tropospheric ozone distribution and budgets, in: C. Varotsos (ed), *Global Environmental Change*, NATO ASI Series, Subseries I, Vol. **53**, Springer Verlag, Heidelberg, pp. 39–59.

Ebel A., H. Elbern, H. Feldmann, H.J. Jakobs, C. Kessler, M. Memmesheimer, A. Oberreuter, G. Piekorz; 1997d, Air pollution studies with the EURAD Model System (3): EURAD - European Air Pollution Dispersion Model System, Mitteilungen aus dem Institut für Geophysik und Meteorologie der Universität zu Köln, No. 120.

Ebel A., M. Memmesheimer, H.J. Jakobs, E. Lippert, J. Kowol-Santen, R. Meyer, H.J. Bock; 1998, Vertical air mass flux in the troposphere and lowest stratosphere and their role for regional ozone budgets at mid-latitudes, in: R.D. Bojkov, G. Visconti (eds.), *Atmospheric Ozone, Proc. XVIII Quadr. Ozone Symp., L'Aquila 1996*, Intern. Ozone Commision, pp. 347–350.

Ehhalt D.H., F. Rohrer, A. Wahner; 1992, Sources and distribution of NO_x in the upper troposphere at northern mid-latitudes, *J. Geophys. Res.* **97**, 3725–3738.

Elbern H., J. Hendricks, A. Ebel; 1998, A climatology of tropopause folds by global analyses, *Theor. Appl. Climat.* **59**, 181–200.

Elbern H., J. Kowol, R. Sládkovic, A. Ebel; 1997, Deep stratospheric intrusions: A statistical assessment with model guided analysis, *Atmos. Environ.* **31**, 3207–3226.

Elbern H., S. Tilmes, H. Schmidt; 1996, Data assimilation for chemistry transport models, a contribution to subproject EUMAC; in: P.M. Borrell, P. Borrell, T. Cvitaš, K. Kelly, W. Seiler (eds), *Proc. of EUROTRAC Symp. '96*, Computational Mechanics Publ., Southampton, pp. 681–685.

Erisman J.W., A. van Pul, G.P. Weyers; 1994, Parametrization of surface resistance for the quantification of acidifying pollutants and ozone, *Atmos. Environ.* **28**, 2595–2607.

Ertel H.; 1944, Wettervorhersagen als Randproblem, *Meteor. Zeitschrift* **61**, 181–190.

Feichter J., P.J. Crutzen; 1990, Parameterization of vertical tracer transport due to deep cumulus convection in a global transport model and its evaluation with ^{222}Radon measurements, *Tellus*, **42B**, 100–117.

Feichter J., E. Kjellstrom, H. Rhode, F. Dentener, J. Lelieveld, G.J. Roelofs; 1996, Simulation of the tropospheric sulfur cycle in a global climate model, *Atmos. Environ.* **30**, 1693–1707.

Fischer M., D.J. Lary; 1995, Lagrangian four-dimensional variational data assimilation of chemical species, *Q.J.R. Meteorol. Soc.* **12**, 1681–1704.

Flatøy F., Hov Ø.; 1996, Model studies of the effects of aircraft emissions in the north Atlantic flight corridor during October 26-November 13, 1994, and June 18 to July 5, 1995, in: U. Schumann (ed), *Contribution to the Final Report of the POLINAT Project*, Commission of the European Communities.

Flatøy F., Ø. Hov; 1997, NO_x from lightning and the calculated chemical composition of the free troposphere, *J. Geophys. Res.* **102**, 21373–21382.

Flatøy F., Ø. Hov, H. Smit; 1995, 3-D model studies of exchange processes of ozone in the troposphere over Europe, *J. Geophys. Res.* **100**, 11465–11481.

Flossmann A., T. Cvitaš, D. Möller, G. Mauersbeger; 1995, *Clouds: Models and Mechanisms,* EUROTRAC ISS, Garmisch-Partenkirchen, 93 pp.

Forkel R., W. Seidel, R. Dlugi, E. Deigele; 1990, A one-dimensional numerical model to simulate formation and balance of sulfate during radiation fog events, *Geophys. Res.* **95**, 18501–18515.

Forkel R., W. Seidl, A. Ruggaber, R. Dlugi; 1995, Fog chemistry during EUMAC joint cases: analysis of routine measurements in southern Germany and model calculations, *Meteor. Appl. Phys.* **57**, 61–86.

Forkel R., A. Ruggaber, W. Seidl, R. Dlugi; 1997, Modelling and parameterisation of chemical reactions in connection with fog events, in: A. Ebel, R. Friedrich, H. Rodhe eds), *Tropospheric Modelling and Emission Estimation,* Springer Verlag, Heidelberg, pp. 139–145.

Friedrich R.; 1997, GENEMIS: Assessment, improvement, and temporal and spatial disaggregation of European emission data, in: A. Ebel, R. Friedrich, H. Rodhe (eds), *Tropospheric Modelling and Emission Estimation,* Springer Verlag, Heidelberg, pp. 181–214.

Gantner L., J. Egger; 1997, Nesting of mesoscale models in complex terrain, in: A. Ebel, R. Friedrich, H. Rhode (eds), *Tropospheric Modelling and Emission Estimation,* Springer Verlag, Heildelberg, pp. 97–103.

Gardner R.M. *et al.*; 1997, The ANCAT/EC global inventory of NO_x emissions from aircraft, *Atmos. Environ.* **31**, 1751–1766.

Giovannoni, J.-M., A. Russell; 1995, Impact of using prognostic and objective wind fields on the photochemical modeling of Athens Greece, *Atmos. Environ.* **29**, 3633–3654.

Giovannoni J.M., A. Clappier, A. Russell; 1995, Ozone Control Strategy Modeling and Evaluation for Athens, Greece: ROG vs. NO_x Effectiveness and the Impact of Using Different Wind Field Preparation Techniques, *Meteor. Appl. Phys.* **57**, 3–20.

Giovannoni J.M., F. Müller, A. Clappier, A.G. Russell; 1997, An air quality model that incorporates aqueous-phase chemistry for calculating concentrations of ozone, as well as nitrogen and sulfur containing pollutants in Europe, in: A. Ebel, R. Friedrich, H. Rodhe (eds), *Tropospheric Modelling and Emission Estimation,* Springer Verlag, Heidelberg, pp. 111–119.

Grell G.A., J. Dudhia, D.R. Stauffer; 1993, A description of the fifth-generation PENNSTATE/NCAR mesoscale model (MM5). NCAR technical note TN-398+IA.

Gronas S., A. Foss, M. Lystad; 1987, Numerical simulations on polar lows in the Norwegian Sea, *Tellus* **39A**, 224–353.

Grosch N.; 1996, Emission inventory of the motor vehicle traffic in Northrhine-Westfalia, *Proc. 29th Intern. Symp. on Automotive Technology and Automation,* Florence, Italy, pp. 157–165.

Hantel M., Y. Wang; 1997, Diagnostics of vertical subscale fluxes (SUBFLUX), in: A. Ebel, R. Friedrich, H. Rodhe (eds), 1997; *Tropospheric Modelling and Emission Estimation,* Springer Verlag, Heidelberg, pp. 65–72.

Hantel M., H.J. Jakobs, Y. Wang; 1995, Validation of parameterized convective fluxes with DIAMOD, *Meteor. Appl. Phys.* **57**, 201–227.

Harley R.A., A.G. Russell, G. McRae, J. Cass, J.H. Seinfeld; 1993, Photochemical air quality modeling of the Southern California air quality study, *Env. Sci. & Tech.* **27**, 378–388.

Hass H.; 1991, Description of the EURAD Chemistry-Transport-Model version 2 (CTM2), Mitteilungen aus dem Institut für Geophysik und Meteorologie der Universität zu Köln, No. 83.

Hass H., E. Berge; 1996, A diagnostic comparison of EMEP and EUARD model results for a wet deposition episode in July 1990, Bericht des EMEP/MSC-W, Note 4/96.

Hass H., A. Ruggaber; 1995, Comparison of two Algorithms for Calculating Photolysis Frequencies including the Effects of Clouds, *Meteor. Appl. Phys.* **57**, 87–100.

Hass H., M. Memmesheimer, H. Geiss, H.J. Jakobs, M. Laube, A. Ebel; 1990, Simulation of the Chernobyl radioactive cloud over Europe using the EURAD model, *Atmos. Environ.* **24A**, 673–692.

Hass H., A. Ebel, H. Feldmann, H.J. Jakobs, M. Memmesheimer; 1993, Evaluation studies with a regional chemical transport model (EURAD) using air quality data from the EMEP monitoring network, *Atmos. Environ.* **27A**, 867–887.

Hass H., H.J. Jakobs, A. Ebel; 1995a, Dispersionsrechnungen für Radioisotope in der Folge des Reaktorunfalls von Tschernobyl, Bericht zum Vorhaben: Wissenschaftler helfen Tschernobyl-Kindern, FuE-Vorhaben 76 293 und 77 551 im Auftrag des GSF-Forschungszentrum für Umwelt und Gesundheit GmbH, 60 S.

Hass H., H.J. Jakobs, M. Memmesheimer; 1995b, Analysis of a regional model (EURAD) near surface gas concentration predictions using observation from networks, *Meteor. Appl. Phys.* **57**, 173–200.

Hass H., P.J.H. Builtjes, D. Simpson, R. Stern, H.J. Jakobs, M. Memmesheimer, G. Piekorz, M. Roemer, P. Esser, E. Reimer; 1996, Comparison of photo-oxidant dispersion model results, EUROTRAC ISS, Garmisch Partenkirchen, 1996.

Hass H., P.J.H. Builtjes, D. Simpson, R. Stern; 1997, Comparison of model results obtained with several European regional air quality models, *Atmos. Environ.* **31**, 3259–3279.

Heimann M.; 1995, The TM2 tracer model, model description and user manual. Technical Report, No. 10, ISSN 0940-9327, Deutsches Klimarechenzentrum, Hamburg, 47 pp.

Heupel M.K.; 1995, Ozone creating potentials of different hydrocarbons as calculated with a one-dimensional version of EURAD model, in: Air Pollution Studies with the EURAD Model System (2), Mitteilungen aus dem Institut für Geophysik und Meteorologie der Universität zu Köln, No. 105, pp. 29–56.

Heymann M.; 1997, Effects of the economic crisis on atmospheric emissions in central and eastern europe 1990–1992: Results of the CERES project, in: A. Ebel, R. Friedrich, H. Rodhe (eds), *Tropospheric Modelling and Emission Estimation*, Springer Verlag, Heidelberg, pp. 269–278.

Hongisto M., S.M. Joffre; 1997, Transport Modelling over Sea Areas, in: A. Ebel, R. Friedrich, H. Rodhe (eds), *Tropospheric Modelling and Emission Estimation*, Springer Verlag, Heidelberg, pp. 52–58.

Hov Ø., F. Flatøy; 1997, Convective redistribution of ozone and oxides of nitrogen in the troposphere over Europe in summer and fall, *J. Atmos. Chem.* **28**, 319–337.

Isaksen I.S.A., J.E. Jonson, T. Berntsen; 1997, Model studies of ozone on regional scales in the troposphere, in: Ø. Hov (ed), *Tropospheric Ozone Research*, Springer Verlag, Heildelberg, pp. 455–460.

Jakobs H.J., H. Feldmann, H. Hass, M. Memmesheimer; 1995, The use of nested models for air pollution studies: an application of the EURAD model to a SANA episode, *J. Appl. Met.* **34**, 1301–1319.

Kanakidou M., P. David, N. Poisson; 1997, , Impact of VOCs of natural and anthropogenic origin on the oxidising capacity of the atmosphere, in: A. Ebel, R. Friedrich, H. Rodhe (eds), *Tropospheric Modelling and Emission Estimation*, Springer Verlag, Heidelberg, pp. 392–400.

Kelder H., M. Allaart, J.P. Beck, R. van Dorland, P. Fortuin, L. Heijboer, A, Jeuken, M. Krol, T.H. The, P. van Velthoven, G. Verver; 1997, Contributions to global modelling of transport, atmospheric composition and radiation, in: A. Ebel, R. Friedrich, H. Rodhe (eds), *Tropospheric Modelling and Emission Estimation*, Springer Verlag, Heidelberg, pp. 405–426.

Kessler CH.; 1995, Entwicklung eines effizienten Lösungsverfahrens zur modellmäßigen Beschreibung der Ausbreitung und chemischen Umwandlung reaktiver Luftschadstoffe, PhD thesis, Karlsruhe, Verlag Shaker, Aachen.

Kley D., A. Volz, F. Mülheims; 1988, Ozone measurements in historic perspective, in: I.S.A. Isaksen (ed), *Tropospheric Ozone*, Reidel Verlag, pp.63–72.

Klug W., G. Graziani, G. Grippa, D. Pierce, C. Tassone (eds); 1991, *Evaluation of Long Range Atmospheric Transport Models Using Environmental Radioactivity Data from the Chernobyl Accident, The ATMES Report*, Elsevier Appl. Sci., London and New York.

Köble R., G. Smiatek; 1997, Mapping land cover for Europe, in: A. Ebel, R. Friedrich, H. Rodhe (eds), *Tropospheric Modelling and Emission Estimation*, Springer Verlag, Heidelberg, pp. 261–268.

Kowol-Santen J.; 1998, Nummerische Analysen von Transport- und Austauschprozessen in der Tropopausenregion der mittleren Breiten, Mitteilungen aus dem Institut für Geophysik und Meteorologie der Universität zu Köln, No. 123.

Kromp-Kolb H., P. Seibert, W. Schöner; 1996, Meteorological Support Study, in: S. Fuzzi, D. Wagenbach (eds), *Cloud Multi-phase Processes and High Alpine Air and Snow Chemistry*, Springer Verlag, Heidelberg, pp. 263–270.

Krüger, B.C., J. Kübler, V. Sathya, A. Clappier, F. Kirchner, A. Martilli; 1998, Modelling of the Geneva area in summer 1996, in: H. Hass, I.J. Ackermann (eds), *Global and Regional Atmospheric Modelling*, Proc. 1st GLOREAM Workshop, Aachen September 1997, Ford Forschungszentrum Aachen, pp. 79–88.

Kuhn M., P.J.H. Builtjes, D. Poppe, D. Simpson, W.R. Stockwell, Y. Andersson-Sköld, A. Baart, M. Das, F. Fiedler, O. Hov, F. Kirchner, P.A. Makar, J.B. Milford, M.G.M. Roemer, R. Ruhnke, A. Strand, B. Vogel, H. Vogel; 1998, Intercomparison of the gas-phase chemistry in several chemistry and transport models, *Atmos. Environ.* **32**, 693–709.

Kunz R., N. Moussiopoulos; 1995, Simulation of the wind field in Athen using refined boundary conditions, *Atmos. Environ.* **29**, 3575–3592.

Kunz R., N. Moussiopoulos; 1997, Implementation and assessment of a one-way nesting technique for high resolution wind flow simulations, *Atmos. Environ.* **31**, 3167–3176.

Lamarque J.-F., P.G. Hess; 1994, Cross-tropopause mass exchange and potential vorticity budget in a simulated tropopause folding, *J. Atmos. Sci.* **51**, 2246–2269.

Langmann B.; 1995, Einbindung der regionalen troposphärischen Chemie in die Hamburger Klimamodellumgebung: Modellrechnungen und Vergleich mit Beobachtung. PhD thesis (ISSN 0938–5177), University of Hamburg.

Langmann B., H. Graf; 1994, Another meteorological driver (HIRLAM) for the chemistry transport model (CTM) of the EURAD system: first simulations, in: P.M. Borrell, P. Borrell, T. Cvitaš, W. Seiler (eds), *Proc. EUROTRAC Symp. '94*, SPB Academic Publishing bv, The Hague, pp. 830–833.

Langmann B., H.-F. Graf; 1997, The chemistry of the polluted atmosphere over Europe: Simulations and sensitivity studies with a regional chemistry-transport-model, *Atmos.Environ.* **31**, 239–3257.

Langner J., H. Rhode, P.J. Crutzen, P. Zimmermann; 1992, Anthropogenic influence on the distribution of tropospheric sulfate aerosol, *Nature* **359**, 712–715.

Lelieveld J., P.J. Crutzen; 1990, Influences of cloud photochemical processes on tropospheric ozone, *Nature* **343**, 227–233.

Lelieveld J., P.J. Crutzen; 1991, The role of clouds in tropospheric photochemistry, *J. Atmos. Chem.* **12**, 229–267.

Lelieveld J., P.J. Crutzen, F.J. Dentener;1998, Changing concentration, lifetime and climate forcing of atmospheric methane. *Tellus*, **50B**, 128–150.

Lenhart L., T. Heck, R. Friedrich; 1997, The GENEMIS inventory: European emission data with high temporal and spatial resolution, in: A. Ebel, R. Friedrich, H. Rodhe (eds), *Tropospheric Modelling and Emission Estimation*, Springer Verlag, Heidelberg, pp. 217–222.

Lenz C.J.;1996, Energieumsetzungen an der Erdoberfläche in gegliedertem Gelände. Dissertation, Fakultät für Physik der Universität Karlsruhe, Institut für Meteorologie und Klimaforschung, 249 pp.

Liu X., G. Mauersberger, D. Möller; 1997, The effects of cloud processes on the tropospheric photo-chemistry: an improvement of the EURAD model with a coupled gaseous and aqueous chemical mechanism, *Atmos. Environ.* **31**, 3119–3135.

Lübkert B., M. Schöpp; 1989, A model to calculate natural VOC emissions from forests in Europe, IIASA, WP–89–082.

Lurman, F.W., W.P.L. Carter, L.A. Coyner; 1987, A surrogate species chemical reaction mechanism for urban-scale air quality simulation models. I. Adaptation of the Mechanism. II. Guidelines for using the mechanism, Report to the U.S. Environmental Protection Agency, EPA/600/3-87/014.

Madronich S.; 1987, Photodissociation in the atmosphere, 1. actinic flux and the effects of ground reflections and clouds, *J. Geophys. Res.* **92,D8**, 9740–9752.

Matthijsen J., P.J.H. Builtjes, D.L. Sedlak; 1995, Cloud model experiments of the effect of eron and copper on tropospheric ozone under marine and continental conditions, *Meteor. Appl. Phys.* **57**, 43–60.

Matthijsen J., P.J.H. Builtjes, E.W. Meijer, G. Boersen; 1997, Modelling cloud effects on ozone on a regional scale: a case study, *Atmos.Environ.* **31**, 3227–3238.

Mauersberger G.; 1997, Cloud chemistry modelling, in: A. Ebel, R. Friedrich, H. Rodhe (eds), *Tropospheric Modelling and Emission Estimation*, Springer Verlag, Heidelberg, pp. 121–126.

Memmesheimer M., H.J. Bock; 1995, Deposition of nitrogen into the North Sea for air pollution episodes calculated with the EURAD model, Mitteilungen aus dem Institut für Geophysik und Meteorologie der Universität zu Köln, No. 105, pp. 7–70.

Memmesheimer M., J. Tippke, A. Ebel, H. Hass, H.J. Jakobs, M. Laube; 1991, On the use of EMEP emision inventories for European scale air pollution modelling with the EURAD model, *Proc. EMEP Workshop on Photo-oxidant Modelling for Long-Range Transport in Relation to Abatement Strategies*, Berlin, pp. 307–324.

Memmesheimer M., H. Hass, H.J. Jakobs, A. Ebel; 1994, Simulation of photo-smog episode in summer 1990, in: P.M. Borrell, P. Borrell, T. Cvitaš, W. Seiler (eds), *Proc. EUROTRAC Symp. '94*, SPB Academic Publishing bv, The Hague, pp. 858–861.

Memmesheimer M., H.J. Bock, A. Ebel, M. Roemer; 1997a, Ozone and its precursors in Europe: photochemical production and transport across regional boundaries, in: Ø. Hov (ed), *Tropospheric Ozone Research*, Springer Verlag, Heidelberg, pp. 153–203.

Memmesheimer M., A. Ebel, M. Roemer; 1997b, Budget calculations for ozone and its precursors: seasonal and episodic features based on model simulations, *J. Atmos. Chem.* **28** (1997b) 283–317.

Memmesheimer M., A. Ebel, H.J. Bock, H. Elbern, H. Hass, H.J. Jakobs, G. Piekorz; 1997c, EURAD in TOR: Simulation and analysis of a photo-smog episode, in: Ø. Hov (ed), *Tropospheric Ozone Research*, Springer Verlag, Heidelberg, pp. 441–447.

Memmesheimer M., H.J. Jakobs, A. Oberreuter, H.J. Bock, G. Piekorz, A. Ebel, H. Hass; 1997d, Application of the EURAD model to a summersmog episode using GENEMIS emission data, in: A. Ebel, R. Friedrich, H. Rodhe (eds), *Tropospheric Modelling and Emission Estimation*, Springer Verlag, Heidelberg, pp. 243–248.

Meyer R.; 1997, Quantitative Analysen von Luftmassenflüssen durch die Tropopause in mittleren Breiten mit einem mesoskaligen Modell, Mitteilungen aus dem Institut für Geophysik und Meteorologie der Universität zu Köln, No. 121.

Mölders N., H. Hass, H.J. Jakobs, M. Laube, A. Ebel; 1994, Some effects of different cloud parametrizations in a mesoscale model and a chemistry transport model, *J. Appl. Met.* **33S**, 845.

Moussiopoulos N.; 1989, Mathematische Modellierung mesoskaliger Ausbreitung in der Atmosphäre, Fortschr. Ber., VDI Reihe 15, Nr. 64.

Moussiopoulos N.; 1993, Athenian photochemical smog: intercomparison of simulations (APSIS), background and objectives, *Environ. Software* **8**, 3–8.

Moussiopoulos N. (ed); 1994, *The EUMAC Zooming Model*, EUROTRAC ISS, Garmisch Partenkirchen, 266 pp.

Moussiopoulos N.; 1995, The EUMAC Zooming Model, a tool for local-to-regional air quality studies, *Meteor. Atmos. Phys.* **57**, 115–134.

Moussiopoulos N., Th. Flassak; 1989, A fully vectorized fast direct solver of the Helmholtz equation; in: C.A. Brebbia, A. Peters (eds) *Applications of supercomputers in engineering. Algorithms, computer systems and user experience*, Elsevier, Amsterdam, 67–77.

Moussiopoulos N., G. Ernst, Th. Flassak, Ch. Kessler, P. Sahm, R. Kunz, Ch. Schneider, T. Voegele, K. Karatzas, V. Megaritti, S. Papalexiou; 1997a, The EUMAC zooming model, a tool supporting environmental policy decisions in the local-to-regional scale, in: A. Ebel, R. Friedrich, H. Rodhe (eds), *Tropospheric Modelling and Emission Estimation*, Springer Verlag, Heidelberg, pp. 81–96.

Moussiopoulos N., P. Sahm, R. Kunz, T. Vögele, Ch. Schneider, Ch. Kessler; 1997b, High-resolution simulations of the wind flow and the ozone formation during the Heilbronn ozone experiment, *Atmos. Environ.* **31**, 3177–3186.

Müller, F., A. Clappier, B. Krüger; 1996, On the consideration of clouds in Eulerian chemistry transport models, in: P.M. Borrell, P. Borrell, T. Cvitaš, K. Kelly, W. Seiler (eds) *Proc. EUROTRAC Symp. '96,* Computational Mechanics Publications, Southampton, pp. 131–136.

Nester K., F. Fiedler; 1991, Comparison of measured and simulated SO_2, NO, NO_2 and ozon concentrarions for an episode of the TULLA experiment, in : H. van Dop, G. Kallos (eds), *Air Pollution Modelling and its Application*, Proc. 19th NATO/CCMS ITM, Ierapetra, Crete, Greece, Plenum Press, pp. 259–266.

Nester A., F. Fiedler; 1992, Modelling of the diurnal variation of air pollutants in a mesoscale area, *Proc. 9th World Clean air Congress,* Montreal, Vol., Paper-No. IU-16C.02.

Nester K., H.J. Panitz, F. Fiedler; 1995, Comparison of the DRAIS and EURAD model simulation of air pollution in a mesoscale area, *Meteor. Atmos. Phys.* **57**, 135–158.

Nester K., F. Fiedler, H.J. Panitz; 1997, Coupling of the DRAIS model with the EURAD model and analysis of subscale phenomena, in: A. Ebel, R. Friedrich, H. Rodhe (eds), 1997; *Tropospheric Modelling and Emission Estimation,* Springer Verlag, Heidelberg, pp. 73–80.

Oort A.H.; 1983, Global atmospheric circulation statistics, 1958–1973, NOAA Professional paper, No. 14, US Government Printing Office, Washington DC.

Panitz H.J., K. Nester, F. Fiedler; 1997, Determination of mass balances of chemically reactive air pollutants over Baden-Württemberg (FRG) - Study for the regions around the cities of Stuttgart and Freudenstadt, in: H. Power, T. Tirabasse, C.A. Brebbia (eds), *Air Pollution V,* Computational Mechanics Publications, Southampton, pp. 413–422.

Petry H.; 1993, Zur Wahl der Anfangskonzentrationen für die numerische Modellierung regionaler troposphärischer Schadstofffelder, Mitteilungen aus dem Institut für Geophysik und Meteorologie der Universität zu Köln, No. 89, 1993.

Petry H., H. Elbern, E. Lippert, R. Meyer; 1994, Three dimensional mesoscale simulations of airplane exhaust impact in a flight corridor, in: U. Schumann, D.Wurzel (eds*), Impact of Emissions From Aircraft,* Proc. Intern. Sci. Colloquium, Köln (Cologne), Germany, 1994; Mitteilung 94-06, DLR, pp. 329–335.

Poppe D., J. Zimmermann, M. Kuhn; 1997, Validation of the gas-phase chemistry of the EURAD model by comparison with field measurements, in: A. Ebel, R. Friedrich, H. Rodhe (eds), *Tropospheric Modelling and Emission Estimation,* Springer Verlag, Heidelberg, pp. 59–64.

Preisendorfer R.W.; 1988, in: C.D. Mobley (ed), *Principal Component Analysis in Meteorology and Oceanography,* Elsevier, Amsterdam, pp. 425.

Raes F., J. Wilson, W. Cooke; 1997, Implementation of aerosol processes in global transport models, in: A. Ebel, R. Friedrich, H. Rodhe (eds), *Tropospheric Modelling and Emission Estimation,* Springer Verlag, Heidelberg, pp, 401–404.

Rodhe H.; 1997, An overview of global atmospheric chemistry modelling, in: A. Ebel, R. Friedrich, H. Rodhe (eds), *Tropospheric Modelling and Emission Estimation,* Springer Verlag, Heidelberg, pp. 361–372.

Rodhe H., P. Crutzen; 1995, Climate and CCN, *Nature* **375**, 111.

Rodhe H., L. Gallardo, U. Hansson, E. Kjellström, J. Langer; 1997, Modelling global distribution of sulfur and nitrogen compounds, in: A. Ebel, R. Friedrich, H. Rodhe (eds), *Tropospheric Modelling and Emission Estimation,* Springer Verlag, Heidelberg, pp. 373–379.

Roeckner E., K. Arpe, L. Bengtsson, M. Christoph, M. Claussen, L. Dümenil, M. Esch, M. Giorgetta, U. Schlese, U. Schulzweida; 1996a, The atmospheric general circulation model ECHAM-4: Model description and simulation of present-day climate, Report Max-Planck-Institute for Meteorology, Hamburg, Germany.

Roeckner E., T. Siebert, J. Feichter; 1996b, Climatic response to anthropogenic sulfate forcing simulated with a general circulation model, in: R.J. Charlson and J. Heintzenberg (eds.), *Aerosol Forcing of Climate,* John Wiley and Sons.

Roelofs G.J., J. Lelieveld; 1995, Distribution and budget of O_3 in the troposphere calculated with a chemistry-general circulation model, *J. Geophys. Res.* **100**, 20983–20998.

Roemer M.G.M., R. Bosman, T. Thijsse, P.J.H. Builtjes, J.P. Beck, M. Vosbeek, P. Esser; 1997, Budget of ozone and precursors over Europe, in: Ø. Hov (edr), *Tropospheric Ozone Research,* Springer Verlag, Heidelberg, pp. 461–468.

Ruggaber A., R. Dlugi, T. Nakajima; 1994, Modelling of Radiation quantities and photolysis frequencies in the troposphere, *J. Atmos. Chem.* **18**, 171–210.

Ruggaber A., R. Dlugi, A. Bott, R. Forkel, H. Herrmann, H.-W. Jacobi; 1997, Modelling of radiation quantities and photolysis frequencies in the aqueous phase in the troposphere, in: EUMAC: European Modelling of Atmospheric Constituents, *Atmos. Environ.* **31**, 3137–3150.

Sandu A., J.G. Verwer, M. van Loon, G.R. Carmichael, F.A. Potra, D. Dabdub, J.H. Seinfeld; 1997, Benchmarking stiff ODE solvers for atmospheric chemistry problems - I. Implicit vs explizit, *Atmos. Environ.* **31**, 3151–3166.

Schädler G., N. Kalthoff, F. Fiedler; 1990, Validation of a model for heat, mass and momentum exchange over vegetated surfaces using LOTREX-10E/HIBE88 data, *Contr. Atmos. Phys.* **63**, 85–100.

Schneider Ch., Ch. Kessler, N. Moussiopoulos; 1997, Influence of emission input data on ozone level prediction for the Upper Rhine Valley, *Atmos. Environ.* **31**, 3187–3205.

Simpson D.; 1992a, Long-period modelling of photochemical oxidants in Europe: A) Hydrocarbon reactivity and ozone formation in Europe, B) On the linearity of country-to-country ozone calculations in Europe, *EMEP MSC-W Note 1/92,* Norw. Meteorol. Inst. Oslo.

Simpson D.; 1992b, Long-period modelling of photochemical oxidants in Europe, Model calculation for July 1985, *Atmos. Environ.* **26A**, 1609–1634.

Simpson D.; 1993, Photochemical model calculations over Europe for two extended summer periods:1985 and 1989, Model results and comparison with observations, *Atmos. Environ.* **27A**, 921–943.

Slanina J. (ed); 1997, *Biosphere-Atmosphere Exchange of Pollutants and Trace Substances,* Springer Verlag, Heidelberg.

Spaete P., D.R. Johnson, T.K. Schaak; 1994, Stratospheric-tropospheric mass exchange during the President's Day Storm, *Mon. Weather Rev.* **122**, 424–439.

Staehelin J., K. Schläpfer, 1997, Emission factors from road traffic from a tunnel study (Gubrist Tunnel, Switzerland), in: A. Ebel, R. Friedrich, H. Rodhe (eds), *Tropospheric Modelling and Emission Estimation*, Springer Verlag, Heidelberg, pp. 249–259.

Staehelin J., J. Thudium, R. Buehler, A. Volz-Thomas, W. Graber, 1994, Trends in surface ozone concentrations at Arosa (Switzerland*)*, *Atmos. Environ.* **28**, 75–87.

Steil B.; 1997, Modellierung der Chemie der Strato- und Troposphäre mit einem dreidimensionalen Zirkulationsmodell. Ph.D. Thesis, Institut für Meteorologie, Universität Hamburg.

Steil B., M. Dameris, C. Brühl, P.J. Crutzen, V. Grewe, M. Ponater, R. Sausen; 1998, Development of a chemistry module for GCMs: first results of a multi-annual integration. *Ann. Geophysicae* **16**, 205–228.

Stern R.; 1994, Entwicklung und Anwendung eines dreidimensionalen photochemischen Ausbreitungsmodells mit verschiedenen chemischen Mechanismen, *Report 8/1*, Serie A, Institute for Meteorology, Free University Berlin.

Stockwell W.R.; 1995, Effects of turbulence on gas-phase atmospheric chemistry: calculation of the relationship between time scales for diffusion and chemical reaction, *Meteor. Appl. Phys.* **57**,) 159–171.

Stockwell W.R., T. Schönemeyer, 1997, Modelling cloud chemical processes, in: A. Ebel, R. Friedrich, H. Rodhe (eds), *Tropospheric Modelling and Emission Estimation*, Springer Verlag, Heidelberg, pp. 134-138.

Stockwell W.R., P. Middleton, J.S. Chang; 1990, The second generation regional acid deposition model chemical mechanism for regional air quality modelling, *J. Geophys. Res.* **95**, 16343–16367.

Strand A., Ø. Hov; 1994, A two-dimensional global study of the tropospheric ozone production, *J. Geophys. Res.* **99**, 22877–22895.

Stull R.B.; 1988, An Introduction to Boundary Layer Meteorology, Kluwer Acad. Publ, Dordrecht, Boston, London.

Tippke J.; 1997, Die Behandlung von Nebelereignissen in einem mesoskaligen Chemietransportmodell, Mitteilungen aus dem Institut für Geophysik und Meteorologie der Universität zu Köln, No. 1177.

Tuovinen J.P., K. Barrett, H. Styve; 1994, Transboundary acidifying pollution in Europe: Calculated fields and budgets 1985–1993, EMEP/MSC-W Report 1/94, The Norwegian Meteorological Institute, Oslo, Norway.

van Velthoven P., H. Kelder, 1996, Estimates of stratosphere–troposphere exchange: sensitivity to model formulation and horizontal resolution, *J. Geophys. Res.* **101**, 1429–1434.

Velders G.J.M., L.C. Heijboer, H. Kelder, 1994, The simulation of the transport of aircraft emissions by a three-dimensional global model, *Ann. Geophys.* **12**, 385–393.

Volz A., D. Kley; 1988, Evaluation of the Montsouris series of ozone measurements made in the nineteenth century, *Nature* **332**, 240–242.

Walcek C.J., G.R. Taylor; 1986, A theoretical method for computing vertical distributions of acidity and sulfate production within cumulus clouds, *J. Atmos. Sci.* **43**, 339–355.

Walmsley J.L., M.L. Wesely; 1995, Modification of code parametrizations of surface resistance to gaseous dry deposition, *Atmos. Environ.* **30**, 1181–1188.

Ziegenbein C., I.J. Ackermann, A. Ebel; 1994a, The treatment of aerosols in the EURAD model: Results from recent developments, in: J.M. Baldasono, C.A. Brebbia, H. Power, P. Zanetti (eds.), *Air Pollution II*, Computational Mechanics Publ. 1, Southampton, pp. 229–237.

Ziegenbein C., I.J. Ackermann, H. Feldmann, H. Hass, M. Memmesheimer, A. Ebel; 1994b, Aerosol treatment in the EURAD model: recent developments and first results, in: P.M. Borrell, P. Borrell, T. Cvitaš, W. Seiler (eds), *Proc. EUROTRAC Symp. '94*, SPB Academic Publishing bv, The Hague, pp. 1176–1179.

Zimmermann P.H.; 1988, MOGUNTIA: A handy global tracer model, in: H. van Dop (ed), *Air Pollution Modelling and its Applications VI*, NATO/CCMWS, Plenum Press, New York.

Zimmermann P.H.; 1994, The Impact of aircraft released NOx to the tropospheric ozone budget: Sensitivity studies with a 3-D global transport/photochemistry model, in: U. Schumann, D. Wurz (eds.), *Impact of Emissions from Aircraft*, Proc. Intern. Sci. Colloquium, Köln (Cologne), Germany, 1994; Mitteilung 94-06, DLR, pp. 211–216.

Chapter 6

Instrumentation

Reginald Colin[1], Gérard Mégie[2] and Peter Borrell[3]

[1]Université Libre de Bruxelles, Laboratoire de Photophysique Moleculaire,
50 avenue F.D. Roosevelt, B-1050 Bruxelles, Belgium
[2]CNRS, Service d'Aeronomie, B.P. 3, F-91371 Verrières Le Buisson Cedex, France
[3]P&PMB Consultants, Ehrwalder Straße 9, D-82467 Garmisch-Partenkirchen,
Germany

6.1   Introduction: requirements for instrument development

The development of new instrumentation and the continual assessment of existing experimental techniques are an essential part of any on-going scientific effort; this is particularly so in the atmospheric sciences which seek to understand the behaviour and global budgets of trace constituents in the lower atmosphere and the effect on them of anthropogenic activities. Instrument development is thus a necessary prerequisite for any quantitative study of photo-oxidation and acidification in the troposphere. in addition to the particular case of laboratory measurements of individual atmospheric processes, the atmosphere itself is the only experimental laboratory where the interactions between the various processes, which influence the emission, transformation and deposition of trace constituents, can be fully quantified taking into account the various chemical, dynamical and biological processes. Therefore, instrument development is of paramount importance in developing a scientific strategy for any environmental project, the objectives being the use of such instruments in co-ordinated field experiments and their deployment in well-assessed scientific networks. Such instruments will then allow the quantification of the natural variability of the atmosphere at various temporal and spatial scales, the assessment of the long-term changes in the physical and chemical state of the atmosphere under man-made influences and the validation of multi-dimensional interactive simulation models, be they global, regional or local. Furthermore, without such field measurements based on reliable instrumentation, no prediction of the further evolution of the atmosphere can be made, since the accuracy and precision of the instruments is a condition for a large part the extrapolations made with prognostic models and thus the usefulness of policy decisions. Moreover, the results of such policy decisions require continuous assessment and control. On global, regional and local scales, the effects of national and international agreements must be thus monitored by instruments in order to judge the effectiveness of the measures decided.

When the continuously evolving state of knowledge on atmospheric processes is taken into account, several objectives have to be addressed in instrument developments including:

*   the development of new instrumentation to measure atmospheric variables or trace species constituents that are not yet accessible;

*   the improvement of the sensitivity of existing instruments to increase their potential applications;

*   the combination of instruments to quantify specific processes and, as such, the development of versatile instruments able to measure several variables of interest simultaneously;

*   the development of mobile, airborne, and balloon-borne instruments to address the problem of temporal and spatial scales integration;

*   the assessment of instrument accuracy and precision through constant quality control and well-organised intercomparison campaigns;

*   the deployment of carefully calibrated and inter-compared instruments in ground-based networks;

*   the transfer of technology from the research laboratories to industry to develop a strong European component in instrumentation for environmental studies.

The EUROTRAC project has been part of this general effort over the past 10 years. At its start in 1987, the project specifically included the promotion of technological development of instrumentation for environmental research and monitoring, as one of its three main objectives, together with the improvement of the basic understanding of atmospheric science and providing a scientific basis for political decision making in environmental management within Europe. As already stated, this requirement for instrument development should be understood not only as an objective *per se*, but as a necessary component of the general scientific strategy developed within EUROTRAC, which is beneficial, not only for the project itself, but also for any on-going or forthcoming environmental projects.

Instrument developments within EUROTRAC were made using two different routes. Three instrument subprojects were established, as part of the initial project in 1987, to address the development of instrumentation for field measurements of atmospheric variables within the scientific objectives of EUROTRAC. These subprojects, TESLAS, TOPAS and JETDLAG, were accepted after consideration by the EUROTRAC Science Steering Committee and International Executive Committee on the basis of their potential for further application to the other science-oriented EUROTRAC subprojects. The initial objective was thus not only to stimulate the development and the assessment at the European level of new and promising instrumentation, but also to use them, before the end of the EUROTRAC project itself, in field campaigns and/or in ground-based networks.

The second route, of equal importance, was the development of instrumentation within the other experimental subprojects, to address their specific objectives. This led to a large number of technical and methodological applications, which include the development and assessment of new instruments, leading for some particular case to technology transfer to the industry. It is worth emphasising that these developments were undertaken in the planning and experimental phase of all the experimental EUROTRAC subprojects, *i.e.* ALPTRAC, ASE, BIATEX, GCE, TOR and TRACT, as well as the two subprojects, LACTOZ and HALIPP, devoted to laboratory measurements. The specific developments of new techniques for such laboratory measurements (Warneck, 1996; Le Bras, 1997; Borrell *et al.*, 1997, chapter 5) will not be considered in this overview which is devoted to field instruments and their components.

A complete description of all achievements in instrument development within the EUROTRAC project is clearly beyond the scope of this section, as such a description can be found in the various EUROTRAC final reports (volumes 2 to 10). Its main objective is rather to give a short summary of some of the work performed . This will be first done by examining the results obtained in the subprojects devoted to instrument development (section 6.2), followed by a short review of the instrument developments accomplished in other EUROTRAC subprojects (section 6.3). A brief examination of the positive and negative outcomes of the instrument subprojects will be given in section 6.4, before ending with some conclusions, recommendations and future needs are examined in section 6.5.

6.2   The instrument subprojects within EUROTRAC

For a very large part, atmospheric measurements make use of spectroscopic techniques based on the specific absorption of electromagnetic radiation by the molecules present in the earth's atmosphere. These spectroscopic techniques are generally considered as powerful, as they can be used to measure simultaneously or sequentially, using the same instrument, several species of interest, as they can often function unattended, as they do not perturb the state of the atmosphere, and they are characterised by low detection limits. Spectroscopic measurements are all based on the Beer-Lambert absorption law, which requires an *a priori* knowledge of the absorption properties of the constituents of interest. Thus, work on spectroscopic techniques also has strong connections with laboratory studies, as one of the prerequisite for field measurements is the establishment of accurate spectroscopic data bases, in the various spectral ranges of interest.

The emphasis was thus given in the three EUROTRAC instrument subprojects (JETDLAG, TESLAS and TOPAS) to active and passive optical spectroscopic methods. The rationale was the potential for further application of such systems to the measurement of many constituents in the atmosphere. In addition, the differential optical absorption spectroscopy (DOAS) method, which was developed in TOPAS, and the tuneable diode laser spectroscopy method of the JETDLAG have also proven to be valuable tools for laboratory studies .

It is worth emphasising that these subprojects were not devoted to the development of entirely new techniques. The differential absorption lidar (DIAL) method, as developed in TESLAS, was already well established, and lidar instruments were already recognised tools for the measurements of stratospheric variables, *i.e.* ozone, temperature and aerosol vertical distributions. Also, research instruments were already available to measure the ozone vertical distribution in the troposphere from the ground, from mobile vehicles or from airborne platforms. However the deployment of such instruments in field campaigns and in monitoring networks was still not very reliable, due to the lack of assessment of the method in tropospheric applications.

Similarly, the DOAS technique (TOPAS) was already in field operation, using either external sources of radiation such as the sun, the moon or the daylight sky for measurements in the stratosphere, or using the long path absorption technique for measurements close to the ground. In the case of JETDLAG, tunable diode laser absorption instruments were already available commercially, although again at an early stage of development. A major objective of the JETDLAG subproject was thus to increase the sensitivity of such instruments, and to develop fast-response instruments to provide direct measurements of the emission and deposition fluxes of trace species using the eddy-correlation technique.

In each case, the challenge within EUROTRAC was to bring these techniques to a degree of maturity where such instruments could be used by scientists other than the ones directly involved in their development, and be operated by trained technicians in the field. Also, the techniques were to be pushed beyond their current limits in order to increase their range of application, in particular so that tropospheric species could be detected and measured in the ppt (part per trillion) to sub-ppt range. The potential applications to other EUROTRAC subprojects were also identified with direct links envisaged with the TOR (TESLAS, TOPAS), BIATEX, ALPTRAC (JETDLAG), TRACT (TESLAS, JETDLAG), ACE and GCE (TOPAS) subprojects.

a.   The subproject TOPAS (Tropospheric OPtical Absorption Spectroscopy)

When the EUROTRAC project started in 1987, several instruments using long-path absorption spectroscopy for the simultaneous measurement of atmospheric constituents had already been in operation for several years. These instruments, essentially laboratory prototypes and not easily mobile, had shown the feasibility and advantages of the technique based on the differential absorption spectroscopy (DOAS). But the technique, which makes use of a spectral analysis of the light transmitted through the atmosphere from a reference light source, was by no means a standard and well-established method for the detection of pollutants and the measurement of their concentrations. There were many experimental problems remaining and no carefully assessed commercial instruments were readily available.

The subproject TOPAS was thus established with the aim of developing high performance instruments based on the DOAS method, to measure minor tropospheric constituents using their absorption properties in the ultra-violet, visible and near-infrared wavelength ranges. Several groups were involved in this project, with previous expertise either in stratospheric and tropospheric measurements. Furthermore, three of these groups had already undertaken preliminary industrial developments of long-path absorption systems for pollution monitoring, and established commercial companies (ATMOS in France, OPSIS in Sweden and Hoffmann Messtechnik in Germany) and one Belgian group was involved in applying a Bruker Fourier transform spectrometer to atmospheric measurements in the visible and ultra-violet regions. A number of improvements were accomplished by TOPAS: the development of the systems using grating spectrographs, reliable long path absorption arrangements, new algorithms for the retrieval of species concentrations and the implementation of UV-visible Fourier transform spectrometers in atmospheric measurements. The number of species which could be measured and detected using the DOAS technique was increased substantially; these include NO, NO_2, NO_3, O_3, SO_2, CS_2, CH_2O, HNO_2, NH_3, ClO, ClO_2, BrO, IO, toluene , naphthalene and xylene, the limit of detection for each of these species being clearly instrument dependent. The routine measurement of the OH radical, which was an objective of the subproject, was achieved outside EUROTRAC.

It was necessary to re-organise TOPAS subproject in 1992, and it was decided to further the work of the subproject by running field intercomparison campaigns. The first took place in 1992 at an urban site in Brussels and involved 8 research groups from 5 nations. From the various blind comparisons made during the campaign, it was clear that the discrepancies between the various instruments which had been developed were not random; they were closely linked to each instruments principally through their spectral resolution and the different algorithms being used. Significant improvements were then made by all participants, since the campaign encouraged the exchange and sharing of various instrument and software improvements between the participating groups. The campaign also emphasised the need for further laboratory work in order to establish a common data base of high spectral resolution absorption cross-sections. This task was undertaken by the Belgian Groups (Camy-Peyret et al., 1996)

A second TOPAS campaign was conducted in 1994 at the Weybourne Atmospheric Observatory in Norfolk, UK with the aim of addressing a number of areas where the DOAS technique might be improved. These were essentially: a comparison of the results of the DOAS measurements with other techniques, in particular with commercial instruments, whose characteristics were already well documented, measurements of cells containing known amounts of absorbers, measurements in a rural area, away from local sources of pollution in order to assess the detection limits of each instrument, and measurements of other important trace gas species. The results of this ten day campaign were encouraging and the full analysis was published recently (Coe et al., 1996).

From the point of view of commercial involvement, the companies which took part in the campaigns (ATMOS (France), Hoffmann Messtechnik and Bruker Analytic Messtechnik (Germany)) certainly gained much precious expertise. Two instruments have been fully commercialised, and DOAS-based systems are now established in several cities for pollution monitoring. DOAS instrumentation has also been assessed by operational agencies, such as INERIS in France for the ATMOS system.

b.   The subproject TESLAS
(<u>T</u>ropospheric <u>E</u>nvironmental <u>S</u>tudies by <u>LA</u>ser <u>S</u>ounding)

Since a number of vertical transfer processes influence the global budget of ozone in the troposphere (stratosphere-troposphere exchange of air masses, ozone formation in the planetary boundary layer and its exchange with the free troposphere, direct production of ozone in the free troposphere and decomposition at the surface), the continuous monitoring of the ozone vertical distribution in the depth of the troposphere would be a powerful tool to identify the altitude-dependent signature of such processes. This is also true for the various temporal and spatial scales involved, from convective processes within the boundary layer to long-term evolution in the global troposphere.

The established technique is based on regular (daily or weekly) *in-situ* measurements with balloon-borne ozone sondes. Now remote-sensing instruments based on the differential absorption lidar technique (DIAL) are proving to be useful in providing continuous measurements of the ozone vertical distribution. Developed first for the measurement of ozone in the stratosphere, the technique has rapidly evolved over the past ten years, to provide accurate measurements of the ozone vertical distribution in the troposphere under cloud-free conditions. Based on the simultaneous emission of two laser lines, one being absorbed by ozone with the second, free of absorption, used as a reference, the DIAL technique provides high vertical (few tens of meters) and temporal (few seconds) resolution measurements from the ground up to the lower stratosphere.

The main objective of TESLAS was to provide reliable laser-based remote-sensing instrumentation for routine measurements of the ozone vertical distribution in the troposphere, and also for detailed studies of atmospheric processes involving transport, production and destruction of ozone. The technical development of lidars requires a large amount of laboratory development work on the various sub-systems of the integrated lidar system (laser, emitting and receiving optics, time-resolving electronics, real-time data acquisition and processing), and the development of careful data analysis using appropriate algorithms to invert the lidar retrieved signals.

Some of the improvements made within the project include: emission sources using stimulated Raman scattering to provide the appropriate wavelengths for differential absorption measurements, grating polychromators to resolve spectrally the scattered signals, laboratory studies of UV spectra of interfering species (SO_2, H_2O), new data acquisition electronics, system semi-automation. An area of particular interest was the addition of Raman scattering channels to improve the retrieval of the ozone concentration in presence of heavy aerosol loads, such as are often observed in the boundary layer. This can truly be considered as a breakthrough, in relation with the availability of powerful laser sources which allow reduced integration time for the measurements. In addition, airborne systems are now available as a result of the technical developments performed in TESLAS.

The principal achievement of TESLAS was its capacity to bring together research groups with complementary expertise in the field of lidar development and to organise, on this basis, instrument and methodology intercomparisons. Such exercises are necessary to quantify the experimental uncertainties and to detect potential biases in experimental system. The intercomparison campaign held in Bilthoven in 1991 (TROLIX) and the further algorithm intercomparison undertaken as part of the 1993 activities in the subproject, are exemplary in this respect.

The campaigns have led to activities concerning the reassessment of existing methodologies and resulted in a narrowing of the uncertainty margins. They are clearly part of the necessary assessment of new instrumentation in order to convince the outside community of the potential of such experimental developments. So a wealth of different techniques was investigated with good success, each of them having advantages for special applications. Of the twelve groups, nine finally succeeded in setting up at least one complete system for use in atmospheric studies, compared with just one in existence in Europe before TESLAS was started. One group also finished the construction of an airborne system.

The TESLAS project was also a success, when one considers that several groups in Europe are now in a position to operate lidar systems for ozone monitoring, or to take part in field campaigns using either ground-based or airborne systems. Groups have participated in field campaigns inside and outside of EUROTRAC, for example the TOR, OCTA and POLLUMET field programmes. The range of application was broad, from high resolution measurements in the lower troposphere to determine transport processes in the boundary layer, between boundary layer and free troposphere, and at the tropopause level, to long-term monitoring on a routine basis up to the lower stratosphere. The European scientists have clearly taken a leading position in this field.

In terms of industrial development, several groups have obtained patents at the subsystem level (optics, electronics) as a result of the developments made within TESLAS. Although the range of application of the DIAL system has been broadened considerably during the developments within TESLAS, the necessary specialisation is probably one of the reasons, why the development of a commercial instrument has not been fully successful, although complete systems can now be purchased from the industry.

c.   *The subproject JETDLAG*
      (Joint European Development of Tuneable Diode Laser Absorption
      Spectroscopy for Measurements of Atmospheric Trace Gases

Most of the constituents of interest in the troposphere have absorption features in the infra-red wavelength range and infrared absorption spectroscopy is now a recognised tool for the monitoring of atmospheric constituents. When high accuracy and sensitivity are required, the measurement of single absorption lines using high resolution spectroscopy is of particular interest.

The development of narrow linewidth emission sources (IR. lasers) opened the path to the development of a new field in atmospheric spectroscopy based on tunable diode lasers. The objective of the JETDLAG subproject was the development and use of the tunable diode laser absorption spectroscopy (TDLAS) for the measurement of concentrations of tropospheric constituents and also their fluxes, using the eddy-correlation technique.

A major requirement for such a development was the availability of reliable laser diodes which constitute the active part of the system. These were available only from a few companies and laboratories in Europe, and a large part of the earlier work performed within JETDLAG was thus devoted, to the development and assessment of the TDLAS components and techniques. This led in 1992 to the development of a new process for manufacturing heterostructure diodes ($PbSnSe$, $PbSrSe$), to the characterisation of the TDLAS spectrometers, and to the development of infrared optical fibre components. Several new methods were also explored, including photo-acoustic spectroscopy for fine spectral tuning of the emissions lines, differential techniques and intra-cavity absorption spectroscopy.

The provision of spectral data on the absorption features of many constituents in the infrared wavelength range was also a prerequisite for further applications of TDLAS spectroscopy in the field. High resolution spectroscopic measurements including line position, and line broadening were thus undertaken as part of the project for such species as NH_3, HNO_2, C_2H_2, C_2D_2, CH_3CHO, C_3O_2, C_2H_6, NO, NO_2, N_2O_4, CH_3Cl and N_2O.

It was necessary to reorganise JETDLAG in 1992. The objectives were re-defined to emphasise field measurements using the five existing TDLAS instruments. One instrument, using a FM technique, was used to estimate background levels of pollutants on the MV Polarstern during a 1994 Atlantic cruise and carried out intercomparison studies between TDLAS measurements of NO_2 and HCHO with those obtained by more conventional methods.

A four laser system (FLAIR) was used on a DC3 for airborne measurements in the SAFARI 92 campaign which studied biomass burning in Africa and the South Atlantic. A series of ground based campaigns were also undertaken with this instrument including two FIELDVOC campaigns and an OCTA campaign at the Izaña TOR station.

Another instrument, developed specifically for HCHO measurements as part of the TOR project, was later housed in a van in order to undertaken a range of field studies with particular emphasis on the influence of HCHO from traffic emissions on photo-chemical ozone production.

A fourth instrument was developed with industrial support for eddy-correlation measurements of O_3, CO and CH_4 on board a Fokker 27 aircraft.

Finally, a simple single-species instrument was modified to measure two species using time division multiplexing and in this form was used for simultaneous measurements of HNO_3 and NH_3.

The JETDLAG subproject made major contributions to the advance of the TDLAS technique for measuring trace gas concentrations in the troposphere. Unfortunately it was not possible to develop the technique sufficiently so that it could be used as a routine method which can easily be operated by non-specialists. This is mainly because the available lasers still have poor reproducibility.

6.3   Instrument development as part of other subprojects

Obviously, the field of instrument techniques to be used in such a wide project as EUROTRAC was not be limited to the three instrument subprojects. As already mentioned, other instrument developments were performed as part of all the EUROTRAC science-oriented experimental subprojects. They result from the co-ordinated action of research groups directly interested in field measurements in relation to the scientific objectives of the various subprojects, and are concerned with the various aspects of instrument developments already outlined in the introduction, *i.e.* new instruments and methods, instrument assessments and intercomparisons, deployment of instruments in co-ordinated campaigns and ground-based networks, technology transfer to the industry. A complete description of all the instrument achievements made within the other EUROTRAC subprojects cannot be given in this overview. The principal areas of progress will be summarised here and the technical details can be found in the annual or final reports of the appropriate subprojects.

Within the subproject ALPTRAC (Fuzzi and Wagenbach, 1997), an intercomparison of analysis methods for snow samples was conducted in Davos (Switzerland) in 1993. Common samples were distributed to the various research groups involved for analysis. The results were quite satisfactory for Cl^-, NO_2^-, SO_4^{2-}, NH_4^+ and reasonable for K^+, Mg^{2+} and H_3O^+, while some difficulties were identified for common ions such as Na^+ and Ca^{2+}. Such intercomparisons have proved to be essential to quantify the differences which can appear in measurements made at widely separated sites.

In the ASE subproject (Larsen *et al.*, 2000), analytical methods have been developed for the analysis of biochemical precursors of dimethylsulfide (DMS), *i.e.* dymethylsulfopropionate (DMSP) and dimethylsuphoxide (DMSO). Such measurements are of particular importance in assessing the complex processes by which DMS is formed. The deployment of reliable instrumentation in ship-borne campaigns, for example on board the Polarstern, is an important outcome of the instrument assessments conducted in this subproject.

As part of the BIATEX subproject, (Slanina, 1996) several instrument field intercomparisons were conducted. A small, lightweight and fast-response ozone sensor for direct eddy flux measurements has been built and tested overt a prolonged period in continuous measurements of ozone fluxes and deposition velocities over different croplands. A similar system for the measurement of total sulfur using a fast-response flame photometer was also developed.

Another instrument has been developed for the measurement of fluxes by the eddy-correlation technique which allows in situ detection of ethylene and ammonia using the photothermal deflection (PTD) , which is one of several methods based on the detection of heat produced in a gas after absorption of laser light by the molecule of interest. A closed cell is not employed thus avoiding the problems created by the absorption of polar gases on the walls of the cell.

A third instrument developed in BIATEX is an aerosol sampler, which samples aerosols by addition of steam and the formation of droplets. Also, droplet samplers for the analysis of NH_3 and organic compounds, and samplers for reduced sulfur compounds have been devised, and a gas chromatography system for the measurement of biogenic emissions, i.e. terpene and isoprene, has been automated. Of particular importance in assessing these various instruments for flux measurements were the intercomparisons held during several field experiments in Leende, Halvergate, Manndorf and in the Bayerische Wald, which have helped to establish the limits of accuracy on concentration and flux measurements, thus increasing the confidence in regional and global budgets derived from such experiments.

In the GCE subproject (Fuzzi and Wagenbach, 1997), several campaigns, in particular at Great Dun Fell and Kleiner Feldberg, have again been used for instrument assessment and intercomparisons, including improved collectors for dry and wet deposition and precipitation. Furthermore specific instrument development have been undertaken which have resulted in the development of a new droplet/aerosol analyser to study the activation of particles as a function of particle size and composition.

Although TOR (Hov, 1997) was not initiated as a technical project, several instrument developments were conducted during the planning and experimental phase of this subproject. This has led to a number of applications in terms of technology transfer to industry, in particular with respect to the measurement of hydrocarbons, nitrogen oxides and photolysis frequencies. Before the EUROTRAC project was initiated, commercial instruments for the quasi-continuous measurement of volatile organic compounds (VOC) were not available or did not have the required sensitivity, although there were concepts for automated measurements in a number of research laboratories. Within TOR, several approaches towards automated VOC measurements were made and implemented at field sites. A prototype for automatic sampling of volatile hydrocarbons was built by CHROMPACK in collaboration with TOR investigators. This instrument was used at a number of TOR stations as well as in other subprojects.

The situation was similar for the nitrogen oxides. Although research instruments had been developed and used in scientific studies, commercial instruments for monitoring of NO_x and NO_y compounds did not have the required sensitivity and accuracy. Technology transfer during the TOR experiments led to the manufacture of sensitive and accurate NO_x analysers which were later implemented at many TOR stations and have been sold to other research groups around the world . Automated instruments for the measurement of peroxides and formaldehyde were also developed, based on existing analytical techniques, and made available commercially through technology transfer to leading companies. Photolysis rates of important molecules such as ozone or NO_2 were usually derived from measurements of the irradiance or from radiative transfer models. In TOR, a photoelectric sensor was improved and made commercially available. In a follow-up project, companies are now building a similar instrument for the photolysis rate of ozone, which is the primary process in formation of OH radicals. A sonde was also developed for the measurement of hydrogen peroxide. Finally, the instrument developments conducted within TOR have greatly improved the measurement capabilities in other projects outside EUROTRAC, such as those sponsored by the European Union (*i.e.* OCTA, FIELDVOC, STREAM, *etc.*).

Quality assurance and quality control were important parts of the TRACT campaigns (Larsen *et al.*, 2000) and have resulted in the comparison of gas sensors using global facilities, and airborne platforms.

Finally, facilities were established that can now be used by European research groups, such as the ozone sonde calibration facility as part of TOR, or the cloud wind tunnel as part of BIATEX, which can be used for calibration of instruments measuring aerosol and droplets size distributions.

6.4   Positive and negative outcomes of instrument development

Instrument developments has a direct influence on environmental policy development since it improves the field measurements required to verify the models used, and is necessary for the instruments required to monitor the effectiveness of policy measures. Development of advanced and reliable instrumentation is required for the implementation of monitoring networks, the organisation of large scale campaigns to validate models, and the assessment of policy measures. It is in this context that the outcome of the EUROTRAC instrument subprojects and instrument developments within science oriented subprojects should be judged

a.   Positive outcomes of the EUROTRAC instrument subprojects

Although some of the instruments which have been developed in the frame of the EUROTRAC subprojects were already in existence at the beginning of the project, it is quite clear from the results, that appreciable progress was made in the development of the instrument techniques both at the subsystem and system levels. All these instruments are complex and thus require a lot of expertise in various fields of instrument research (active and passive spectroscopic methods, chemical analysis, fast time-response electronics, signal processing, real time data computing).

Furthermore, one of the most important contribution of the co-operative work which has been implemented through the establishment of EUROTRAC is certainly the assessment of the methodologies and the instrument measurements through the organisation of intercomparison exercises and field validation experiments, concerning both the methodologies (algorithms, input data) and the measurements themselves. These exercises, performed on an European basis, have definitely increased our confidence in the measurement techniques to a level where the instruments can now be used in field campaigns and monitoring networks with a clear appreciation of their potentialities and uncertainties. As part of this general assessment and quality control of the instrument, the improvement and development of common data bases (spectroscopic data and parameters required for signal inversion) and European facilities (ozone sonde calibration, cloud wind tunnel), should also be considered as a positive outcome with future potential applications to laboratory studies and field measurements.

As a result of these technical developments, an overall improvement in the instrument reliability has been achieved, which for some of the instruments, has led to the development of transportable and/or airborne versions. The link between chemical and transport processes, which characterise atmospheric processes at all temporal and spatial scales, makes the construction of such mobile, especially airborne instrumentation very useful. Although such instrumentation for chemical measurements is not at the same level of development in Europe as compared to the United States, partly due to the lack of availability of appropriate aircraft, the potential expertise has now been built which should contribute to closing this gap.

As emphasised in the rationale for the implementation of instrument developments, some of the instruments which have been developed and assessed have, were used in the other EUROTRAC subprojects. This holds particularly true for the lidar instruments (TESLAS) which contributed to the on-going monitoring effort in the frame of the TOR stations, and as tools in extended field experiments (ground-based, mobile, airborne). Similar applications are foreseen for the TDLAS instruments developed within JETDLAG. The DOAS instruments developed within TOPAS are already in use for pollution monitoring at the local scale. Similarly, the instruments developed within BIATEX (ozone sensor, ethylene and ammonia detectors) and TOR (VOC, nitrogen oxides, photolysis rates) have also been used for field measurements outside EUROTRAC.

A definitive proof of the quantitative impact of these instrument developments on future policy decisions, is certainly the increasing role played by the European research groups in the implementation of research networks and large scale campaigns based on innovative instrumentation. This is clearly recognised at the international level, and corresponds to a drastic change in comparison with the situation at the outset of EUROTRAC. In this respect, the European co-ordination efforts which have been fostered by EUROTRAC, and also by the European Union environmental programmes, have certainly been essential. This new position will have an important impact on policy decisions, as the potential for organising field campaigns on an European basis, in connection with regional environmental objectives has increased considerably, thus allowing a more accurate validation of integrated models, and their adaptation to local and regional European conditions.

b.   Negative outcomes of the EUROTRAC instrument subprojects

This sub-section tries to identify what has not been achieved in the instrument developments. In doing so, it is not intended to give a negative impression of these developments, but rather to identify the present shortcomings in order to assess the potential for instrument projects in future European co-ordinated projects.

It is quite clear from the various reports delivered within EUROTRAC, that each research group involved in the subprojects conducted its own instrument development with reference to its own scientific objectives. Although some expertise has certainly been gained through the discussions within the subprojects, one cannot consider that common developments of instrumentation at the European level has been performed. These would have required a much stronger co-ordination, as done for example in the European Union environmental programmes or European Space Agency research and development programmes, which require a prompt development of the "work packages" as identified in the initial projects. However, this is not a great disadvantage provided that the individual instruments have undergone a common assessment.

In this respect, the reorganisation of the TOPAS and JETDLAG subprojects which have occurred in 1991 and 1993 respectively, were positive in giving more emphasis on the measurement intercomparisons. Furthermore, the structure of EUROTRAC itself did not allow the strong co-ordination required for a common development. The lack of centralised co-ordination of funds, required by EUREKA, and the differences in the timing of funds availability in each participating country were important factors inhibiting the international co-ordination of the instrument subprojects in particular.

A consequence was the lack of full participation by some of the research groups in the instrument subprojects and the loss of some of the initial objectives (for example the measurement of the water vapour vertical distribution within TESLAS and the OH concentration measurement in TOPAS). A further consequence was a general slowness of the projects. In instrument development, there are given technical milestones which have to be achieved before a new part of the development can be undertaken efficiently. As a result, the speed is determined by the slowest of the participating research groups, which might have been determined by local financial restrictions. Even after reassessments of some subprojects were made taking these deficiencies into account, there was still some delay in the achievement of the project. This was particularly true for JETDLAG, and to a lesser extent for TOPAS.

Despite the appreciable progresses made both in terms of instrument techniques and instrument validation, one should also recognise that only a few of the instruments which have been developed within the EUROTRAC subprojects were fully automated and commercialised at the system level. Some exceptions do exist however, as for example the DOAS instruments developed within TOPAS by the ATMOS and Hoffmann Messtechnik companies, which were already in existence at the initial stage of TOPAS, the fast-response ozone sensor developed within BIATEX, some sub-systems of the Differential Absorption Lidars developed within TESLAS, and various sensors developed within the TOR subproject.

These considerations raise an important question with respect to EUROTRAC itself. The project is part of the EUREKA initiative, which, in principle, is directed to technological development. However, the EUREKA connection of the EUROTRAC project was not a driving force and very few projects were able to raise EUREKA funds at the national level, which would have definitely increased the commercial participation in the subprojects. Two reasons can be given. First, the general lack of interest of the scientists involved in these subprojects for a strong industrial involvement in their development, at least at the system level, with some exceptions already mentioned. Second, the lack of interest of the industry in the proposed developments in environmental technology, which, as always do not have a large market in Europe. As the EUREKA rule requires a 50% participation of the industrial companies on their own funds in the project, this is perhaps understandable.

In some cases (e.g. TOPAS) the initial involvement of several competing companies led to unforeseen problems: Some were reluctant to reveal their plans or to participate in intercomparison campaigns. The problem of the development of an European industrial component in environmental monitoring technology is still open. Furthermore, once an instrument has been built on an commercial basis, it still requires assessment as an operational instrument. This part of the development cannot be made by the research laboratories which do not have the required infrastructure and manpower, or by the commercial companies. It requires at some point, the involvement of operational agencies (environmental agencies, meteorological office) which were also not part of the EUROTRAC initiative for this purpose.

6.5   Conclusions, recommendations and future needs

In conclusion, the instrument subprojects (TESLAS, TOPAS, JETDLAG), the science oriented experimental subprojects (ALPTRAC, ASE, BIATEX, GCE, TOR, TRACT) and the laboratory measurement subprojects (LACTOZ, HALIPP) within EUROTRAC developed numerous innovative and reliable instruments. One point of particular importance is that these instruments have undergone thorough validation and intercomparison exercises, and have already been used in field experiments and implemented in scientific monitoring networks and other environmental projects within Europe. They have also led to a large increase in the international role of European research groups in measurement related activities, such as still larger scale field campaigns or tropospheric networks. The instrument developments have also had an effect on policy development by improving the measurements required to validate models and developing the instrumentation required for monitoring the progress of environmental policy measures.

For the future, one of the main question to be addressed, is to understand if projects entirely devoted to innovative development of instrumentation, are the optimal way to achieve such developments. The conclusion which can be drawn from the EUROTRAC experience, is that such projects can only be developed efficiently if centralised funds are provided to ensure the timeliness of the various contributions. The mechanism will then be similar to the one used by other European agencies for their research and development activities (European Commission environmental programmes or European Space Agency space projects).

In a structure of the EUROTRAC type, two routes could then be explored. If the EUREKA umbrella is used to provide from the early stage of the project a strong industrial involvement, which requires strong co-ordination and detailed planing, specific instrument projects might be valuable. The objectives will then rather be commercial development of existing instrumentation, including the assessment of their operational character, rather than true innovative developments.

The second route is to develop new instruments as part of the science driven projects. The results obtained within EUROTRAC clearly demonstrate that this route can be appropriate, as long as the validation and quality control phases of the instrument development are still considered as a primary requirement, before any use in field experiments or science driven networks is made. As the implementation of fully equipped research stations or the organisation of field campaigns are important milestones in such projects, these will constitute the driving force for the instrument development. This condition finally implies that the instrument developments have to be considered as an important part of the science driven projects, and thus the constraint on a timely development is still directly linked to the funds and manpower availability.

6.6   References

Borrell, P., Ø. Hov, P. Grennfelt, P. Builtjes (eds); 1997; *Photo-oxidants, Acidification and Tools: Policy Applications of EUROTRAC Results,* Springer Verlag, Heidelberg; Vol. 10 of the EUROTRAC final report.

Bösenberg, J., D. Brassington, P. Simon (eds); 1997; *Instrument Development for Atmospheric Research and Monitoring,* Springer Verlag, Heidelberg; Vol. 8 of the EUROTRAC final report.

Camy-Peyret, C., B. Bergvist, B. Galle, M. Carleer, C. Clerbaux, R. Colin, C. Fayt, F. Goutail, M. Nunes-Pinharanda, J.P. Pommereau, M. Hausman, F. Hertz, U. Platt, I. Pundt, T. Rudolph, C. Hermans, P.C. Simon, A.C. Vandaele, J.M.C. Plane, N. Smith; 1996, Intercomparison of Instruments for Tropospheric Measurements using Differential Optical Absorption Spectroscopy, *J. Atmos. Chem.* **23**, 51-80.

Coe, H., R.L.Jones, R.Colin, M.Carleer, R.M.Harrison, J.Peak, J.M.C.Plane, N.Smith, B.Allen, K.C.Clemitshaw, R.A.Burgen, U.Platt, T.Etzkorn, J.Stutz, J.P.Pommereau, F.Goutail, N.Nunes-Pinharanda, P.Simon, C.Hermans, and A.C. Vandaele; 1996, A Comparison of Differential Optical Absorption Spectrometres for Measurements of NO_2, O_3, SO_2 and HONO in: P.M. Borrell, P.Borrell, T.Cvitaš, K. Kelly and W.Seiler (eds), *Proc. EUROTRAC Symp. '96, Vol. 2*, Computational Mechanics Publications, Southampton, pp. 757-762.

Fuzzi, S., D. Wagenbach (eds); 1997, *Cloud Multi-phase Processes and High Alpine Air and Snow Chemistry*, Springer Verlag, Heidelberg; Vol. 5 of the EUROTRAC final report.

Hov, Ø. (ed); 1997; *Tropospheric Ozone Research*, Springer Verlag, Heidelberg; Vol. 6 of the EUROTRAC final report.

Larsen, S., F. Fiedler, P. Borrell (eds), 2000; *Exchange and Transport of Air Pollutants over Complex Terrain and the Sea*, Springer Verlag, Heidelberg (in preparation); Vol. 9 of the EUROTRAC final report.

Le Bras, G. (ed); 1997; *Chemical Processes in Atmospheric Oxidation*, Springer Verlag, Heidelberg; Vol. 3 of the EUROTRAC final report.

Slanina, J. (ed); 1997, *Biosphere-Atmosphere Exchange of Pollutants and Trace Substances*, Springer Verlag, Heidelberg; Vol. 4 of the EUROTRAC final report.

Warneck, P. (ed); 1996; *Heterogeneous and Liquid Phase Processes, Laboratory studies related to aerosols and clouds*, Springer Verlag, Heidelberg; Vol. 2 of the EUROTRAC final report.

Chapter 7

Policy Applications of the EUROTRAC Results:
The Application Project

Peter Borrell[1], Peter Builtjes[2], Peringe Grennfelt[3], Øystein Hov[4]
(Coordinator and Conveners)

Roel van Aalst[5], David Fowler[6], Gérard Mégie[7], Nicolas Moussiopoulos[8],
Peter Warneck[9], Andreas Volz-Thomas[10,] Richard Wayne[11] (Members)

[1]P&PMB Consultants, D-82467 Garmisch-Partenkirchen, Germany
[2]TNO-MEP, NL-7300 AH Apeldoorn, Netherlands
[3]Swedish Environmental Research Institute (IVL), S-40258 Göteborg, Sweden
[4]NILU, Norwegian Institute for Air Research, N-2007 Kjeller, Norway
[5]RIVM, (EEA Topic Centre: Air Quality), NL-3720 BA Bilthoven, Netherlands
[6]Institute of Terrestrial Ecology, Penicuik Midlothian E26 0QB, UK
[7]CNRS Service d'Aeronomie F-91371 Verrières Le Buisson France
[8]Aristotle University GR-54006 Thessaloniki, Greece
[9]Max-Planck-Institut für Chemie, D-55020 Mainz, Germany
[10]Forschungzentrum Jülich (FZJ), D-52425 Jülich, Germany
[11]Physical Chemistry Laboratory, Oxford OX1 3QZ, UK

7.1 The origin of the Application Project

The Application Project found its origin in the third aim of EUROTRAC (chapter 1.1): *to improve the scientific basis for taking future political decisions on environmental management in the European countries.* It was not clear at the outset of the project how this aim would be achieved. It was realised that, with so many of the scientists responsible for advising their respective governments on policy participating in EUROTRAC, the scientific consensus necessary to the development of European policy would be attained, but nothing specific had been planned to convey the results and conclusions of the project to those responsible for policy development.

In 1991 the IEC decided to seek an outside evaluation of EUROTRAC before approving the continuation into the second half. The task was undertaken by a UK consultancy, Serco Space Ltd which invited seven scientific experts from North America to advise it on the scientific aspects of the programme, while it concentrated on the organisational aspects and the profile of EUROTRAC, both within the project and in the community outside.

The principal recommendation of the favourable report, (Serco Space, 1992) was that a special effort be made to bring together the scientific results of the project and make them available in a form suitable for those concerned with environmental management and policy development in Europe. The IEC accepted the recommendation and formed the Application Project. Ten scientists, the authors of this chapter, were selected, principally funded via special contributions made by the participating countries to the International Scientific Secretariat. Two countries, Norway and the Netherlands, funded their participating scientists directly.

The Application Project was divided into three groups, each headed by a convenor and each covering a particular theme.. There was a coordinator for the whole project. As part of the project, each of the 250 EUROTRAC Principal Investigators was requested to say which of their results they believed to be of potential importance, directly or indirectly to policy development. Some 120 principal investigators responded and thus made a direct contribution to the policy-oriented evaluation of the project. The major part of the evaluation was made by reviewing the published work and reports of the principal investigators and subprojects. The draft report was reviewed by the IEC, SSC, the subproject coordinators and the principal investigators and the many comments taken into account in preparing the final draft which was finally accepted by the IEC. In this way the AP report represented an overall consensus of the whole project on the issues discussed. The report is volume 10 of this series (Borrell *et al.*, 1997) and the remainder of this chapter is taken from the Executive Summary.

7.2 EUROTRAC and the Application Project:
An Introduction to the Executive Summary

The section numbers refer to volume 10 of this series (Borrell *et al.*, 1997).

In this report, the scientific results from EUROTRAC have been assimilated by the Application Project, and the principle findings are presented in a condensed form, suitable for use by those responsible for environmental planning and management in Europe.

In most European countries, air pollution exceeds the acceptable national and international standards but, as many pollutants are transported great distances in the air, only international measures will be successful in control and abatement. The initial measures taken under the Convention on the Long Range Transport of Air Pollution (LRTAP) are bringing some benefits, but it is clear that further reductions in emissions will probably be very expensive to implement. Future cost effective abatement strategies will only be successful if they are underpinned by a thorough understanding of the detailed atmospheric processes in which the pollutants are produced, transformed, consumed, transported and deposited from the air. EUROTRAC, the policy-relevant findings of which are discussed in this report, was set up in 1985 to help provide the improved scientific understanding necessary for future policy development in the field of air pollution control and

EUROTRAC and the
Application Project

abatement. It was recognised that the scientific problems associated with trans-boundary air pollution could only be solved by an international high level state-of-the-art interdisciplinary project.

EUROTRAC is a co-ordinated environmental research programme, within the EUREKA initiative, studying the transport and chemical transformation of trace substances in the troposphere over Europe. The project consists of more than 250 research groups in 24 European countries and is organised into 14 subprojects. (*section 2.4.1)

The EUROTRAC Application Project (AP) was set up

> "to assimilate the scientific results from EUROTRAC and present them in a condensed form, together with recommendations where appropriate, so that they are suitable for use by those responsible for environmental planning and management in Europe".

Three themes were addressed by the AP and are presented in this report:

- Photo-oxidants in Europe; in the free troposphere and in rural and urban atmospheres;

- Acidification of soil and water, and the atmospheric contribution to nutrient inputs;

- The contribution of EUROTRAC to the development of tools for the study of tropospheric pollution; in particular, tropospheric modelling, the development of new or improved instrumentation and the provision of laboratory data.

The first two themes bear directly on issues of environmental concern in Europe. The third, "tools", is a recognition that, while some of the work done during the project has found use immediately, much will find its application in the longer term either directly, by serving applications to policy, or indirectly, by incorporation into the general understanding (*section 2.4.2).

1.   The research accomplished in EUROTRAC provides substantial scientific support for the negotiations in the second generation of abatement strategy protocols under the UN-ECE Convention on the Long Range Transport of Air Pollution (LRTAP), in particular the second revised NO_x protocol and the revisions of the recently signed sulfur protocol. It also finds application for work within the European Environmental Agency, the EU Framework and ozone directives, and in the development of national strategies. EUROTRAC models and measurements are also used by WMO/UNEP and the IPCC for assessment of the current ozone budget and its sensitivity to changes in precursor concentrations.

| Policy applications of EUROTRAC scientific results |

The application of effects-based control strategies by European governments, in which the maximum environmental benefit is being sought for the investment in control technology, places great demands on our knowledge and understanding of the links between sources, deposition and effects. Within EUROTRAC, major developments in the science for developing these links have been made (chapter 1).

7.3 Photo-oxidants in Europe: in the free troposphere, in rural and in urban atmospheres

2.   From experimental studies it is concluded that the natural background of ozone over Europe at the turn of the century, within the atmospheric boundary layer, was about 10 to 15 ppb at ground level and 20 to 30 ppb one to two kilometres above the ground. Today, the concentration of ozone near the sea surface is 30 to 35 ppb before air masses move into Europe from the west. On a seasonal basis, photochemical processes over western and central Europe add about 30 to 40 % to this background in summer, and subtract about 10 % in winter.

> The concentration of photo-oxidants in Europe is strongly influenced by photochemical production from man-made precursors that are emitted within the region

Within Europe very high concentrations of more than 100 ppb are observed during photochemical episodes under unfavourable meteorological conditions, *i.e.* high solar radiation combined with stagnant air or circulating wind systems.

In the free troposphere, that is from the top of the atmospheric boundary layer (1 to 2 km above the ground) to the tropopause which constitutes the boundary with the stratosphere (10 to 12 km above the ground), the background concentration before the air masses pass over Europe is higher than in the atmospheric boundary layer, being about 40 to 50 ppb in winter and autumn and 50 to 70 ppb in spring and summer.

The concentration of free tropospheric ozone over Europe is influenced not only by European emissions but also by North American and Asian emissions (*sections 3.2.1, 3.2.5 and 3.3.2).

3.   The concentration of ozone in the troposphere north of $20°$ N has increased since the beginning of modern measurements. This increase was larger at northern mid-latitudes than in the tropics, and larger over Europe and Japan than over North America. Comparison with historical data suggests that ozone in the troposphere over Europe has doubled since the turn of the century and that most of the increase has occurred since the 1950s. Measurements of nitrate in ice cores from Alpine glaciers provide strong circumstantial evidence for man-made emissions being responsible for the observed ozone trend (*section 3.2.1).

> Tropospheric ozone in the northern hemisphere has increased since the 1950s

4.   Long-term observations show that the increase of ozone in the free troposphere was smaller in the eighties than in the seventies. The average ozone concentrations in the boundary layer near the ground have even decreased at some locations, for example at Garmisch-Partenkirchen in

> **Tropospheric ozone increase slowed down in the 1980s**

Germany and at Delft, a polluted site in the Netherlands. The concentration of peroxyacetylnitrate (PAN), a photo-oxidant like ozone, increased by a factor of three at Delft in the 1970s and stabilised in the 1980s (*section 3.2.1).

5.   The understanding of the temporal and spatial resolution of the emissions of NO_x and VOC has improved considerably. Results show for example that emissions are approximately 30 % lower at weekends than during the week. Such variations provide a regular "natural

> **Weekday/weekend differences in the emissions of ozone precursors**

experiment" for examining the effects of the short-term reductions of emissions. The effects of the reductions on the photo-oxidant concentrations in this case appear to be rather small and indeed may increase ozone levels in areas of high pollution.

6.   Recent model simulations have suggested that the effective abatement of elevated ozone concentrations in Europe require the reduction of the emissions of both NO_x and VOC, with more emphasis being put on VOCs, especially in north-west and central Europe. However some field experiments in EUROTRAC have identified possible shortcomings in the models that are presently used for quantifying ozone/VOC and ozone/NO_x relationships. These experiments emphasise the greater importance of NO_x emissions in controlling the photochemical ozone balance. The reasons are (*sections 3.2.3, 3.2.5, 3.2.6 and 3.3.2):

-       There is an indication, based on ambient measurements, that biogenic VOC emissions, which cannot be subjected to abatement, form a base level of VOCs which is higher than that assumed previously.

> **Should NO_x or VOC be controlled, or both?**

-       The photochemistry in urban plumes seems to proceed faster than is assumed in models. The results suggest that the oxidation of VOCs leads to more peroxy radicals and, hence, more ozone over a shorter time than predicted by the photochemical schemes currently used in atmospheric models, and to a faster removal of NO_x, the catalyst in ozone formation.

Based on today's knowledge, the following picture emerges:

-       An effective way to reduce ozone concentrations on urban and suburban scales appears to be to reduce VOC emissions. However, NO_x reductions are required to reduce the concentrations of other oxidants such as NO_2 and PAN.

- Both NO_x and VOC reductions are required in order to reduce ozone levels on a European scale. NO_x reductions are essential to reduce ozone concentrations on a global scale.

- The emission reductions need to be substantial (40 to 60 %) to obtain noticeable reductions in ozone concentrations.

7.   The pre-industrial concentration of ozone of about 10 to 15 ppb at ground level resulted from the approximate balance between the transfer of ozone from the stratosphere to the troposphere, the destruction in the troposphere by photochemical reactions and by deposition to the ground. The 15 to 20 ppb difference between today's ozone levels, near the sea surface before air masses pass over Europe, and the pre-industrial ozone concentration is probably due to photochemical formation from precursors such as VOC and NO_x, emitted in other parts of the northern hemisphere, in particular North America. Reduction in the background tropospheric level will require agreement on a hemispheric scale. In addition, precursors emitted from biomass burning have a large impact on ozone concentrations in the tropics and in the southern hemisphere (*section 3.2.6).

> **Are photo-oxidants a home-made or trans-boundary problem?**

The very high ozone, NO_2 and PAN concentrations that are observed in some urban and suburban areas (photochemical smog) are due to photochemical production in the atmospheric boundary layer from precursors that are mostly emitted within the area (*sections 32.2.4 and 3.2.5).

On a continental scale, the enhanced photo-oxidant concentrations observed are a consequence of both *in situ* chemistry and transport from regions with higher emissions. Studies in EUROTRAC have greatly added to our understanding of the relevant processes for quantifying ozone/precursor relationships and, hence, provide a better basis for determining how reductions in precursor emissions in one region would reduce the photo-oxidant levels in regions downwind, especially in moderately populated and rural areas (scales >50 km). However, source-receptor relationships are still difficult to assign because the chemistry is non-linear and there are large differences in the residence times of photo-oxidants and precursors in the atmospheric boundary layer close to the ground, compared with the residence times in the free troposphere (*sections 3.2.5, 3.2.6, 3.3.1, 3.3.2 and 3.3.3).

While the highest photo-oxidant levels can be counteracted by local pollution control measures, abatement of enhanced ozone formation on a European scale requires a co-ordinated abatement strategy (*sections 3.2.4, 3.2.5, 3.3.1, 3.3.2).

8.   Strategies to abate photochemical air pollution at any relevant scale may be assessed with models developed within EUROTRAC.

On the local scale, a zooming model (EZM) has already been successfully utilised to optimise the air pollution abatement strategy for Athens, to support the decisions taken with regard to traffic regulations in Barcelona during the 1992 Olympics and to interpret the observations during measuring campaigns in the Upper Rhine Valley.

Models have also been developed to describe the distribution of ozone in the global troposphere. The results show that the concentrations of ozone throughout large parts of the northern hemisphere have been substantially increased by anthropogenic

> **Practical applications of photo-oxidant models on all scales**

emissions of nitrogen oxides, VOCs and carbon monoxide. Since ozone is a greenhouse gas, these elevated levels could be making an appreciable contribution to global warming (*section 3.3.3).

The EURAD model as well as global models have been used to calculate the influence of aircraft emissions on upper tropospheric ozone levels (*section 3.3.2).

7.4 Acidification of soil and water and the atmospheric contribution to nutrient inputs

9.   There is clear evidence that deposition of anthropogenic sulfur and nitrogen compounds over central and northern Europe has caused severe changes in the composition and functioning of many ecosystems. The deposition of these compounds has also been an important factor in the deterioration of materials and of our cultural heritage of ancient buildings. These effects have to a large extent occurred during the second half of the twentieth century and are mainly caused by emissions of sulfur dioxide, nitrogen oxides and ammonia *within* Europe (*sections 2.2.3 and 4.1).

> **Sulfur and nitrogen deposition is still causing severe damage to ecosystems in Europe**

Annual budget calculations on a national scale and for Europe as a whole show that 50 to 70 % of the emissions of sulfur and oxidised nitrogen and approximately 80 % of the emissions of reduced nitrogen are deposited within Europe. These inputs exceed the critical loads for soils and for freshwater acidification over large areas of Europe. A protocol to the LRTAP for reducing the sulfur deposition over the next decade has recently been agreed. For nitrogen, protocols are being developed.

SO_2 emissions vary markedly with the time of day, the day of the week and the season, the variations being explicable in terms of working hours and the influence of temperature on energy demand (*section 4.2.1).

10.   The inputs of dimethylsulfide (DMS) to the atmosphere are important on a global scale for the atmospheric sulfur budget. The emissions from oceans contribute to the background 'natural' sulfur, and sulfur compounds are deposited together with anthropogenically derived sulfur. However, except in small coastal areas of

> **The natural sources of sulfur are unimportant relative to the anthropogenic sources in Europe**

northern and western Europe, the contributions from natural sources to annual sulfur inputs is negligible (*section 4.3.1).

11.   The protocols agreed in 1994 should substantially reduce sulfur emissions and deposition throughout Europe by 2005, and a decrease in sulfur emission and deposition in Europe will reduce the scale of environmental damage due to acidification. Sulfur is however not the only contributor. Deposited nitrogen (both oxidised and reduced) contributes to the acidification problem and in many areas the deposition of nitrogen alone exceeds critical loads to ecosystems. The downward trends in sulfur emission over the last decade, and those expected in the next, are rapidly increasing the relative importance of nitrogen in acidification and eutrophication (*section 4.3.3).

| The contribution of deposited nitrogen to acidification and eutrophication is increasing |

12.   Cloud chemical processes have been shown to be at least as important as gas-phase processes for the oxidation of sulfur dioxide. New pathways for the aqueous-phase oxidation have been discovered and quantified. Incorporation of transfer and reaction mechanisms in simulation models is in progress and it should soon be possible to include more realistic cloud modules in source-receptor models (*section 4.2.2).

| Most of the sulfur dioxide oxidation in Europe occurs in clouds |

13.   Field experiments performed in recent years have led to a better understanding of the formation of cloud droplets and of how particulate matter is incorporated in the droplets. In these experiments two groups of aerosols were observed with different hygroscopic properties. These findings show that the different chemical species in aerosols are scavenged by clouds and therefore by wet deposition processes at different rates, which will lead to different residence times in the atmosphere (*section 4.2.3).

| Cloud chemical processes cause non-linearity |

The reduction of sulfur emissions in western Europe after 1980 is not reflected in atmospheric concentrations in a simple way. Measurements show that sulfur dioxide concentrations are reduced faster than expected, while concentrations of sulfur in precipitation show a slower decrease than expected. Aqueous-phase chemistry in clouds may be a key factor contributing to the observed anomalies (*section 4.2.3).

14.   The application of approaches to develop emission controls, based on critical loads, provides a means of maximising the environmental benefits for the investments in controls. These approaches require the ecosystem sensitivity and the actual deposition input to be quantified on the same scale. This scale of variability in ecosystem sensitivity to acidification varies with land use and vegetation, with the landscape scale being mostly between 1 and 10 km (*section 4.3.4).

| Effects-based emission controls demand great spatial detail for ecosystem sensitivity and deposition |

15.   Models and data bases have been developed to provide the fine scale (1 to 10 km) resolution in the land-use-specific deposition maps that are required to calculate exceedances of critical loads. These methods have been developed from field and laboratory based studies of deposition processes and validated by long-term flux measurements. The new equipment developed to make the measurements, the collaborative field campaigns and the long-term flux studies have provided

> **Fine-scale deposition maps of the key acidifying species have been provided by EUROTRAC work**

the science which underpins the deposition maps. The work has been developed for both national and international approaches (*section 4.3.3).

16.   The dry deposition of sulfur dioxide to vegetation has been shown to be influenced by the presence of surface water (rain, dew *etc.*). These effects lead to larger rates of SO_2 deposition in the presence of ammonia. Atmospheric ammonia may therefore reduce the atmospheric lifetime of sulfur dioxide. Similar effects of SO_2 on rates of ammonia exchange over vegetation are expected. The quantitative detail has yet to

> **The elimination of ammonia emissions in Europe would change the trans-boundary fluxes of sulfur**

be provided, but it is clear that emission controls of either of these gases will influence the lifetime and the deposition 'footprint' of both gases (*section 4.2.3).

17.   Up to now the control of emissions of nitrogen compounds has been largely on emissions of NO_x from vehicles and large combustion plants. However some of the major environmental effects (acidification and eutrophication) are caused by *both* oxidised and reduced nitrogen. Moreover, deposition and the effects of deposited nitrogen over large

> **Reduced as well as oxidised nitrogen emissions need to be controlled**

areas of Europe are dominated by inputs of reduced nitrogen as NH_3 in dry deposition and NH_4^+ in wet deposition. To reduce the effects or eliminate exceedances of critical loads for nitrogen, it is necessary to control emissions of both NO_x and NH_3 (*sections 4.2.1 and 4.3.1).

18.   Models have been developed to describe the global transport of sulfur and nitrogen compounds. These models show that sulfate aerosols (originating from anthropogenic sulfur emissions) reflect sunlight back into space and thereby cool the surface. In heavily industrialised regions, where the concentration of sulfate aerosols is high, this cooling

> **Anthropogenic sulfate aerosols in the free troposphere counteract global warming**

effect may mask a major fraction of the warming due to greenhouse gases (*section 4.3.5).

7.5 The contribution of EUROTRAC to the development of tools for the study of tropospheric pollution

19.   Methods and tools, in the form of simulation models, instrumentation and the results of laboratory studies, have been developed which are being used and will be used in the future in the evaluation of abatement

> **Tools now available for the evaluation of abatement strategies**

strategies on continental and urban scales, for air pollution episodes and long-term averaged situations (*section 5.1).

20.   Major improvements and new developments in instrumentation have been made. These include long-path absorption techniques, tunable-diode laser spectrometers, tropospheric lidar techniques, automated VOC measuring systems and a sensitive NO_x analyser. Numerous instrument intercomparisons have led to a better assessment of the accuracy of the instruments.

> **New and improved instrumentation has been developed within this EUREKA project**

Several of these techniques are now in general use and have been commercialised; as expected in a EUREKA project. The EUROTRAC experience indicates however that if the *separate* development of innovative instrumentation is required, precisely defined objectives and centralised funding are needed (*section 5.2.5).

21.   Laboratory studies of free radical processes, of the oxidation reactions of aromatic and biogenic compounds and of individual reactions of many chemical intermediates have considerably improved the knowledge of tropospheric ozone chemistry. Laboratory studies of chemistry in cloud droplets and aerosol particles have added to the understanding of the elementary reaction steps of the oxidation of sulfate in

> **Laboratory results are improving our knowledge of atmospheric chemistry**

clouds, of the absorption process of gases by cloud droplets and of aerosol surface reactions. These new findings have been incorporated into simulation models, so improving their reliability (*sections 5.3.2, 5.3.3 and 5.3.4).

22.   Simulation models on urban, continental and global scales have been developed, improved and validated. These models, which integrate the results of process-oriented studies, describe in a quantitative manner the relationship between emissions and atmospheric concentrations or deposition. As such models are based on state-of-the-art knowledge and experience, and they should be viewed as adequate and reliable tools which have been

> **Simulation models required to address policy-oriented questions**

and will be asked to answer policy-oriented questions. However simulation models require continuous improvement to include new findings and understanding, and frequent validation to demonstrate their reliability (*section 5.4.7).

23.   Emission data bases, including the procedures developed for temporal and spatial disaggregation of the data, are an essential tool since both accurate emission data and optimal meteorological input is required as input for simulation models in order to provide quantitative source/receptor relationships. Only with good and accurate

> **Good emission data essential**

emission data can reliable results be expected from simulation models. Data bases of field observations and field campaigns collected within EUROTRAC are also valuable tools for future studies (*sections 3.2.2, 5.5 and 5.5.2).

24.   The network of scientists, created by EUROTRAC, has and will lead to more synergistic scientific activities and a coherent environmental policy in Europe. The network provides an efficient way of transferring knowledge and provides a major source of expert advice to the participating countries, to the EU and the UN-ECE (*section 5.5.5).

> **Networks of scientists are sources of expert advice**

7.6 Uncertainties in our present knowledge

25.   While much has been achieved within the present project a number of uncertainties remain. These will have to be resolved in the future if the source/receptor relationships, quantitative enough to support the present trends in policy development, are to be produced. The following headings mention the major uncertainties and lists some of the work needed to reduce them.

> **Uncertainties requiring resolution**

For photo-oxidants, there are uncertainties in estimates of precursor emissions and in the detailed chemical and meteorological mechanisms by which photo-oxidants are processed. Studies are necessary to:

- develop better emission estimates for nitrogen oxides and speciated volatile organic compounds including biogenic species;

- validate chemical transport models on regional and sub-regional scales;

- resolve the current disagreements between field experiments and model calculations on the role of NO_x versus VOC limitation of photo-oxidant formation.

For acidification and deposition of nutrients, the principle uncertainties are in emission estimates, the effects of clouds and aerosols on the production of acidity and the effects of complex terrain on deposition. Studies are necessary to:

- develop better validated emission estimates for nitrogen oxides and ammonia;

- develop a more complete understanding, and hence parameterisation of aerosol processes governing their life cycle from their formation and transport to their transformation and deposition in the atmosphere, and incorporate this into source-receptor models;

- develop validated models for local-scale deposition of reactive species;

- develop methods and models to quantify deposition in complex terrain (hills and forests).

For the development of tools, improvements are needed in the models themselves, in their validation, in instrumentation for monitoring and research and in the mechanisms of the fundamental reactions involved in the formation of photo-oxidants and acidity. Studies are necessary to:

- validate models with more dedicated field measurements and specifically designed field campaigns;

- validate emission data using an integrated approach that includes ambient concentration measurements and the use of models;

- increase and improve scientific monitoring for process studies, trend analysis and the evaluation of policy measures;

- develop improved instrumentation for monitoring and field measurements within any future scientific programmes;

- elucidate the mechanisms for the oxidation of aromatic and biogenic VOCs.

- complete the chemical mechanisms used in models and test them in chemical simulation experiments; particularly requiring attention are the mechanisms describing the effect of aromatic hydrocarbons on photo-oxidant formation, the oxidation of biogenic VOCs, and nitrogen chemistry, including heterogeneous processes;

- incorporate validated and more realistic mechanisms for aqueous oxidation in cloud droplets into simulation models.

7.7 Further information

The full report of the Application Project is given in volume 10 of this series (Borrell *et al.*, 1997). An indication of the way in which the scientific uncertainties are being addressed is shown in the project description for EUROTRAC-2 (ISS 1999).

7.8 References

Borrell, P., Ø. Hov, P. Grennfelt and P. Builtjes (eds); 1997; *Photo-oxidants, Acidification and Tools: Policy Applications of EUROTRAC Results,* Springer Verlag, Heidelberg; Vol. 10 of the EUROTRAC final report.

ISS; 1999; *EUROTRAC-2: Project Description and Handbook,* EUROTRAC-2 ISS, München, 1998. See also: http://www.gsf.de/eurotrac

Serco Space, 1992, *Review of EUROTRAC,* Serco Space Ltd, Sunbury on Thames.

Part II

An Overview of the Scientific Work of the Subprojects

Chapter 8

Field Measurements

Field work is required in atmospheric chemistry to gain direct experimental evidence about the processes in the atmosphere. The understanding obtained provides a guide to modellers on what should be included in the models and the results provide the ultimate yardstick against which chemical transport models must be measured.

There were seven subprojects principally concerned with field measurements within EUROTRAC. The four dealt with in this chapter are:

8.1 **ALPTRAC** High Alpine Aerosol and Snow Chemistry Study

8.2 **GCE** Ground-based Cloud Experiments

8.3 **TOR** Tropospheric Ozone Research

8.4 **TRACT** Transport of Pollutants over Complex Terrain

Two subprojects, ASE and BIATEX, concerned with the exchange between the biosphere and the atmosphere are dealt with in chapter 8.

At the outset there was further subproject, ACE (Acidity in Cloud Experiments) which was closed at the end of 1991; most of the principal investigators were able to transfer their contributions to GCE.

8.1 ALPTRAC: High Alpine Air and Snow Chemistry

Dietmar Wagenbach (Coordinator)

Universität Heidelberg, Institut für Umweltphysik, Im Neuenheimer Feld 366, D-69120 Heidelberg, Germany

8.1.1 Aims

The general objective of ALPTRAC was to investigate the occurrence, transformation and deposition of atmospheric constituents (primary and secondary aerosol species and associated precursor gases) within the high altitude environment of the Alps. To achieve this, four different but closely connected observational programs were placed in operation.

SNOSP: *The alpine wide investigation of the winter snow pack chemistry above 3000 m, mainly devoted to the seasonal and geographical distribution of snow concentrations and deposition fluxes of ionic species.*

The SNOSP data set, reflecting on a large spatial scale the present day anthropogenic impact on the alpine snow chemistry, will be fundamental to recognise any future changes of the precipitation chemistry in this ecologically sensitive area.

CORE: *The retrospective ice core study on the long term chemical changes of high elevation snow fields, covering the decadal as well as the centennial time scale.*

This is essentially an attempt to investigate total deposition samples as presumably archived in non-temperated glaciers of the highest alpine summit ranges. Perhaps most important here (although somewhat beyond the more 'secular' EUROTRAC scope), are the expected insights into the natural properties of atmospheric pollutant cycles, which are difficult to obtain from model calculations in view of the tremendous uncertainties involved about the natural fluxes.

ALASS: *The process oriented field study on the various interrelations between the air, aerosol and cloud chemistry, which finally determine the pollutant concentrations of alpine precipitation at high elevation sites.*

This part of ALPTRAC performed at the Sonnblick and Jungfraujoch observatories as well as on Mt. Rigi focused on three different topics, distinguished by their characteristic time scale:

a) the chemical climatology of high alpine sites, which is closely related to the *seasonal* snow pack chemistry studied under SNOSP,

b) the reactive aerosol chemistry connecting atmospheric records of gaseous and particulate species to the alpine precipitation chemistry on the *synoptical time scale* and,

c) the transfer of various atmospheric species to super-cooled cloud droplets and to ice crystals as observed during *single precipitation events.*

SNOWMET: *The meteorological study providing support regarding basic information on the climatology of all alpine sites under investigation, specific meteorological situations during individual events and interpretation of measured aerosol data.*

One may conclude, that the characteristic feature of ALPTRAC within the concert of the EUROTRAC research activities, is to address air chemical issues, which are logically associated with the high alpine environment, as in particular:

❋   the air mass transport to, and the vertical exchange within the high alpine terrain,

❋   the chemistry of the free troposphere,

❋   the scavenging of pollutants by solid hydrometeors,

❋   the deposition of pollutants onto snow covered receptor areas and,

❋   the long term glacio-chemical records.

8.1.2 Principal scientific results

a.   Alpine aerosol and trace gas levels

Monthly averages of selected gases (nitric acid, sulfur dioxide, ammonia) and major aerosol constituents (particulate nitrate, sulfate and ammonium) determined at the 3 km level in the eastern part of the Alps (Mt. Sonnblick, Austria) ranged from 0.8–12 nmol/m^3 for nitric acid, 1.1–23 nmol/m^3 for sulfur dioxide, 2.4–23 nmol/m^3 for ammonia and 0.3–12 nmol/m^3, 1.7–33 nmol/m^3 and 1.9–62 nmol/m^3 for particulate nitrate, sulfate and ammonium, respectively. On a daily basis, even stronger variations could be observed (Kasper and Puxbaum, 1994). Maximum values reach up to concentrations typical for measurements in remote areas within the planetary boundary layer. Annual average concentrations at the Sonnblick observatory are one order of magnitude lower than at a rural site in Eastern Austria (Wolkersdorf, 240 m a.s.l.). Similarly low concentrations were measured in winter (October through February) and spring (March to May) at Jungfraujoch (3450 m a.s.l., Swiss Alps, Schwikowski *et al.*, 1995). Here, observations of Radon decay products (Gäggeler *et al.*, 1995) corroborate the findings of chemical species.

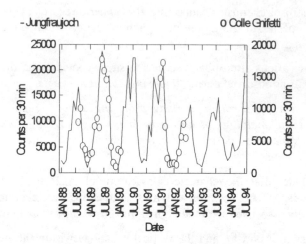

Fig. 8.1.1:   Monthly means of the epiphaniometer signals at Jungfraujoch (3450 m a.s.l.) and Colle Gnifetti (4450 m a.s.l.). A count rate of 1000 counts/30 min corresponds roughly to an aerosol mass concentration of 0.5 µg/m³ (Baltensperger *et al.*, 1991).

A strong seasonality of aerosol species is found in the Alps, as exemplified by the monthly means of the epiphaniometer signals from Jungfraujoch and Colle Gnifetti, a glacier in the Monte Rosa region, Switzerland (4450 m a.s.l.). Fig. 8.1.1 shows that at both sites the average summer aerosol signal was a factor of ten higher than the average winter signal. At Sonnblick observatory (SBO), Kasper and Puxbaum (1994) found a similar seasonality for a variety of atmospheric constituents, with summer/winter ratios of 8 for nitrate, 11 for sulfate, 17 for ammonium, and 5 for both nitric acid and ammonia. The general conclusion from the strong seasonalities is that different vertical exchange conditions are responsible for the different concentrations in the various seasons. During winter free tropospheric conditions prevail, whereas during summer boundary layer air is transported to the highest sites in the Alps. This fact is also reflected in the clear diurnal variations of the aerosol signal emerging only during the summer half year (Baltensperger *et al.*, 1991).

b.   Cloud chemistry and scavenging of pollutants

Cloud water chemistry was studied at the Sonnblick and Jungfraujoch sites, indicating a seasonality of cloud water concentrations comparable to the conditions observed for the wet precipitation data (Brantner *et al.*, 1994; Kalina and Puxbaum, 1994b; Schwikowski *et al.*, 1994). Cloud to snow ratios showed high variability for the single events and different compounds with average ratios between 2 and 10 for sulfate, nitrate, ammonium and hydrogen ions, being comparable to data reported for Mount Rigi, Switzerland (Collet and Steiner, 1992). The cloud to snow ratios for sulfate were used to reconstruct the mixing ratios of sublimation grown ice phase and cloud water droplets (Brantner *et al.*, 1994) indicating that an attachment of 70 to 30 % of cloud water to the ice phase is necessary to explain the sulfate concentrations in precipitating snow (Kalina and Puxbaum, 1994a).

Scavenging ratios have been determined at three ALPTRAC sites. Results are summarised in Table 8.1.1. The Weißfluhjoch data are from a campaign from winter/spring 1988 (Baltensperger *et al.*, 1993). The Jungfraujoch data are from 4 campaigns in the years 1990–1993 (Baltensperger *et al.*, 1992; Schwikowski *et al.*, 1994; Schwikowski *et al.*, 1995). The Sonnblick data are from a two year time series of aerosol and precipitation measurements (Kasper, unpublished results).

Table 8.1.1:   Scavenging ratios at three ALPTRAC sites. Volume based ratios $\times 10^6$

	Sulfate	Nitrate + HNO$_3$	Ammonium
Weißfluhjoch	0.3	0.7	0.2
Jungfraujoch	1.0	3.3	0.6
Sonnblick	1.7	2.5	1.6

Scavenging ratios appear to exhibit a seasonality with higher values in winter and lower values in summer. The seasonality has been observed at Jungfraujoch as well as at Sonnblick. The Weißfluhjoch data are from one late winter campaign and not representative for one year.

The aerosol scavenging efficiency (aerosol concentration incorporated into cloud water divided by the pre-cloud aerosol concentration) shows a strong increase with the liquid water content (LWC). As the LWC exceeds 2 g/m^3 the scavenging efficiencies remain rather constant, around 0.85 (Brantner, 1994).

The role of riming was investigated at both alpine observatories and at Mt. Rigi. According to a scale running from 0 (unrimed) to 5 (graupel) (Mosimann *et al.*, 1993) the degree of riming ranged from 0.5 to 4.5 during all campaigns at Sonnblick. At the onset of precipitation higher values could be observed which decreased during the snowfall (Kalina and Puxbaum, 1994a). A similar result was presented by Mosimann *et al.* (1993) for a campaign conducted at Mt. Rigi, Switzerland. Time weighted averages of the degree of riming are found to differ not substantially between summer and winter conditions. At Jungfraujoch, for three case studies in November, the degree of riming varied between 1 and 4. During single snow fall periods, constant, increasing and decreasing degrees of riming were found (Poulida, 1994).

Under the assumption that riming is the dominant process for the incorporation of impurities into the ice crystal, the concentration in snow can be estimated from the cloudwater concentrations of these impurities (Baltensperger *et al.*, 1994, see Fig. 8.1.2). As a general observation, the degree of riming was found to be higher at Mount Rigi than at Sonnblick and at Jungfraujoch which are at higher elevations.

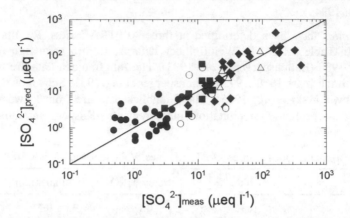

Fig. 8.1.2:   Measured and predicted (see text) concentrations of sulfate in fresh snow from Mount Rigi (♦); summer (Δ) and winter (O) data from Sonnblick and spring (■) and winter (●) data from Jungfraujoch (compiled by Baltensperger *et al.*,1994).

In most cases, at Sonnblick and Jungfraujoch a simultaneous increase of the degree of riming with the LWC (Kalina and Puxbaum, 1994a, Poulida *et al.*, 1994) as well as with the concentration of the wet precipitation samples (Kalina and Puxbaum, 1994a) could be found. The poorest correlation was determined for nitrate, which also showed the largest fluctuation of cloud water concentrations. These findings confirm that riming is the predominant process determining the ion composition of the precipitating snow.

c.   Ion concentration in wet deposition and snow packs

Wet deposition chemistry

Wet deposition measurements conducted with a WADOS (Wet And Dry Only Sampler) at the Sonnblick observatory showed low concentration levels of ionic constituents but high deposition loads. During 1991 and 1994 the annual wet flux was in the range of 22–38 meq/m^2 for ammonium, 27–38 meq/m^2 for sulfate and 18–24 meq/m^2 for nitrate. Average concentrations ranged between 11–22 μeq/l for ammonium, 17–25 μeq/l for sulfate and 11–16 μeq/l for nitrate (Kalina and Puxbaum, 1994a; 1994b). While ionic concentrations are up to a factor of 5 lower at the high alpine site than at lower level sites, the annual flux of sulfate, ammonium and nitrate is equal at both regions or tends to be even higher in the alpine environment.

The seasonal variation of aerosol components was found to control the respective change of the wet deposition concentrations observed earlier (Puxbaum *et al.*, 1991). Fig. 8.1.3 compares the annual cycles for particulate sulfate and for sulfate in precipitation determined at the Sonnblick observatory (Kasper, 1994).

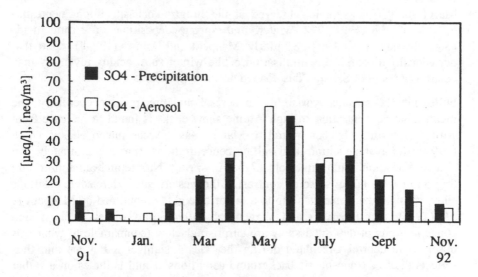

Fig. 8.1.3:   Seasonal cycle for atmospheric sulfate (neq/m³) and for sulfate in precipitation (µeq/L) determined at the Sonnblick observatory (Kasper, 1994).

Spatial and seasonal variability of snow pack chemistry (SNOSP)

Regular snow pit studies on selected glaciers are performed to investigate total pollutant deposition. At up to 17 sites across the whole range of the Alps in the years 1991 through 1994. A detailed description of the sites, being mainly close to the 3000 m level, is given in the ALPTRAC Data Catalogue (1993). The sampling and analytical skill of the participating groups and their laboratories was tested repeatedly and was found to be reproducible or interchangeable to between 10 % and 20 % for chloride, nitrate, sulfate, ammonium, sodium and calcium(Schwikowski *et al.*, 1997). The results attained for magnesium, pottasium and pH were less satisfying, however.

An overview on the temporal-spatial variability of the SNOSP data is given by Nickus *et al.* (1997). Furthermore, at the French glaciers sites Maupetit and Delmas (1994a; 1994b) provided for the first time a comprehensive study on the ionic snow pack composition including all relevant species (major ions, organic acids and halogens)

West-East transect: According to Nickus *et al.* (1997), Puxbaum and Wagenbach (1994), Schöner *et al.* (1994), and illustrated in Fig. 8.1.2, the volume weighted average concentrations of sulfate, ammonium and nitrate observed for the accumulation period (October-May) exhibit a distinct geographical pattern with a generally increasing trend of nitrate from west to east. The sulfate concentration increases are much weaker or even absent.

Mineral dust events: Several mineral dust events have been observed at the time of their occurrence or have clearly been identified in dated snow layers at a number of sites (Maupetit and Delmas, 1994b; Schwikowski *et al.*, 1995) and were attributed to long range transport of Saharan dust. One single event in

March 1990, for example, observed at the Jungfraujoch site might neutralise 50 % of the 1991/1992 winter and spring deposition of strong acids (Schwikowski *et al.*, 1997). Similarly, Maupetit and Delmas (1994b) report that occasionally a complete neutralisation of the winter snow acidity inventory may occur as a result of Saharan dust loaded air mass intrusions.

Following the seasonal variability in aerosol and precipitation chemistry, the mean ionic concentration in high Alpine snow packs is found to increase from winter to spring (October–March to March–May). According to Nickus *et al.* (1997), the average nitrate and sulfate concentrations were approximately two and that of ammonium three fold higher in spring. Nitrate and ammonium are found to have a larger seasonal contrast at the western sites. Here some positions show spring enhancements of up to a factor of eight. As outlined by Maupetit *et al.* (1995), and Puxbaum and Wagenbach (1994), mid winter levels of ion concentrations in deposited snow are surprisingly low (comparable to what was observed in central Greenland during the "clean" summer half year) and thus reflects the free tropospheric background conditions at mid latitudes on a rather large spatial scale.

d.   Transport processes and source areas

There are three main source areas for aerosols as measured by the epiphaniometer at Jungfraujoch: north-east central Europe, the Iberian and the Italian Peninsula. For the Sonnblick site, the Iberian Peninsula is of lesser importance, while the Balkans come second after north-east Europe (Kromp-Kolb in: Fuzzi and Waggenbach, 1997).

Low aerosol concentrations occur in air masses arriving from the north west, usually of maritime origin, even though they crossed the British Isles, or, in the case of the eastern Alps, NW Europe. Thus long range advection from south west and south east seems be more important for the Alpine ridge than from the north west. These unexpectedly high contributions from southern Europe are attributed to reservoir layers, which build up due to lack of rain. Transport of the high pollutant levels to mid troposphere is enhanced by mountains.

The low contributions from the high emission regions in north western Europe seem to be a consequence of the high likelihood of precipitation in northwesterly flow before these air masses reach the Alps. Some episodes of extremely high sulfate and nitrate concentrations have been traced to transport from the Po Valley. Transport of air from the Po Valley below 1000 m (mixing layer) to the Sonnblick occurs in about 16 % of the cases with trajectories across the Po Valley and 29 % at the Jungfraujoch.

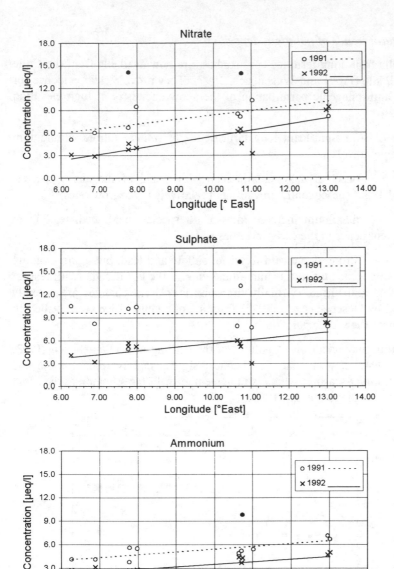

Fig. 8.1.4: Precipitation weighted mean nitrate, sulfate and ammonium concentrations of the alpine snow cover observed at the end of the winter and spring accumulation period in 1991 and 1992. The years shown were selected in view of their most complete geographical coverage of the Alpine ridge. The SNOSP sampling sites were arranged according to their longitudinal position. Outliers, displayed as full circles, are not used for linear regression. (adapted from Schöner *et al.*, 1994; Nickus *et al.*, 1997)

e.   Long term trends of pollutant deposition

CORE results from long term ice core studies, are now available from the Colle Gnifetti drill site (Fuzzi and Wagenbach, 1997). They provide clear evidence for a dramatic anthropogenic perturbation of the snow chemistry at high elevation alpine sites, as:

≈ 1750–1850:   no significant long term trends of sulfate, nitrate, ammonium and heavy metals,

1850–1950:   onset of a systematic increasing trend of the above species, around 1860, subsequently prevailing of relatively weak grow rates,

1950–1970:   maximum increase rates of all species mentioned, leading to prominent peaks in the early seventies and

1970–present: strong downward trends of sulfate and lead, but again ongoing increasing trends for nitrate and ammonium in the last decade (with respect to sulfate and nitrate, virtually the same long term changes are recently derived by Döscher et al., (1995) from a Colle Gnifetti ice core, drilled in 1982 near the saddle point).

Isotopical and nivological investigations of the drilling area (Wagenbach, 1993; Preunkert, 1994) showed, that the chemical signals mainly represent the late spring and summer seasons, which is in agreement with the situation at the Mont Blanc drill site (Maupetit et al., 1995).

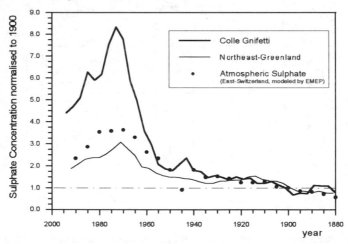

Fig. 8.1.5:   The 20th-century change of mean sulfate concentrations in alpine and Greenland snow and calculated data. To facilitate comparison between alpine and Greenland snow, both records are smoothed identically by a decadel running mean (according to respective the low time resolution of the Greenland data) and subsequently normalised to the level at turn of the century. Calculated dates are adapted from model calculation of near surface atmospheric sulfate by Molyna (1993), referring to the EMEP rid 24,14. (Greenland data by Fischer et al., unpublished results, University of Heidelberg).

A unique finding of the 20th-century change of the ionic snow composition in the Alps appears to be the steady increase of ammonium to a recent concentration level, exceeding that at turn of the century by roughly a factor of 5 (Wagenbach, 1993). Surprisingly, no significantly increasing ammonium trends have been detected in various Greenland ice cores (Legrand *et al.*, 1995). Both, the historical ammonia emission inventories in western Europe as well as the atmospheric NH_x records deduced from them exhibit over the last 100 years only an increase not much more than a factor two (Asman and Dukker, 1988).

Important implications arising from the strong perturbation of the atmospheric load of sulfate and nitrate are, among others:

❋ the relatively large nitrate grow-rates between 1950 and 1970, which clearly control the overall change since turn of the century. This temporal pattern is closely paralleled by lead (but not by other heavy metals) indicating a substantial contribution of non stationary NO emissions to the observed nitrate record. As outlined in the EUROTRAC Application Project, the above finding is believed to confirm the evaluations of past European ozone measurements which suggest a two-fold ozone increase, mainly confined to the 1950–1970 period as well (Staehelin *et al.*, 1994) (see NO_x limited ozone production rate).

❋ the dramatic change of the free tropospheric sulfate load, exceeding during the early seventies the 1900 level by roughly a factor of eight. As illustrated in Fig. 8.1.4, this relative change is considerable larger than in Greenland snow (although the temporal pattern is surprisingly well mimicked there). It is also significantly higher during the post 1950 era, than anything predicted by Molyna, 1993, from a EMEP/MSC-W model calculation, based on historical sulfur dioxide emission inventories. The most simple explanation for the latter discrepancy, would be a significant underestimation of the (quasi-)natural background sulfate level, used in the EMEP model. Taking the Colle Gnifetti records, at least, representative on the regional scale, the extreme increase in (summer time) sulfate may be interesting as entry in models on the regional radiative forcing by the anthropogenic (sub-micron) aerosol body (IPCC, 1994). Indeed, there is apparently a striking anti-correlation between the alpine sulfate record and the surface air temperature deviation synthesised by Engardt and Rodhe (1993) from globally distributed regions, significantly affected by anthropogenic emissions.

❋ the clearly declining sulfate trend in the past 1970 period (see Fig. 8.1.5) which is in contrast to an ongoing strong increase of nitrate and ammonium confirms the growing importance of the N species in the acidification and enthrophication of natural ecosystem (as presented by the high alpine region). Only in a quantitative sense, similar trends were identified during the last decade at a SNOSP glacier site in eastern Austria (Winiwarter, 1994).

8.1.3 Conclusions

The three observational ALPTRAC subprograms backed up by the meteorological support study, already reached a good deal of the original ALPTRAC aims (see below). The wealth of new field data on various alpine pollution aspects, spanning from the very local to the continental scale, but also from the diurnal to the centennial time scale deserves, however, further evaluations.

a.   High Alpine Aerosol and Snow Chemistry (ALASS)

A strong seasonal cycle for concentrations of major ions (sulfate, nitrate, ammonium) in high alpine wet deposition was observed at Sonnblick. In the ALASS study concurrent observations of atmospheric gas, aerosol and wet precipitation measurements have revealed, that the low concentration of sulfate and nitrate in snow during the cold season is not primarily a result of less efficient scavenging during colder conditions but a result of cleaner air masses ingested into the precipitating clouds at that time of the year. A two year time series of SO_2, HNO_3, sulfate and nitrate at Sonnblick (3106 m) reflects the aerosol climatology at the 3 km level in the European Alps.

Scavenging ratios appear to exhibit a seasonality with higher values in the cold and lower values in the warm season. The average volume based scavenging ratio for sulfate was 1.0×10^6 at Jungfraujoch and 1.7×10^6 at Sonnblick.

The aerosol scavenging efficiency was found to be a function of the liquid water content (LWC) of the cloud. At LWCs above 2 g/m^3 the fine particle scavenging efficiency was relatively constant around 0,85.

Investigations of the mass fraction of droplets (b) attached to the mass fraction of ice (a) were performed microscopically and by comparing the composition of snow with the composition of cloud water. Both methods resulted in an average range of 0.25–0.50 for the attached mass fraction b of super-cooled droplets.

b.   High Alpine Snow Chemistry (SNOSP)

A five year data set from alpine wide snow sampling campaigns from up to 17 glacier sites supplemental by a 10 year time series at Sonnblick is available now allowing to evaluate geographical trends which appear more pronounced for nitrate than for sulfate. The temporal trends observed in the 10 year study point towards a decrease in sulfate and an increase in ammonium which is consistent with results from the CORE study. With the aid of intra-site variability studies it was possible to verify the representativity of the geographical trends.

Large seasonal variations of the concentrations of major ions (sulfate, nitrate, ammonium) in wet precipitation were found to be consistent with the high resolution snow pack studies. There is now evidence from different reports from the ALASS study that the striking seasonality of the ionic concentration levels, is predominantly originating from a seasonal variation of the vertical mixing intensity during the precipitation events. A decoupling of air masses in the boundary layer from those at cloud level in the mid troposphere during the cold season leads to concentration levels of major ions in snow which can be extremely low (comparing to concentrations observed on the Greenland ice cap). During the warm season vertical exchange processes lead to ionic concentrations in the high alpine wet deposition which can come close to levels observed at lower sites situated within the boundary layer.

c.   Long term trends (CORE)

Alpine ice core studies, performed in ALPTRAC, allowed for the first time, to make a comprehensive sketch on the chemical long term changes, having occurred in the free atmosphere, outside the polar ice sheets. Among the wealth of essentially new information, following findings appear to be specifically important with respect to other EUROTRAC research activities:

❋   the univocal trends, exhibited by the above species (including lead) during the last 20 years, that may now be reviewed in consideration of the relatively well known emission records available for that period,

❋   the synchronous relative increases of sulfate, nitrate and ammonium in the past 1950 period, which are much larger than those observed in Greenland or (in the case of sulfate and ammonium) predicted for ground level positions by simple modelling studies. This finding calls for a closer look to the interplay of these and related species, when dealing with their source-remote receptor relationships,

❋   the relatively clear picture emerging now about the continental clean air chemistry during the pre-industrial era providing an unique observational approach to the natural baseline levels of atmospheric pollutants.

In making full profit of the glacio-chemical long term records, it will be crucial to correct them for site specific glacio-meteorological changes, affecting mainly the modern, as well as the pre-industrial periods. Equally important will be to place some reliable limits on the spatial scales being represented by the individual Colle Gnifetti records. This is expected to be provided by the total of four new ice cores drilled to bedrock there and in the Mont Blanc region, respectively. Thus an immediate validation of the existing records, will be possible over several hundred years within the drill site area and over the last 50 years, approximately, within the interregional scale.

8.1.4 Further information

The ALPTRAC contribution is shown in context in chapters 2, 3 and 5 of this volume.

A complete account of the subproject can be found in volume 5 of this series (Fuzzi and Wagenbach, 1997) which includes the following reviews by the subproject coordinator:

An Introduction to ALPTRAC Research
Aims and Internal Structure of ALPTRAC

Principal Scientific Results

and contributions from the individual principal investigators, the titles of which are given in the next section.

8.1.5 ALPTRAC: Steering group and principal investigators

Steering group

Dietmar Wagenbach (Coordinator)	Heidelberg
Heinz Gäggler	PSI, Villigen
Helga Kromp-Kolb	Vienna
Michael Kuhn	Innsbruck
Hans Puxbaum	Vienna

Principal investigators

Here is a list of principal investigators (*) and some of their co-workers together with the title of their contribution to the ALPTRAC final report (Fuzzi and Wagenbach, 1997).

High Alpine Snow Pack Chemistry

M. Kuhn and U. Nickus,*
   Accumulation of Pollutants and Nutrients in the Snow Pack at High Altitudes along a North-South Transect in the Eastern Alps

F. Pichlmayer and K. Blochberger,*
   Stable Isotope Ratio of Sulfur, Nitrogen and Carbon as Pollution Tracers for Atmospheric Constituents

M. Staudinger, W. Schöner, H. Puxbaum and R. Böhm,*
   Accumulation of Acidic Components in Two Snowfields in the Sonnblick Region

Long-Term Pollution Trends

D. Wagenbach, K. Geis, K. Hebestreit, S. Preunkert, J. Schäfer, R. Schajor, V. Ulshöfer and P. Weddeling,*
   Retrospective and Present State of Anthropogenic Aerosol Deposition at a High Altitude Alpine Glacier (Colle Gnifetti, 4450 m a.s.l.)

High Alpine Air, Aerosol and Cloud Chemistry

A. Berner and C. Kruisz,*
   Segregation of Hydrometeors

H.W. Gäggeler, M. Schwikowski*, U. Baltensperger* and D.T. Jost,*
   Transport, Scavenging and Deposition Studies of Air Pollutants at High-Alpine Sites

H. Puxbaum, B. Brantner, H. Fierlinger, M. F. Kalina, A. Kasper*, S. Paleczek
and W. Winiwarter*,*
Alpine Aerosol and Snow Chemistry Study at the Sonnblick Observatory
(Austria, 3106 m a.s.l.)

J. Staehelin* and A. Waldvogel,
Aerosol and Hydrometeor Concentrations and their Chemical Composition
during Winter Precipitation along a Mountain Slope (Mt. Rigi, Switzerland)

Meteorological Support to ALPTRAC

H. Kromp-Kolb*, P. Seibert* and W. Schöner,
Meteorological Support Study

8.1.6 References

ALPTRAC Data Catalogue, Kromb-Kolb H., W. Schöner, P. Seibert; 1993, *EUROTRAC Special Publication*, EUROTRAC ISS, Garmisch-Partenkirchen, 137 pp.

Asman W., B. Drukker; 1988, Modelled historical concentrations and depositions of ammonia and ammonium in Europe. *Atmos. Environ.* **22**, 725–735

Baltensperger U., H.W. Gäggeler, D.T. Jost, M. Emenegger, W. Nägeli; 1991, Continuous background aerosol monitoring with the Epiphaniometer. *Atmos. Environ.* **25A**, 629–634.

Baltensperger U., M. Schwikowski, H.W. Gäggeler; 1994, Scavenging of atmospheric constitutents by snow; in: P.M. Borrell, P. Borrell, T. Cvitaš, W. Seiler (eds) *Proc. EUROTRAC Symp. '94.* SPB Academic Publishing bv, The Hague, pp. 973–977

Baltensperger U., M. Schwikowski, H.W. Gäggeler, D.T. Jost; 1992, The scavenging of atmospheric constitutents by Alpine snow, in: S.E. Schwartz, W.G.N. Slinn (eds), *Precipitation Scavenging and Atmosphere-Surface Exchange*, Hemisphere Publishing Corporation, Washington, pp. 483–493

Brantner B.; 1994, Multiphase chemistry in the subcooled droplet regime at Mt. Sonnblick (3106 m a.s.l.). Thesis, Vienna University of Technology.

Brantner B., H. Fierlinger, A. Berner, H. Puxbaum; 1994, Cloudwater chemistry in the subcooled droplet regime at Mount Sonnblick (Austrian Alps), *Water Air and Soil Pollut.* **74**, 363–384.

Collett J.L., M. Steiner; 1992, Investigations of the relationship between cloudwater and precipitation chemistry using Doppler radar; in: S.E. Schwartz, W.G.N. Slinn (eds), *Precipitation Scavenging and Atmosphere-Surface Exchange*, Hemisphere Publishing Corporation, Washington, pp. 381–392.

Döscher A., H.W. Gäggeler, U. Schotterer, M. Schwikowski; 1995, A 130 years deposition record of sulfate, nitrate and chloride from a high alpine glacier, *Water, Air and Soil Pollut.* **85**, 603–609.

Enghardt M., H. Rodhe; 1993, A comparison between patterns of temperature trends and sulfate aerosol pollution. *Geophys. Res. Lett.* **20**, 117–120

Fuzzi, S, D. Wagenbach (eds); 1997, *Cloud Multi-phase Processes and High Alpine Air and Snow Chemistry*, Springer Verlag, Heidelberg,. Vol. 5 of the EUROTRAC final report.

Gäggeler H.W., D.T. Jost, U. Baltensperger, M. Schwikowski, P. Seibert; 1995, Radon and thoron decay products and ^{210}Pb measurements at Jungfraujoch, Switzerland. *Atmos. Environ.* **29**, 606–616.

Kalina M.F., H. Puxbaum; 1994a, A study of the influence of riming of ice crystals on snow chemistry during different seasons in precipitating continental clouds. *Atmos. Environ.* **28**, 3311–3328.

Kalina M.F., H. Puxbaum; 1994b, Riming - an explanation for the strong seasonality of major ions in wet precipitation at a high alpine site (Sonnblick Observatory, Austria)? in: P.M. Borrell, P. Borrell, T. Cvitaš, W. Seiler (eds) *Proc. EUROTRAC Symp. '94.* SPB Academic Publishing bv, The Hague, pp. 1235–1239.

Kasper A.; 1994, Saisonale Variation ausgewählter Spurenstoffe sowie deren Auswaschverhalten an der Hintergrundmeßstelle Hoher Sonnblick. Thesis, Vienna University of Technology, Vienna.

Kasper A., H. Puxbaum; 1994, Determination of SO_2, HNO_3, NH_3 and aerosol components at a high alpine background site with a filter pack method. *Anal. Chim. Acta* **291**, 297–304.

Maupetit F., D. Wagenbach, P. Weddeling, R. Delmas; 1995, Recent chemical and isotopic properties of high altitude cold Alpine glaciers *Atmos. Environ.* **29**, 1–9

Maupetit F., R. Delmas; 1994a, Carboxylic acids in high-elevation Alpine glacier snow; *J. Geophys. Res.* **99**, 16491–16500

Maupetit F., R. Delmas; 1994b, Snow chemistry of high altitude glaciers in the French Alps *Tellus* **46B**, 304–324

Mosimann L., M. Steiner, J.L. Collett, W. Henrich, W. Schmid, A. Waldvogel; 1993, Ice crystal observations and the degree of riming in winter precepitation. *Water, Air and Soil Pollut.* **68**, 29–42

Molyna S.; 1993, Trends of sulfur dioxide emissions, air concentrations and depositions of sulfur in Europe since 1880. *EMEP/MSC-W Report 2/93.*

Nickus, U., M. Kuhn, F. Pichlmayer, U. Baltensperger, R. Delmas, H.W. Gäggeler, A. Kasper, H. Kromp-Kolb, F. Maupetit, A. Novo, S. Preunkert, H. Puxbaum, G. Rossi, W. Schöner, M. Schwikowski, P. Seibert, M. Staudinger, V. Trockner, D. Wagenbach; 1997, SNOSP: Ion deposition and concentration in high alpine snow packs. *Tellus* **49B**, 56–71.

Poulida O., M. Schwikowski, U. Baltensperger, J. Staehelin, W. Henrich; 1994, Physical processes affecting the chemical composition of precipitation at the high Alpine site Jungfraujoch, Switzerland; in: P.M. Borrell, P. Borrell, T. Cvitaš, W. Seiler (eds.) *Proc. EUROTRAC Symp. '94.* SPB Academic Publishing bv, The Hague, pp. 1226–1230

Preunkert S.; 1994, Glazio-chemische Verhältnisse des Colle Gnifetti im Vergleich zu seiner regionalen Umgebung, MS-Thesis, Institut für Umweltphysik der Universität Heidelberg.

Puxbaum H., D. Wagenbach; 1994, High alpine precipitation chemistry. in: P.M. Borrell, P. Borrell, T. Cvitaš, W. Seiler (eds) *Proc. EUROTRAC Symp. '94.* SPB Academic Publishing bv, The Hague, pp. 597–605.

Schöner W., H. Puxbaum, M. Staudinger, F. Maupetit; 1994, Geographical patterns in the chemical composition of Alpine glacier snow, 1973–1993. in: P.M. Borrell, P. Borrell, T. Cvitaš, W. Seiler (eds) *Proc. EUROTRAC Symp. '94.* SPB Academic Publishing bv, The Hague, pp. 721–724

Schwikowski M., O. Poulida, U. Baltensperger, H.W. Gäggeler; 1994, In-cloud scavenging of aerosol particles by cloud droplets and ice crystals during precipitation at the high alpine site Jungfraujoch. in: P.M. Borrell, P. Borrell, T. Cvitaš, W. Seiler (eds) *Proc. EUROTRAC Symp. '94.* SPB Academic Publishing bv, The Hague, pp. 1221–1224.

Schwikowski M., P. Seibert, U. Baltensperger, H.W. Gäggeler; 1995, A study of an outstanding Saharan dust event at the high-alpine site Jungfraujoch, Switzerland, *Atmos. Environ.* **29**, 1829–1842.

Schwikowski M., A. Novo, U. Baltensperger, R. Delmas, H.W. Gäggeler, A. Kasper, M. Kuhn, F. Maupetit, U. Nickus, S. Preunkert, H. Puxbaum, G.C. Rossi, W. Schöner, D. Wagenbach; 1997, Intercomparison of snow sampling and analysis within the alpine wide snowpack investigation (SNOSP), *Water, Air, and Soil Pollut.* **93**, 67–91.

Staehelin J., J. Thudium, A. Bühler, A. Volz-Thomas, W. Graber; 1994, Trends in surface ozone concentrations at Arosa (Switzerland). *Atmos. Environ.* **28**, 75–87

Wagenbach D.; 1993, Special problems of mid latitude glacier ice core research, *Proc. ESF/EPC workshop: Greenhouse gases, isotopes and trace elements in glaciers as climatic evidence for Holocene time.* Zürich 27.–28.10.1992 VAW–ETH-Zürich.

Winiwarter W., W. Schöner, H. Puxbaum; 1994, Ionic contents of wintertime deposition at an Alpine glacier: results from a decade of sampling. in: P.M. Borrell, P. Borrell, T. Cvitas, W. Seiler (eds) *Proc. EUROTRAC Symp. '94.* SPB Academic Publishing bv, The Hague, pp. 731–734.

8.2 GCE: An Overview of Ground Based Cloud Experiments

Sandro Fuzzi (Coordinator)

Institute of Atmospheric and Oceanic Sciences, ISAO - C.N.R., Via Gobetti 101, I-40129 Bologna, Italy

8.2.1 Introduction

The need for a cloud project within EUROTRAC derived from an increasing awareness of the importance of multi-phase cloud processes in tropospheric chemistry as a whole and of the fate of several key tropospheric trace components, processes that were poorly understood when EUROTRAC commenced. The interaction of trace gases and atmospheric particles with cloud droplets, liquid phase chemical reactions, and the chemistry of the cloud ice phase were subjects whose understanding was (and in part still is) inadequate in order to provide a framework for atmospheric and transport models that would allow a reasonably accurate description of the role of clouds in tropospheric chemistry.

Cloud multi-phase processes act on a wide variety of spatial (and therefore temporal) scales, ranging from molecular processes at a spatial scale of a few Ångström (corresponding to a microseconds temporal scale) to synoptic meteorology characterised by a spatial scale of thousands of km (temporal scale of the order of days). The wide range of spatial and temporal scales involved in cloud processes (more than 15 orders of magnitude) is one of the main problems encountered in cloud modelling, since processes at the smallest scale cannot always be neglected in describing larger scale processes. A graphic representation of the processes occurring in the multiphase atmospheric system is reported in the overall flow diagram of Fig. 8.2.1 (Fuzzi, 1994). The processes evidenced in the figure are interdependent and may proceed simultaneously. Ice-phase processes are not represented in the figure. It should also be noted that the listed processes are essentially all of a physical nature; chemical transformations also occur within the system, greatly increasing its overall complexity.

The original general aims of GCE were stated at the beginning of the project as:

* to determine which factors control the rate of acid formation in cloud droplets;

* to determine which factors control the concentration of oxidants and catalysts in cloud droplets;

*   to determine how the relative importance to cloudwater composition of incorporation of the pre-existing aerosol versus in-cloud reactions depends on the physical and chemical environment of the cloud;

*   to ascertain how the efficiency of nucleation scavenging depends on the aerosol size distribution and type of cloud.

During the project, further issues were faced as they emerged from the work in progress. The most relevant issue which was addressed within GCE that was not among the original aims was the effect of cloud cycling on aerosol particles, *i.e.* how an aerosol population is modified after passing through a cloud.

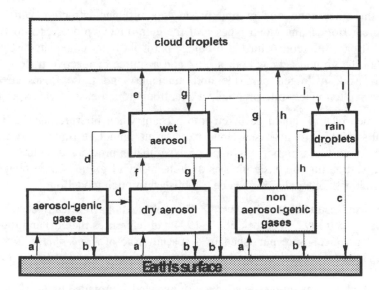

Fig. 8.2.1:   Graphic representation of the multiphase atmospheric system. The numbered arrows represent the physical and chemical processes leading to mass exchange between the different reservoirs (boxes): a) emission, b) dry deposition, c) wet deposition, d) gas-to-particle conversion, e) nucleation, f) condensation, g) evaporation, h) dissolution, i) aerosol capture by falling droplets, l) formation of precipitation.

With the above objectives in mind the GCE community, that comprises nearly all European groups working on cloud research, decided on a research approach based on three joint field experiments at selected locations in Europe, characterised by different cloud systems and different climatic and pollution conditions. The experiments were organised and carried out, not merely as a series of individual activities at the same site, but rather as a coordinated effort in which each participating group would focus on a particular aspect(s) according to their own expertise and instrumental capabilities. This choice was made taking into account the wide range of skills and interdisciplinarity involved in cloud studies, as well as the wide range of instrumentation needed to fully characterise the complex multiphase system of clouds. Scientists of one single group or experts in one discipline alone could never have conducted experiments of such complexity as the GCE field campaigns.

Ground-based clouds offer a distinct advantage in that large amounts of equipment can be brought together at a single site so as to allow a much more comprehensive set of measurements than would be possible on an aircraft. In addition, in a ground-based experiment the cloud advects past the instruments at the local wind-speed. This is, of course, a much lower velocity than on an aircraft and thus allows a much better spatial resolution to be achieved. Moreover, the simple dynamic structure of orographic clouds considerably facilitates the interpretation of the data.

On the other hand, the ground produces a turbulent structure which is very different from many cloud types and the entrainment processes into the cloud may also be different. A further problem, particularly for cloud chemistry studies, is that the ground acts as both a sink and a source of reactive trace gases (*e.g.* nitric acid, hydrogen peroxide and ammonia) and these fluxes need to be measured to reliably interpret budget data from ground-based field experiments.

All such factors should be taken into account when interpreting the data. The results from a ground-based cloud experiment cannot be transferred directly to other cloud types; rather, this type of experiments provides a natural laboratory for investigating the basic physics and chemistry of ground-based clouds, while the link with other clouds must be established through modelling.

The main results of the first two GCE experiments have already been published (Fuzzi *et al.*, 1992; Wobrock *et al.*, 1994 and references therein) and those of the last experiment were published in a special issue of *Atmospheric Environment* (Fuzzi *et al.*, 1997). In this paper we aim to summarise and critically review the work accomplished during the GCE project, to provide an integrated view of the overall project, examine the advance in knowledge provided by GCE in the field of cloud research, and discuss the major issues which, in our opinion, need to be addressed by future cloud projects.

8.2.2 Overview of the three GCE joint field experiments

The three GCE experiments took place in the Po Valley (PV '89), Italy (Fuzzi *et al.*, 1992), Kleiner Feldberg (KF '90), Germany (Wobrock *et al.*, 1994) and Great Dun Fell (GDF '93), United Kingdom (Choularton *et al.*, 1997). In all, fifteen research groups participated in GCE over the entire period of the project (Table 8.2.1); other groups, not formally GCE participants, joined the different experiments upon invitation of the groups hosting the experiments.

As reported above, the three sites were chosen as examples of cloud systems characterised by different dynamical structures (different cloud formation and evolution mechanisms) and also of different levels of pollution (different chemical systems). The choice of the three sites was also based on previous cloud research carried out prior to GCE and on the body of information on cloud properties obtained from previous work at the sites themselves, which was of great help in planning the experiments and in the interpretation of data.

Table 8.2.1:   Alphabetical list of research groups participating in the different phases of the GCE project, giving the names of principal investigators. Other groups who were not formally GCE members also took part in the various field campaigns (see Fuzzi *et al.*, 1992; Wobrock *et al.*, 1994; Choularton *et al.*, 1997 for a complete list).

Research Group	Principal Investigator(s)	PV '89	KF '90	GDF '93
Department of Meteorology, Stockholm University, Stockholm (Sweden)	J.A. Ogren (Coordinator, 1988-1991)	yes	yes	
Deutscher Wetterdienst, Meteorologisches Observatorium Hamburg, Hamburg (Germany)	P. Winkler	yes	yes	yes
Division of Nuclear Physics, Lund University, Lund (Sweden)	H-C. Hansson B. Martinsson	yes	yes	yes
Fraunhofer Institut für Atmosphärische Umweltforschung, Aussenstelle für Luftchemie, Berlin (Germany)	D. Möller			yes
Fraunhofer Institut für Toxicologie und Aerosolforschung, Hannover (Germany)	K. Levsen			yes
Institut für Analytische Chemie, Technische Universität Wien, Wien (Austria)	W. Winiwarter	yes	yes	
Institut für Experimentalphysik, Universität Wien, Wien (Austria)	A. Berner	yes	yes	yes
Institut für Meteorologie und Geophysik, Universität Frankfurt, Frankfurt (Germany)	H.W. Georgii	yes	yes	yes
Institut für Physik, Universität Hohenheim, Stuttgart (Germany)	R. Gieray			yes
Istituto FISBAT - C.N.R., Bologna (Italy) (host PV '98)	S. Fuzzi (Coordinator, 1992–1995)	yes	yes	yes
Max Planck Institut für Chemie, Abteilung Biogeochemie, Mainz (Germany)	G. Helas			yes
National Institut of Chemistry, Ljubljana (Slovenia)	M. Bizjak			yes
Netherlands Energy Research Foundation, Petten (the Netherlands)	B. Arends	yes	yes	yes
UMIST, Physics Department, Manchester (United Kingdom) (host GDF '93)	T. Choularton		yes	yes
Zentrum für Umweltforschung, Universität Frankfurt, Frankfurt (Germany) (host KF '90)	W. Jaeschke	yes	yes	yes

As predicted in view of the long series of observations previously performed at the three field sites, high levels of pollutants, particularly NO_X and NH_3, were encountered at the PV station, along with low oxidant levels. Higher concentrations of SO_2, but still with low oxidant levels characterised the atmosphere at the KF station. In contrast to PV and KF, significantly lower pollution levels characterised the remote location of GDF, where high oxidant levels were also encountered (Table 8.2.2).

Table 8.2.2:   Summary of the three joint GCE field campaigns.

Site	Date of the experiment	Geographic coordinates	Height (m a.s.l.)	Characteristics of the site
S. Pietro Capofiume, Po Valley, northern Italy	November 1989	44° 39' N 11° 37' E	10	Rural site in the eastern Po Valley. The entire area of the valley has a very high population density and is also characterised by intensive industrial, trading and agricultural activities. High levels of pollutants are therefore reported for this region, especially during periods of strong temperature inversions when fog forms.
Kleiner Feldberg, Taunus Highlands, central Germany	October-November, 1990	50° 18' N 8° 30' E	826	The Taunus Highlands run SW to NE and form the northern barrier of the Upper Rhine Valley. The Rhine-Main area (west and south of the Taunus mountains) is a large industrial region with a population 2.5 million people, and is characterised by a dense heavy traffic road system. High pollution levels are therefore reported in the region.
Great Dun Fell, Pennines Region, northern England	April-May, 1993	57° 2' N 3° 7' W	850	Rural northern England. The peak is part of a long ridge running NE to SW. For SW, West and NE wind directions, air is clean oceanic. Main pollution sources are Manchester/Liverpool (to the SSW) and Teesside (to the East): road traffic and various types of industrial plants are present in both areas.

The cloud types which typically appeared at the three sites were: ground fogs at PV, stratocumulus or stratus clouds at KF and cap clouds at GDF. The differences in the clouds impinging KF and GDF are mainly due to the fact that KF is a relatively isolated mountain in the Taunus Highlands, while GDF is part of a long mountain ridge of the Northern Pennines, where the orographic effect alone (condensation by adiabatic lifting and subsequent cooling of air) is sufficient for the cloud formation. KF clouds, on the other hand, are mostly associated with the passage of frontal systems: as the front approaches, the height of the associated cloud fields decreases and KF consequently becomes immersed in cloud. The airmasses prevailing on a synoptic scale at the KF and GDF mountain sites can be quite different, depending on the large scale airflow. Atlantic airmasses typically dominate both sites for westerly flow, but continental airmasses arrive at the sites under easterly wind conditions. In contrast to the mountain sites, the airmass origin at PV should be examined at the mesoscale level, *i.e.* the scale of the valley itself. The ground fogs normally evolve within a 400-500 m height, reaching a persistent inversion layer, wherein the exchange with airmasses of the synoptic flow is a minor effect and can thus be neglected.

Table 8.2.3 shows a compilation of some dynamical and microphysical parameters observed during the three GCE experiments. The temperature was in the same range for all sampling campaigns. Wind speed and turbulence, however, differed significantly from site to site. For the surface measurements, at PV wind speed typically ranged between 2–4 m s^{-1}, although wind speed up to 7 m s^{-1} was observed in the first 30 m height. In addition, values of the turbulent kinetic energy (tke) around 0.2 m^2 s^{-2} show that fog formation and persistence is not necessarily associated with a lack of wind and turbulence in the levels near the ground, as often reported in the literature (*e.g.* Brown and Roach, 1976). The dynamical conditions for KF can be divided in two distinct periods. In the first part of the experiment, cumulus clouds evolved in a period with strong horizontal and vertical winds and high values of the turbulent kinetic energy (2–5 m^2 s^{-2}), while in the second observational period stratiform clouds prevailed, vertical wind velocity diminished and turbulent kinetic energy dropped to below 1.5 m^2 s^{-2} and even reached values as low as those measured in the PV fog. This similarity between stratus clouds and fog also becomes obvious when comparing the droplet size during both experiments.

Table 8.2.3:   Typical dynamical and microphyisical conditions for the clouds encountered during the three GCE field experiments, data for KF are split for situations with stratocumulus(Sc) and stratus (St) clouds.

	PV '89	KF '90		GDF '93
horizontal wind (m s^{-1})	2–4	4–8		10–15
vertical wind (m s^{-1})	± 0.3	1–2		2–4
turbulent kinetic energy (m^2 s^{-2})	0.2	2–5 < 1.5	(Sc) (St)	0.6
range of LWC (g m^{-3})	0.2–0.5	0.3–0.6 0.1–0.3	(Sc) (St)	0.4–0.8
VMD (μm)	19–23	12 16–19	(Sc) (St)	15–20

Typically, the PV fog droplet spectra had a volume mean diameter (VMD) in the range of 19 to 23 μm, while the stratus clouds at KF had a VMD in the range of 16–19 μm. In contrast to this, the VMD of the cumulus clouds during the first period of the KF experiment was only around 12 μm.

From Table 8.2.3, we can also see that the cloud liquid water content (LWC) and VMD of the droplet spectra observed for GDF clouds are significantly larger than those for KF. The main reason for this is that most cloud events during the GDF experiment were encountered under north-easterly flow conditions. Hereby, the air parcels had already passed over several ridges east of GDF and clouds with large droplets may have developed. These clouds developed further due to the adiabatic lifting process along the slopes of GDF which increased the LWC. In Fig. 8.2.2, the shape of the droplet spectrum observed at GDF appears much smoother than that observed at KF. One possible explanation may be the different turbulent conditions prevailing at the two sites. KF is forested up to the summit with 15–20 m high conifers, while GDF and its surroundings is covered in hummocky grassland. This is reflected in the different values of the turbulent kinetic energy (see Table 8.2.3). The different degree of turbulence is also an indication of a different mixing of cloudy and clear air (entrainment). This is supported by the strong bimodal shape of the drop spectrum at KF (Arends *et al.*, 1994) which is an indication of dry air entrainment into the cloud.

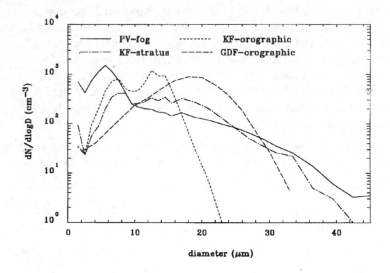

Fig. 8.2.2:   Typical droplet spectra observed for the PV fog (LWC = 0.32 g m^{-3}, 12 Nov. 1989), KF orographic clouds (LWC = 0.26 g m^{-3}, 28 Oct. 1990), KF stratus clouds (LWC = 0.34 g m^{-3}, 13 Nov. 1990), and GDF orographic clouds (LWC = 0.73 g m^{-3}, 11 May 1993). All spectra represent average values over 60 minutes.

8.2.3 Main results obtained within GCE

To describe and summarise the extent to which GCE has contributed to the advancement in knowledge on cloud processes, the following discussion is organised following the most important issues arising in this field of research.

a.   Size distribution, hygroscopic properties and chemical composition of aerosol particles

During the GCE field campaigns, size distributions of atmospheric aerosol particles were recorded using the Differential Mobility Particle Spectrometer (DMPS) which offers high accuracy and size resolution. In both the GDF campaign and in several additional minor experiments, we were able to force the lower detection limit from 20 nm down to 3 nm using a Ultrafine Condensation Particle Counter (UFCPC, TSI Inc.), and we consistently found a size distribution that differed from the above mentioned general view of the atmospheric particle size distribution. Thus, together with other scientists, we have put forward a modified picture of the particle size distribution (Hoppel *et al.*, 1994; Covert *et al.*, 1996; Wiedensohler *et al.*, 1997) (Fig. 8.2.3). A major development is that today we can clearly state that the accumulation mode is actually comprised of two modes, a smaller one with a mean diameter of 20–80 nm, which we suggest should be called "Aitken mode", and a larger one, with a mean size of 150–300 nm, which should still be called accumulation mode as it represents the final size of the particles in the fine mode range, *i.e.* particles less than 1 μm in diameter. Above this size, the efficiency of the growth processes decreases and the deposition processes become effective.

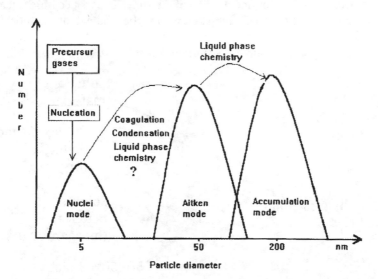

Fig. 8.2.3:   A conceptual description of the submicron atmospheric particle size distribution with an indication of the major processes causing the particles to grow in the different size modes.

The main process transferring the Aitken mode particles into the accumulation mode size range is today believed to be the liquid phase chemistry taking place in cloud droplets.

Within GCE, a new concept of measuring the hygroscopicity of different classes of particles was tested for the first time. By using a coupled Tandem Differential Mobility Analyser (TDMA), with accurately controlled relative humidity (r.h.) in between the two DMAs, it was possible to determine the growth in size of particles exposed to increasing r. h. This allowed the growth in particle size to be measured with a very high precision and accuracy, *i.e.* about 1 %. The size interval investigated was 30–300 nm, which covers the two modes (Aitken and accumulation) involved in cloud processes.

The results of the TDMA measurements were certainly striking as they did not show the expected hygroscopic behaviour. Instead of revealing a broad distribution of hygroscopic properties from completely hydrophobic particles to pure salt particles, as expected from theory, they show a bimodal hygroscopic distribution. In continental aerosol we almost invariably found the particles to grow either ca. 5 % or ca. 45 % in size at 85 % r.h., compared to the size at 20 % r.h. (Svenningsson *et al.*, 1992; 1994; 1997, Hansson and Svenningsson, 1994). The data available today in the world, of which the GCE data constitute a majority, cover so many areas and such long time periods that there remains no doubt that continental aerosol, strongly perturbed by anthropogenic sources, consists of two major groups of particles characterised by different hygroscopicity. At GDF, where the aerosol particles were strongly influenced by marine sources (Swietlicki *et al.*, 1997), only the more hygroscopic type of particles was observed for about 50 % of the time. The reduced presence of less-hygroscopic particles at GDF compared with earlier campaigns is probably due to the strong influence of marine aerosol and to the long transport time from the possible anthropogenic sources (Svenningsson *et al.*, 1997).

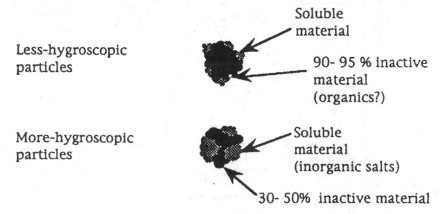

Fig. 8.2.4: A conceptual picture of the chemical composition of the two hygroscopic types of aerosol particles. The soluble/insoluble composition is based on the known hygroscopic behaviour of inorganic salts and comparison with actual measurements.

This new finding was mainly derived from studies within GCE and considerably enhances our understanding of how particles interact with water vapour in the atmosphere, *i.e.* they form cloud droplets, and of how anthropogenic pollution affects the formation of droplets and thus the chemistry and radiative properties of clouds.

Using Raoult's Law corrected for non-ideal solutions also taking the curvature effect into account (Kelvin effect), and using results from laboratory studies on aerosol consisting of pure salts and of internally and externally mixed composition, we have to conclude that the atmospheric particles carry a substantial fraction of insoluble material, in both hygroscopic modes. The less hygroscopic particles carry as much as 90 % insoluble mass, while the more hygroscopic particles carry from 20 % up to perhaps 50 % (Fig. 8.2.4). When integrating the soluble and insoluble fractions found in both hygroscopic modes, a good agreement is found with bulk measurements of inorganic compounds (the soluble fraction), elemental carbon and organic compounds (the insoluble fraction) (Heintzenberg, 1989).

b.   *Nucleation scavenging*

The term nucleation scavenging refers to the processes through which some atmospheric particles, the CCN, grow into cloud droplets in a supersaturated air parcel. Strictly speaking, the term "nucleation" is somewhat inappropriate, as cloud droplet formation is not a real nucleation process but rather a condensation of water vapour on a wet aerosol solution. The actual nucleation occurs in sub-saturated conditions, when the hygroscopic substances within the atmospheric aerosol particles reach their deliquescence point, and the particles grow into concentrated droplets.

The mathematical formulation of how water droplets grow in high relative humidity conditions and supersaturated environments (Köhler equation) has long been known and is used in all models. The equation has been improved by the availability of more accurate measurements and estimates of water activity, allowing a better description of how the water vapour uptake depends on the chemistry of the droplet. However, the major question that remains open is whether the present formulation of the Köhler equation correctly describes the dynamical situation occurring during the formation of cloud droplets.

In view of the rather complex chemical composition of atmospheric particles, their hygroscopic behaviour cannot analytically be described in the Köhler equation. However, the TDMA measurements discussed in the previous paragraph offer a unique opportunity to assess the hygroscopic response of aerosol particles to changing r.h. A simple approach was adopted here: the TDMA measurements were used to describe the aerosol particles in terms of a soluble fraction of inorganic salts (whose chemical composition can be measured), and an insoluble fraction, and the results were introduced in the Köhler equation for comparison with measurements. This proved to be quite successful: the PV experiment, for example, showed the strongest evidence so far of how the droplet size at high relative humidity depends on particle size, by taking into account the above bimodal hygroscopic nature of the aerosol (Svenningsson *et al.*, 1992, Noone *et al.*, 1992). Most results obtained during the GCE experiments clearly show that by introducing the measured hygroscopic properties of the particles into the Köhler formulation it is possible to explain the uptake of particles in fog and cloud (Svenningson *et al.*, 1992; 1994; 1997; Colvile *et al.*, 1994). We now believe there is sufficient evidence of the reliability of the Köhler formulation in describing the nucleation process, provided the appropriate water activity coefficients are introduced.

c.   Other in-cloud particle scavenging processes

The capture of interstitial aerosol by cloud droplets can be an important aerosol removal mechanism in cloud. The series of GCE field experiments enabled this process to be studied in some detail. It was found that in the radiation fogs studied in PV, where air remains within the fog for many hours and the lifetime of an individual droplet is about 1 hour, the process is of negligible significance in the transfer of aerosol mass and accumulation mode particles to the droplets. Instead, it was predicted that it would play an important role in the transfer of Aitken mode aerosol particles to fog droplets (Noone *et al.*, 1992). When the airstream containing the particles remains in cloud for a much shorter time, *e.g.* at GDF, where the characteristic transit time of an air parcel through the cloud is about 15 minutes, Brownian diffusion is a sink for nucleation mode particles of typically 10 nm but has little effect on mass transfer to the droplets, which is totally dominated by nucleation scavenging and take-up of soluble species from the gas phase.

d.   Scavenging of gases

The GCE experiments provided the most extensive data set so far on Henry's Law equilibrium for gases like NH_3, H_2O_2, SO_2, HCOOH and CH_3COOH and HCHO by simultaneous measurements in both liquid and gas phases. Large deviations from Henry's Law equilibrium were encountered in both the PV and KF experiments (Facchini *et al.*, 1992b, Winiwarter *et al.*, 1994). Surprisingly, smaller deviations from Henry's Law equilibrium were detected in the case of GDF clouds (Laj *et al.*, 1997b). The pH dependency of these deviations was clearly highlighted during PV and KF experiments, where formic and acetic acids were generally supersaturated in the liquid phase at low pH and subsaturated at high pH, as opposed to the behaviour of NH_3. The pH dependence of formic acid during GDF contrasted with the results for PV and KF, due to the in-cloud formation of HCOOH (Laj *et al.*, 1997b). Because the reported deviations from equilibrium cannot be fully explained either by bulk sampling and time integration artefacts (which only explain deviations up to a factor of 3), or by the formation of additional compounds, other hypotheses have been advanced such as a shift in equilibrium due to the presence of chemical substances not taken into account in the Henry's Law calculation, or a kinetic inhibition due to mass transfer limitation by an organic film coating cloud droplets. Neither of the two hypotheses were investigated during the GCE experiments, but future research should certainly address them and aim at reducing sampling times and performing droplet size segregated sampling.

e.   Cloud droplet liquid phase chemistry and its dependence on cloud microphysics

In-cloud chemical processes are believed to play a major role in the modification of both the size distribution and composition of atmospheric aerosol. Chemical transformations within the aqueous phase were accurately studied during the GCE field experiments. It was shown that the evolution of fog and clouds can be well described using the concept of atmospheric acidity, defined as the base neutralising capacity of a unit volume of an atmosphere including gas, interstitial aerosol and liquid phase (Facchini *et al.*, 1992a, Fuzzi *et al.*, 1994). In fact, cloud and fog systems at the three experimental sites showed active exchanges of acidic and basic components among the different phases: at both PV and KF, advection of HNO_3-rich air resulted in an acidification of the fog and cloud systems after exhaustion of their neutralising capacity, followed, in the case of the PV fog, by immediate neutralisation due to high concentrations of NH_3 in the atmosphere (Facchini *et al.*, 1992a). A fundamental difference between the three experiments was the lack of major atmospheric oxidants during both the PV and KF experiments, where changes in atmospheric acidity were simply due to advection of acidic air masses to the sampling sites, rather than to S(VI) or HNO_3 production processes. On the contrary, high levels of gas phase O_3 and H_2O_2 were encountered during the GDF experiment (Choularton *et al.*, 1997), leading to an efficient transformation of both sulfur and nitrogen species and in-cloud production of other chemical species, *e.g.* HCOOH (Laj *et al.*, 1997a; 1997b).

Detailed studies were carried out on the transformations of oxidised nitrogen and particulate nitrate at GDF (Cape *et al.*, 1997). During one of the cloud events, significant conversion of NO_x to HONO and HNO_3 in cloud was monitored, followed by degassing of HNO_3 as the cloud dissipated. In fact, degassing of HNO_3 from dissipating clouds was often monitored during other cloud events, but no conclusions could be drawn on the mechanisms by which HNO_3 was formed or absorbed into cloud droplets. This was partly due to difficulty in monitoring transformations of nitrogen species in the gas, liquid and particulate phases through the GDF cloud. Similar problems were encountered in investigating the role of clouds in the transformations of NH_3 at GDF. The fate of NH_3 and its role in determining whether nitrogen species are transferred to the gas phase (HNO_3) or to the particulate phase (NH_4NO_3) appears to be driven by both cloud microphysics and acid-base chemistry (Sutton *et al.*, 1997; Wells *et al.*, 1997).

Similarly, the identification of the different factors involved in the transformation of sulfur species in cloud was not straightforward. Sulfate aerosol loading before and after passage through the cloud at GDF showed significant differences with much larger concentrations in the outflow of the cloud system (Fig. 8.2.5) (Laj *et al.*, 1997a). Although part of the increase in aerosol sulfate mass results clearly from S(IV) to S(VI) oxidation by H_2O_2 in the liquid phase, difficulties in quantifying the mechanisms leading to sulfate production arose from the fact that dynamical mixing in the GDF cloud was also found to be a key process in supplying oxidants (H_2O_2) and possibly particulate sulfate to the system. In addition, liquid phase formation of HCOOH, possibly through HCHO hydrolysis (Laj *et al.*, 1997b), may have interfered in the odd hydrogen cycle leading to liquid phase production of H_2O_2. It was shown that the study of in-cloud chemical conversion requires that both dynamical and chemical aspects of the cloud system be taken into account in interpreting the results.

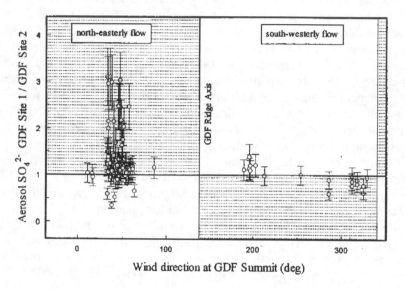

Fig. 8.2.5:   During GDF '93, SO_4^{2-} mass addition to the aerosol population passing through the cap cloud enveloping GDF Summit was measured. The added mass could be (in part) ascribed to in-cloud chemical reactions. GDF site 1 and GDF site 2 are out-of-cloud measuring sites which are either pre- or post-cloud depending on wind direction. GDF site 1 is the pre-cloud site in south-westerly wind conditions and vice-versa. The shaded areas in the figure indicate a post-cloud SO_4^{2-} concentration in the aerosol higher than the pre-cloud one, *i.e.* where SO_4^{2-} in-cloud production occurred. The figure is oriented along the GDF ridge axis (Laj *et al.*, 1997a).

Size dependent cloud/fog water measurements within GCE were performed at the three European sites. The chemical analyses of the samples taken with a two-stage fog water impactor (TFI) (Schell *et al.*, 1994, Schell *et al.*, 1997b) for pH, conductivity, and major ions show some differences in the concentrations as well as in the relative composition of the sampled cloud water between the two size ranges. The differences in concentration were found to vary both in amount and sign among the various types of cloud, providing evidence of the strong influence of microphysical and chemical processes on the size dependence of cloud droplet chemical composition. The results from measurements of the PV '89 experiment, where data from a CVI and a wet aerosol impactor set-up were mainly available (Ogren *et al.*, 1992), show a strong decrease of droplet solute concentrations with increasing diameters, whereas the results from the measurements at the two mountain stations (GDF and KF) show both increasing and decreasing solute concentrations with increasing droplet diameters (Wobrock *et al.*, 1994; Schell *et al.*, 1997a).

Fig. 8.2.6 illustrates the time dependence of solute concentrations vs. cloud droplet size: the curves are the result of sensitivity studies (Schell *et al.*, 1997a) using a diffusional growth model (Wobrock, 1988) and represent the volume weighted mean concentrations for each droplet size class at the reported time steps of the model run. Fig. 8.2.6a is calculated for a marine aerosol number distribution (Jaenicke, 1993), with an upper limit of particles at 10 μm, Fig. 8.2.6b for particles d < 5 μm and Fig. 8.2.6c for particles d < 1 μm. In principle, we always observe decreasing solute concentrations with increasing size up to a droplet diameter of 5 μm, followed by a strong increase of solute concentrations with droplet size which, however, attenuates with increasing growth time. At a certain diameter, the size dependence curves reverse again. The parameter determining both the size and value of the maximum concentration is the width of the CCN number distribution, which is, in turn, determined by the value of the peak supersaturation and by the number of large particles. Limiting the maximum particle size available for condensation, as shown in Figs. 8.2.6b and 8.2.6c, decreases the tendency towards higher solute concentrations at diameters > 5 μm, as does limiting the minimum particle size available by reducing the peak supersaturation. However, in the latter case, we would expect to observe solute concentrations which are on average at a higher level.

Our experimental and model results imply the conclusion that, for droplets with diameters above *ca.* 5 μm, we will observe:

*   increasing solute concentrations with increasing size during the initial stage of a cloud, *e.g.* near the cloud base where the droplets have just formed;

*   decreasing solute concentrations with increasing diameters in aged cloud parcels which can be observed, for example, high above the cloud base in cumuliform clouds or in stratiform clouds advected to the observation point.

Clearly the factors controlling the size dependence of cloud droplet solute concentration can be found in many different combinations within clouds.

During the PV experiment, Ogren *et al.* (1992) found a strong decrease in solute concentration with increasing droplet size in the observed ground fog, as reported above. Several possible reasons for this were given, but ultimately no satisfactory explanation could be found for the presumed contradiction. In view of the above considerations, we are now able to explain the discrepancies of size dependent solute concentration observed at PV which were the consequence of the relatively high droplet age (Wobrock *et al.*, 1992) and the low supersaturation (around 0.04 %) that allowed only particles larger than 0.3 μm to become activated (Noone *et al.*, 1992).

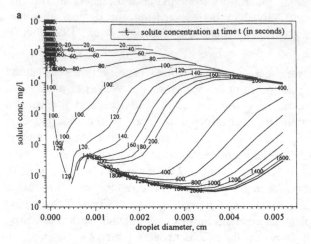

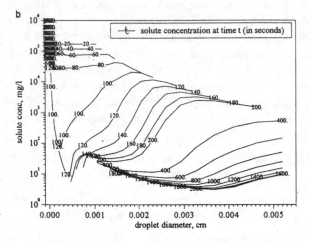

Fig. 8.2.6:   Solute concentrations vs. droplet diameter for droplets grown on Jaenicke-type marine aerosol particle distribution (Jaenicke, 1993) The three figures differ for the upper particle size-cut chosen: a) only particles with diameters < 10 μm; b) only particles with diameters < 5 μm; c) only particles with diameters < 1 μm. Group parameter of the curves is the model time in seconds (Schell *et al.*, 1997a).

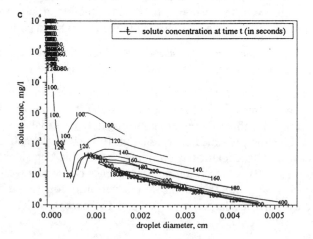

f.   Effect of entrainment on cloud chemistry and microphysics

It is well known that the entrainment of air into clouds from outside their boundaries can have a marked effect on the microphysical development of the clouds. This entrainment tends to have the greatest effect in convective clouds, where vigorous turbulent motions cause the cloud to engulf large volumes of dry outside air. In layer clouds the effects of entrainment are less marked, but may still be very important. It has also been shown that entrainment may have a marked effect on the chemical properties of the cloud by introducing extra oxidants, particularly hydrogen peroxide which is abundant in the free troposphere but is often exhausted within cloud due to its rapid reaction with S(IV) (Gallagher et al., 1991; Bower et al., 1991).

The effects of entrainment on the chemistry of radiation fog were previously unknown and were investigated during the PV experiment. Here, it was shown (Wobrock et al., 1992) that a radiation fog cannot be treated as a closed box. In particular, soluble species are mixed into the fog either by horizontal transport or by vertical mixing with a major effect on the chemistry. It was shown that the effects of the entrained species can be described quantitatively and are therefore now well understood (Facchini et al., 1992a).

Further evidence of the entrainment of chemical species into the cloud, particularly oxidants, but in one case also SO_2, was found also in the hill cap cloud studies. Evidence was found for periods of mixing in hill cap clouds which sometimes followed an inhomogeneous pattern and sometimes a more homogeneous type of mixing. This can be explained by the different scales of mixing associated with different cloud types and turbulent regimes at the hill sites. It was suggested by Baker et al. (1984) that the mixing will follow an extremely inhomogeneous pattern if the time scale for droplet evaporation is short, compared to the mixing time scale of entrained parcels of air. If the reverse is true, the pattern will be homogeneous. This has generally been confirmed by studies in hill cap clouds.

Study of droplet residues from the Droplet Aerosol Analyser (DAA) (Martinnson et al., 1997) showed that on many occasions no strong correlation existed between the size of the droplet residue (the aerosol particle on which the droplet formed) and the size of the droplet, while on other occasions they were strongly correlated. In the former case, this supports the suggestion that droplets with a range of lifetimes were present in the cloud, consistent with several periods of activation rather than a single activation event at cloud base. This means that many of the droplets have ages much lower than the transit time of a parcel of air from cloud base to the observation site.

g.   Evaporating clouds

It is well known that most clouds do not lead to precipitation but evaporate. Upon cloud evaporation, gas and particles are released back to the atmosphere.

During the GCE campaigns studies were performed to investigate outgassing from evaporating cloud and the possible link with new particle formation. A model sensitivity study showed that for an ammonia laden airstream flowing into a cap cloud, the amount of ammonia taken up into the aqueous phase varies according to the acidity present within the cloud, with increasing acidity favouring ammonia uptake. However, the fraction of ammonia remaining fixed as ammonium within the evaporating cloud droplets depends upon the source of cloud acidity, and whether it is capable of forming stable ammonium salts. Hence, for species such as nitric acid, the formation of ammonium nitrate will modify the aerosol population emerging downwind of the cap cloud but, unless oxidised in solution, SO_2 will not lead to a significant conversion of ammonia to ammonium aerosol. It was shown that in the absence of other salt forming species, ammonium aerosol production is directly linked to the in-cloud production of sulfate.

Comparison of these results with the field observations performed during the GCE campaign at GDF (Wells *et al.*, 1997) shows broad qualitative agreement. In general, however, it was found that the concentrations of ammonia in the gas phase downwind of the cloud generally exceeded that predicted by the modelling exercise discussed above. This is because the above model treats the evaporating droplets as ideal solutions. Including the effects of high ionic strength increased the proportion of the ammonia returned to the gas phase, a treatment resulting in a good agreement between observations and predictions on many occasions. It was able to predict that (consistent with observations) a net transfer of ammonium aerosol to ammonia gas sometimes occurred upon passage through the hill cap cloud.

The effect of cloud on the aerosol size distribution and hygroscopic properties was a major focus of the third GCE experiment performed at GDF, and was studied through a combination of modelling and observational work. A modelling study of the effects of aqueous phase cloud chemistry in changing the activated CCN spectrum was performed. Significant modification of the CCN spectrum emerging downwind of the processing cloud was observed. This led to the formation of strongly bimodal aerosol size distributions. The degree of modification was strongly correlated to sulfur dioxide concentration and, in oxidant limited situations, to H_2O_2 concentration. The cloud droplet chemistry was seen to have the largest effect on the smallest activated CCN, enabling them to activate much more readily after cloud processing at critical supersaturation up to 20 times lower than was originally required.

Fig. 8.2.7 shows the data obtained from the three Differential Mobility Particle Sizing (DMPS) instruments operating upwind of, within and downwind of the hill cap cloud. The diagram illustrates features which are found in much of the data set when the airflow between the three sites is connected and aqueous phase oxidation of S(IV) to sulfate occurs in the cloud water. At the largest sizes the aerosol particles are activated to form cloud droplets. The DMPS at the summit only measures interstitial particles. At the downwind site the cloud droplets evaporate, returning the CCN to the aerosol phase with some growth due to sulfate production. At night, the production of nitrate ions in solution can also be significant in adding soluble mass to the activated aerosol particles (Colvile *et al.*, 1994).

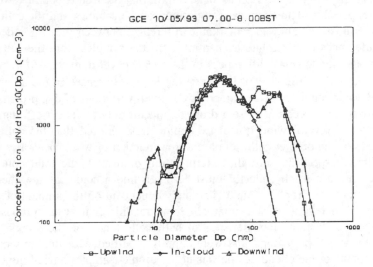

Fig. 8.2.7:   Aerosol size spectra measured by three DMPSs operated simultaneously upwind of the GDF cap cloud, within the cloud and downwind of the cloud. These data were obtained on 10 May, 1993 as part of the GDF '93 field campaign.

h.   Cloud droplet deposition

Chemical deposition onto coniferous forests due to cloud droplet interception was addressed within the KF experiment using a highly time-resolved technique. The results were employed to verify a one-dimensional resistance model for cloud water deposition (Pahl *et al.*, 1994) and the modelled cloud water deposition agreed quite well with observed values. Although the cloud deposition rates were quite low compared to the average precipitation deposition, $(0.3–0.6 \text{ mm h}^{-1})$ the chemical deposition due to cloud turned out to be from 3 to 6 times higher than that due to precipitation. This result offers further evidence of the important role played by cloud water deposition in the damage to subalpine forests in Germany and other parts of Europe.

i.   Instrument development and testing within GCE

Development and testing of cloud droplet collectors

The initial phase of GCE involved discussions as to the type of cloud droplet collector that was most suitable for all the planned field experiments. At the time two devices were available which met the requirements of the project, so these instruments were operated side by side throughout all GCE field experiments. The impactor of the Institute of Experimental Physics, University of Vienna (IEP)

(Berner, 1988) consists of four identical upward facing jet impaction stages. The impactor of the German Weather Service, Meteorological Observatory Hamburg (MOH) (Winkler, 1986) consists of two rectangular U-shaped impaction plates.

The first GCE experiment in PV '89 was used to perform a comparison of the two instruments (Schell *et al.*, 1992) to make certain that the results obtained were as reliable as possible. The chemical composition of the samples and the collection efficiencies observed showed remarkable differences between the two collectors. Some differences in the solute concentration in the samples from the collectors was expected due to small differences in the 50 % cut-off diameters ($d_{50} = 5$ μm (MOH) and $d_{50} = 7$ μm (IEP)). However, the large differences in the amount of collected water could not be explained by the small variations of d_{50}, proving that it is not sufficient to characterise a droplet collector only by its cut-off diameter. The results of several wind tunnel calibration tests showed that the collection efficiencies of the droplet collectors are also a function of wind-speed and shape of the droplet spectrum. For the interpretation of results, the total collection efficiency curve of each collector must be taken into account, as the chemical composition of the droplet sample is strongly dependent on the amount of water collected in each droplet size interval. Some modifications were proposed to improve the collection characteristics of both collectors. For the second GCE experiment at KF both impactors were slightly modified with the consequence that the results of the chemical analysis of the droplet samples agreed quite well (Wobrock *et al.*, 1994).

On the basis of the impactor comparison study, two new cloud water impactors were also developed and operated within GCE:

* the Isokinetic Cloud Probing System (ICPS) (Maser *et al.*, 1994; Schell *et al.*, 1994);

* the Two-stage Fog water Impactor (TFI) (Schell *et al.*, 1994; Schell *et al.*, 1997b).

Intercomparison of LWC measuring techniques

During the PV '89 experiment different methods for the measurement of LWC in fog were employed simultaneously (Arends *et al.*, 1992). It was concluded that the PVM-100 is the best instrument available at the moment for field observations of fog or ground-based clouds. The main advantages are that the instrument has no problems with inlet losses as the LWC is measured in situ without perturbing the system, and that the calibration of the instrument can easily be performed in the field. Although the maximum droplet diameter reliably detected by the PVM-100 is probably less than 30 μm, the error in LWC determination by this instrument is smaller than the errors of the other instruments. For field experiments the combination of a PVM-100 to determine LWC and a FSSP-100 to measure drop size distributions turned out to be the optimal configuration.

Development and testing of the Droplet Aerosol Analyser (DAA)

A new basic approach was developed within GCE for measuring the relationship between droplets and their residues. The idea was to electrically charge the droplets directly, while sampling with an isokinetic inlet. In this way, the droplets acquire a number of elementary charges proportional to their size. Subsequently, the droplets are dried in a diffusion dryer, without loosing the information on the original droplet size, *i.e.* the original charge. By measuring the electrical mobility and the size of the droplet residues with ordinary DMAs, it is possible to obtain directly the relation between the size of the original droplet and its residue. Since the development of the DAA was well integrated within the defined scientific goals of GCE, the time for the development and implementation was very short (4 years).

8.2.4 GCE seen in retrospect

At the end of the project, we can confidently state that the original goals of GCE have been fulfilled and the contribution of GCE to the scientific knowledge on clouds has been widely recognised. New issues have also been highlighted as a result of the GCE work which should be the focus of new studies in the years to come. There are other aspects, both scientific and methodological, in what we call the "GCE approach", which require further retrospective discussion in order to bring them to the attention of the entire atmospheric science community.

a.   The importance of integrated field work, instrumental development and modelling

A particularly important aspect of GCE was the integration of groups with a wide variety of expertise and experimental skills (e.g. cloud microphysics, aerosol physics, cloud chemistry) working together at a number of sites, with the aim of obtaining far more comprehensive data sets than had been possible in the past. The combination of these detailed field projects along with the development of new instrumentation and modelling studies has lead to advances in our understanding of cloud chemistry, aerosol physics and chemistry and cloud-aerosol interactions which would otherwise have been impossible. It should be stressed that the overall strategy of the large collaborative field campaigns was very successful and that during the course of GCE lessons were learned concerning the planning and integration of groups with different expertise. The GCE experience will, in our opinion, be a model for future investigations in the important areas of cloud/aerosol interaction and heterogeneous atmospheric chemistry.

b.   Airborne measurements as a complement to ground-based investigations

Surface cloud fields are of varying structure, *e.g.* in the spatial distribution of liquid water and cloud height. Thus, ground-based cloud experiments, with observations at very few or sometimes only one point in cloud, can provide only limited information on the cloud field in general. In order to resolve this spatial structure, airborne observation platforms and/or remote sensing techniques are required. Aircraft measurements outside the cloud fields are very important for completing ground based cloud experiments. The lack of such observations during all GCE campaigns was certainly a major weakness of the experiments. The deficit is mainly reflected in the uncertainties in interpreting the sudden changes observed in gas and aerosol particle concentrations or in the chemical composition of cloud water and aerosol particle mass.

c.   Joint evaluation and dissemination of GCE results

Bearing in mind the reasons given in the introduction for centring the study of the cloud multiphase system around three large experiments, it should be noted that there are, of course, both advantages and disadvantages in organising such large projects. A major disadvantage is undoubtedly the need for a close coordination of the project, given the fact that no one group is in control of the overall project and all of the pieces are required to reconstruct the overall picture a-posteriori. The failure of one group would therefore most probably imply a general failure of the project. Fortunately, this problem did not arise in the case of GCE.

The same joint approach was also taken for the data evaluation, which was mainly performed with periodic meetings of the whole GCE group. Here again, advantages and disadvantages are inherent in this collective approach; large and frequent meetings are expensive and time-consuming, and the data analysis process may be considerably delayed due to the inter-dependency of the various groups. In the case of GCE, the disadvantages were kept to a minimum, as also testified by the relatively short interval between the experiments and the publication of the results: 3.5 years on average. The idea of publishing the results of the GCE experiments in the form of special issues of international scientific journals was, in our opinion, a means of providing the whole atmospheric science community with a comprehensive and self-consistent set of information which could be used by others in the field. The collaborative nature of the work accomplished within GCE is emphasised by the large authorship of most papers presented in these special issues. This should be regarded as an indication of the interdisciplinary efforts of the GCE scientific community in the accomplishment of these studies.

8.2.5 Further information

The GCE contribution is shown in context in chapter 3 of this volume. A set of papers were also published in a special issue of *Atmospheric Environment* (Fuzzi *et al.*, 1997).

A complete account of the subproject can be found in volume 5 of this series (Fuzzi and Wagenbach, 1996) which includes the following reviews by the subproject coordinators:

> GCE: an Overview
> Principal Results from GCE

GCE in Retrospective
Future Directions for Cloud Research

and contributions from the individual principal investigators, the titles of which are given in the next section.

8.2.6 GCE: Steering group and principal investigators

Steering group

Sandro Fuzzi (Coordinator)	Bologna
Hans-Christen Hansson	Stockholm
Thomas W. Choularton	Manchester
Wolfgang Jaeschke	Frankfurt
Wolfram Wobrock	Aubière
Dieter Schell	Frankfurt

Here is a list of principal investigators (*) and their co-workers together with the title of their contribution to the GCE final report (Fuzzi and Wagenbach, 1996).

B.G. Arends*, G.P.A. Kos and H.J. Möls,
   Field and Wind Tunnel Experiments in Clouds and Fog by ECN

A. Berner*, I. Solly and C. Kruisz,
   Study of Interstitial Aerosol

M. Bizjak*, A. Berner, I. Grgıc, B. Dıvjak, V. Hudnık, Š. Kozak-Legiša, M. Poje,
   Formation and Reactivity of Aerosols

T. Choularton*, M.W. Gallagher, K.N. Bower and R.N Colvile,
   Field and Modelling Studies of Cloud Chemical Processes

S. Fuzzi*, M.C. Facchini, G. Orsi and P. Laj,
   The Multiphase Chemistry of Clouds and Fog

R. Gieray*, T. Engelhardt and P.A. Wieser,
   A Single Particle Approach to Characterising Droplet Residues and Interstitial Particles

A. Hallberg*, K.J. Noone and J.A. Ogren,
   Partitioning of Aerosol Particles Between Droplets and Interstitial Air

H-C Hansson*, I.B. Svenningsson, E. Swietlicki, A. Widensohler, B.G. Martinsson and G. Frank,
   Influence of the Chemical Composition of the Atmospheric Particles on Fog and Cloud Formation

W. Jaeschke*, H.W. Georgii, D. Schell, M. Preiss, R. Maser and W. Wobrock,
   The Dynamic Behaviour of Pollutants in Fog and Captive Clouds

K. Levsen* and J. Lüttke,
   The Role of Nitrophenols in Cloud Chemistry

B.G. Martinsson*, S.I. Cederfelt, H.C. Hansson and G. Frank,
A Droplet Aerosol Analysing System for Cloud Studies

D. Möller*, K. Acker, W. Wieprecht and R. Auel,
Study of the Interaction of Photo-oxidants and Acidic Components between
Gas and Liquid Phase

P. Winkler* and S. Pahl,
Input of Trace Substances to High Elevation Forests by Cloud Water
Interception

8.2.7   References

Arends B.G., G.P.A. Kos, W. Wobrock, D. Schell, K.J. Noone, S. Fuzzi, S. Pahl; 1992, Comparison of techniques for measurements of fog liquid water content, *Tellus* **44B**, 604–611.

Baker M.B., R.E. Breidenthal, T.W. Choularton, J. Latham; 1984, The effect of turbulent mixing in clouds. *J. Atmos. Sci.* **41**, 299–304.

Berner A.; 1988, The collection of fog droplets by a jet impaction stage, *Sci. Total Environ.* **73**, 217–228.

Bower K.N., T.A. Hill, H. Coe, T.W. Choularton; 1991, Sulfur dioxide oxidation in an entraining cloud model with explicit microphysics. *Atmos. Environ.* **25A**, 2401–2418.

Brown R., W.T. Roach; 1976, The physics of radiation fog: II-a numerical study. *Quart. J. Roy. Meteor. Soc.* **106**, 335–354.

Cape, J.N., K.J. Hargreaves, R.L. Storeton-West, B. Jones, T. Davies, R.N. Colvile, M.W. Gallagher, T.W.Choularton, S. Pahl, A. Berner, C. Kruisz, M.Bizjak, P. Laj, M.C. Facchini, S. Fuzzi, B.G. Arends, K. Acker, W. Wieprecht, R.M. Harrison, J.D. Peak; 1997, Budget of oxidized nitrogen species in orographic clouds. *Atmos. Environ.* **31**, 2635–2637.

Choularton T.W., R.N. Colvile, K.N. Bower, M.W. Gallagher, M.Wells, K.M. Beswick, B.G. Arends, J.J. Möls, G.P.A.Kos, S. Fuzzi, J.A. Lind, G. Orsi, M.C. Facchini, P. Laj, R. Gieray, P. Wieser, T. Engelhardt, A. Berner, C. Kruisz, D. Möller, K. Acker, W. Wieprecht, J.Lüttke, K. Levsen, M. Bizjak, H-C. Hansson, S-I.Cederfelt, G. Frank, B. Mentes, B. Martinsson, D. Orsini, B.Svenningsson, E. Swietlicki, A. Wiedensohler, K.J. Noone, S. Pahl, P. Winkler, E. Seyffer, G. Helas, W. Jaeschke, H.W. Georgii, W. Wobrock, M. Preiss, R. Maser, D.Schell, G. Dollard, B. Jones, T. Davies, D.L. Sedlak, M.M. David, M. Wendisch, J.N. Cape, K.J.Hargreaves, M.A. Sutton, R.L. Storeton-West, D. Fowler, A.Hallberg, R.M. Harrison, J.D. Peak; 1997, The Great Dun Fell Cloud Experiment 1993: an overview. *Atmos. Environ.* **31**, 2393–2407.

Colvile R. N., R. Sander, T.W. Choularton, K.N. Bower, D.W.F. Inglis, W. Wobrock, R. Maser, D. Schell, I.B. Svenningsson, A. Wiedensohler, H-C. Hansson, A. Hallberg, J.A. Ogren, K.J. Noone, M.C. Facchini, S. Fuzzi, G. Orsi, B.G. Arends, W. Winiwarter, T. Schneider, A. Berner; 1994, Computer modelling of clouds at Kleiner Feldberg. *J. Atmos. Chem.* **19**, 189–229.

Covert, D.S., A. Wiedensohler, P. Aalto, J. Heintzenberg, P.H. McMurry, C. Leck; 1996, Aerosol number size distribution from 3 to 500 nm diameter in the arctic marine boundary layer during summer and autumn. *Tellus* **48B**, 197–212.

Facchini M.C., S. Fuzzi, M. Kessel, W. Wobrock, W. Jaeschke, B.G. Arends, J.J. Möls, A. Berner, I. Solly, C. Kruisz, G. Reischl, S. Pahl, A. Hallberg, J.A. Ogren, H. Fierlinger-Oberlinninger, A. Marzorati, D. Schell; 1992a, The chemistry of sulfur and nitrogen species in a fog system. A multiphase approach. *Tellus* **44B**, 505–521.

Facchini M.C., S. Fuzzi, J.A. Lind, M. Kessel, H. Fierlinger-Oberlinninger, M. Kalina, H. Puxbaum, W. Winiwarter, B.G. Arends, W. Wobrock, W. Jaeschke, A. Berner, C. Kruisz; 1992b, Phase partitioning and chemical reactions of low molecular weight organic compounds in fog. *Tellus* **44B**, 533–544.

Fuzzi S.; 1994, Clouds in the troposphere; in: C.F. Boutron (ed), *Topics in Atmospheric and Interstellar Physics and Chemistry*, Les Editions de Physique, Les Ulis, pp. 291–308.

Fuzzi S., M. C. Facchini, G. Orsi, J. A. Lind, W. Wobrock, M.Kessel, R. Maser, W. Jaeschke, K.-H. Enderle, B. G. Arends, A. Berner, I. Solly, C. Kruisz, G. Reischl, S. Pahl, U. Kaminski, P. Winkler, J. A. Ogren, K. J. Noone, A. Hallberg, H. Fierlinger-Oberlinninger, H. Puxbaum, A. Marzorati, H.-C. Hansson, A. Wiedensohler, I. B. Svenningsson, B. G. Martinsson, D. Schell, H.-W. Georgii; 1992, The Po Valley Fog Experiment 1989: An Overview, *Tellus* **44B**, 448–468.

Fuzzi S., M.C. Facchini, D. Schell, W. Wobrock, P. Winkler, B.G. Arends, M. Kessel, J.J. Möls, S. Pahl, T. Schneider, A. Berner, I. Solly, C. Kruisz, M. Kalina, H. Fierlinger, A. Hallberg, P. Vitali, L. Santoli, G. Tigli;   1994,   Multiphase chemistry and acidity of clouds at Kleiner Feldberg. *J. Atmos. Chem.* **19**, 87–106.

Fuzzi S., *et al.*; 1997, *Atmos. Environ.* **31**, 2391–2695.

Gallagher M.W., T.W. Choularton, R.M. Downer, B.J. Tyler, I.M. Stromberg, C.S. Mill, S.A. Penkett, B. Bandy, G.J. Dollard, T.J. Davies, B.M.R. Jones; 1991, Measurements of the entrainment of hydrogen peroxide into cloud systems. *Atmos. Environ.* **25A**, 2029–2038.

Heintzenberg J.; 1989, Fine particles in the global troposphere. A review. *Tellus* **41B**, 149–160.

Hoppel W.A., G.M. Frick, J.W. Fitzgerald, R.E. Larson; 1994, Marine boundary layer measurements of new particle formation and the effects nonprecipitating clouds have on aerosol size distribution. *J. Geophys. Res.* **99**, 14443–14459.

Jaenicke R.; 1993, Tropospheric aerosols; in: P.V. Hobbs (ed), *Aerosol-Cloud-Climate Interactions*, Academic Press, San Diego, pp. 1–31.

Laj P., S. Fuzzi, M.C. Facchini, G. Orsi, A. Berner, C. Kruisz,W. Wobrock, A. Hallberg, K.N. Bower, M.W. Gallagher, K.M. Beswick, R.N. Colvile, T.W.Choularton, P. Nason, B. Jones; 1997a, Experimental evidence for in-cloud production of aerosol sulfate. *Atmos. Environ.* **31**, 2503–2515

Laj P., S. Fuzzi, M.C. Facchini, J.A. Lind, G. Orsi, M. Preiss, R. Maser, W. Jaeschke, E. Seyffer, B.G.Arends , J.J. Möls, K. Acker, W. Wieprecht, D. Möller, R.N. Colvile, M.W. Gallagher, K.M. Beswick, K.J.Hargreaves, R.L. Storeton-West, M.A. Sutton; 1997b, Cloud processing of soluble gases. *Atmos. Environ.* **31**, 3589–2599.

Martinsson B.G., S-I. Cederfelt, B. Svenningsson, G. Frank, H-C. Hansson, E.Swietlicki, A. Wiedensohler, M. Wendisch, M.W. Gallagher, R. N. Colvile, K.M. Beswick, T.W. Choularton, K.N. Bower; 1997, Experimental determination of the connection between cloud droplet size and its dry residue size. *Atmos. Environ.* **31**, 2477–2491.

Maser R., H. Franke, M. Preiss, W. Jaeschke; 1994, Methods provided and applied in a research aircraft for the study of cloud physics and chemistry. *Contr. Phys. Atmos.* **67**, 321–334.

Noone K.J., J.A. Ogren, A. Hallberg, J. Heintzenberg, J. Ström, H-C. Hansson, I.B. Svenningsson, A. Wiedensohler, S. Fuzzi, M.C. Facchini, B.G. Arends, A. Berner; 1992, Changes in aerosol size- and phase distributions due to physical and chemical processes in fog. *Tellus* **44B**, 489–504.

Ogren, J.A., K.J. Noone, A. Hallberg, J. Heintzenberg, D. Schell, A. Berner, I. Solly, C. Kruisz, G. Reischl, B.G. Arends, W. Wobrock; 1992, Measurements of the size dependence of the concentration of non-volatile material in fog droplets. *Tellus* **44B**, 570–580.

Pahl S., P. Winkler, T. Schneider, B.G. Arends, D. Schell, R. Maser, W. Wobrock; 1994, Deposition of trace substances via cloud interception to a coniferous forest at Kleiner Feldberg. *J. Atmos. Chem.* **19**, 231–252.

Schell D., H.W. Georgii, R. Maser, W. Jaeschke, B.G. Arends, G.P. A. Kos, P. Winkler, T. Schneider, A. Berner, C. Kruisz; 1992, Intercomparison of fog water samplers. *Tellus* **44B**, 612–631.

Schell D., R. Maser, M. Preiss, W. Wobrock, K. Acker, W. Wieprecht, S. Pahl, H.W. Georgii, W. Jaeschke; 1994, The dependence of cloud water chemical concentrations on drop size and height above cloud base at Great Dun Fell. Instrumentation and first results; in: P.M. Borrell, P. Borrell, T. Cvitaš, W. Seiler (eds), *Proc. EUROTRAC Symp. '94*, SPB Academic Publishing bv, The Hague, pp. 1133–1137.

Schell D., W. Wobrock, R. Maser, M. Preiss, W. Jaeschke, H.W. Georgii, M.W. Gallagher, K.N. Bower, K.M. Beswick, S. Pahl, M.C. Facchini, S. Fuzzi, A. Wiedensohler, H-C. Hansson, M. Wendisch; 1997a, The size-dependent chemical composition of cloud droplets. *Atmos. Environ.* **31**, 2561–2577.

Schell D., R. Maser, W. Wobrock, W. Jaeschke and H.W. Georgii, G.P.A. Kos, B.G. Arends, K.M. Beswick, K.N. Bower, M.W. Gallagher;1997b, A two-stage impactor for fog droplet collection: design and performance. *Atmos. Environ.* **31**, 2671–2675.

Sutton M.A., D. Fowler, R.L. Storeton-West, J.N. Cape, B.G.Arends, J.J. Möls; 1997, Vertical distribution and fluxes of NH₃ at Great Dun Fell. *Atmos. Environ.* **31**, 2615–2625.

Svenningsson I.B., H-C. Hansson, A. Wiedensohler, J.A. Ogren, K.J. Noone, A. Hallberg; 1992, Hygroscopic growth of aerosol particles in the Po Valley. *Tellus* **44B**, 556–569.

Svenningson I.B., H-C. Hansson, A. Wiedensohler, K.J. Noone, J.A. Ogren, A. Hallberg, R.N. Colvile; 1994, Hygroscopic growth of aerosol particles and its influence on nucleation scavenging in cloud: experimental results from Kleiner Feldberg. *J. Atmos. Chem.* **19**, 129-152.

Svenningson I.B, H-C Hansson, B.G. Martinsson, A. Wiedensohler, E. Swietlicki, S.-I. Cederfelt; 1997, Cloud droplet nucleation scavenging in relation to the size and hygroscopic behaviour of aerosol particles. *Atmos. Environ.* **31**, 2463–2477.

Swietlicki E., H-C. Hansson, B. Martinsson, B. Mentes, D. Orsini, B. Svenningsson, A. Wiedensohler, M. Wendisch, S. Pahl, P. Winkler, R.N. Colvile, R. Gieray, J. Luttke, J. Heintzenberg, J.N. Cape, K.J. Hargreaves, R.L. Storeton-West, K. Acker, W. Wieprecht, A. Berner, C. Kruisz, M.C. Facchini, P. Laj, S. Fuzzi, B. Jones, P. Nason; 1997, Source identification during the Great Dun Fell Cloud Experiment 1993. *Atmos. Environ.* **31**, 2441–2453.

Wells M., K.N. Bower, T.W. Choularton, J.N. Cape, M.A.Sutton, R.L. Storeton-West, D. Fowler, A. Wiedensohler H-C. Hansson, B. Svenningsson, E. Swietlicki, M. Wendisch, B. Jones, G. Dollard, K. Acker, W. Wieprecht, M. Preiss, B.G. Arends, S. Pahl, A. Berner, C. Kruisz, P. Laj, M.C. Facchini, S. Fuzzi; 1997, The reduced nitrogen budget of an orographic cloud. *Atmos. Environ.* **31**, 2599–2615.

Wiedensohler A., H-C. Hansson and D. Orsini, M. Wendisch, F. Wagner, K.N. Bower, T.W. Choularton, M. Wells, M.Parkin, K. Acker, W. Wieprecht, M.C. Facchini, J.A. Lind, S. Fuzzi, B.G. Arends; 1997, Night-time formation of new particles associated with orographic clouds. *Atmos. Environ.***31**, 2545–2561.

Winiwarter W., H. Fierlinger, H. Puxbaum, M.C. Facchini, B.G. Arends, S. Fuzzi, D. Schell, U. Kaminski, S. Pahl, T. Schneider, A. Berner, I. Solly, C. Kruisz; 1994, Henry's Law and the behaviour of weak acids and bases in fog and cloud. *J. Atmos. Chem.* **19**, 173-188.

Winkler P.; 1986, Observation of fogwater composition in Hamburg, in: H.-W. Georgii (ed), *Atmospheric Pollutants in Forest Areas*, Reidel, Dordrecht, pp. 143–151.

Wobrock W. (1988) Numerische Modellsimulation von Strahlungsnebelsituationen unter Berücksichtigung spektraler Wolkenmikrophysik. Dissertation, Institute of Meteorology und Geophysics, University Frankfurt.

Wobrock W., D. Schell, R. Maser, M. Kessel, W. Jaeschke, S. Fuzzi, M.C. Facchini, G. Orsi, A. Marzorati, P. Winkler, B.G. Arends, J. Bendix; 1992, Meteorological characteristics of the Po Valley fog. *Tellus* **44B**, 469–488.

Wobrock, W., D. Schell, R. Maser, W. Jaeschke, H.-W. Georgii, W. Wieprecht, B.G. Arends, J.J. Möls, G.P.A. Kos, S. Fuzzi, M.C. Facchini, G. Orsi, A. Berner, I. Solly, C. Kruisz, I.B. Svenningsson, A. Wiedensohler, H.-C. Hansson, J.A. Ogren, K.J. Noone, A. Hallberg, S. Pahl, T. Schneider, P. Winkler, W. Winiwarter, R.N. Colvile, T.W. Choularton, A.I. Flossmann, S. Borrmann; 1994, The Kleiner Feldberg Cloud Experiment 1990. An overview. *J. Atmos. Chem.* **19**, 3–35.

8.3 TOR: An Overview of Tropospheric Ozone Research

Øystein Hov[1], Dieter Kley[2], Andreas Volz-Thomas[2], Jeannette Beck[3], Peringe Grennfelt[4] and Stuart A. Penkett[5]

[1]Norwegian Institute for Air Research (NILU), Kjeller, Norway
[2]Forschungszentrum Jülich GmbH, Institut für Chemie und Dynamik der Geosphäre, Postfach 1913, D-52425 Jülich, Germany
[3]RIVM, National Institute of Public Health and Environmental Protection, Antonie van Leewenhoeklaan 9, P.O.Box 1, NL-3720 BA Bilthoven, Netherlands
[4]Swedish Environmental Research Institute (IVL), Box 47086, S-40258 Göteborg, Sweden
[5]University of East Anglia, School of Environmental Sciences, Norwich NR4 7TJ, UK

8.3.1 Aims of the subproject

In the original plan for the subproject TOR (Kley *et al.*, 1986) the aim was stated to provide scientific information about a number of questions which later were grouped into the following tasks:

Task 1: (Q1) How much higher is the mean ozone concentration in the boundary layer over Europe than that averaged over northern mid-latitudes, and what is the seasonal, latitudinal and vertical variation of ozone within the adjacent troposphere? and (Q4) Is there a secular trend in the concentrations of ozone and precursor molecules in the boundary layer or in the background atmosphere?

Task 2: (Q2) What are the emissions and distributions of the precursors responsible for that excess of ozone?

Task 3: (Q3) Can we measure how much of the excess ozone in the boundary layer over Europe spills over into the background atmosphere? and (Q6) Is it possible to quantify by co-measurements of ozone and other tracers the proportion of ozone produced in the troposphere to that transferred from stratosphere to troposphere at our location?

Task 4: (Q7) How much ozone and precursors are transported across regional boundaries? and design of an optimal network for ozone and precursors in Europe.

In the original plan one further question was defined (Q5): Can we model the observations, and how well do the model calculations agree with the observations? Later in the project, the steering committee recommended that *in-situ* measurements and other experimental data, and numerical and statistical models should be used to analyse the scientific questions within each task. The use of models was then seen as an integral part of the analysis of the scientific questions asked in each task.

8.3.2 Activities

In order to address the scientific questions in TOR, a high quality observation network was designed and implemented. The basic instrumentation was, in part, available commercially or could be adopted from existing networks. Additional instrumentation, in particular for hydrocarbons (VOC) and nitrogen oxides (NO_x, NO_y) was developed especially for TOR. Quality assurance and quality control principles were implemented, a data centre was established at RIVM in the Netherlands, deterministic as well as statistical models were developed, and links were established with the weather services in the different countries as well as at the European Centre for Medium Range Weather Forecast (ECMWF) in the UK, in order to provide the general meteorological information required for the interpretation of the chemical measurements.

a. Observation network

The original idea behind the TOR network was to obtain information about trace gas concentrations that are representative of larger regions within Europe. However, given the relatively short atmospheric life times of ozone and some of its precursors, *e.g.* NO_x and the more reactive VOCs, it was found essential to study the chemical/meteorological regime in the vicinity of urban areas as well.

Vertical soundings of ozone, water vapour and temperature were given a high priority in the TOR plan because it was seen as essential to provide more information on the vertical distribution of ozone throughout the troposphere and its variability with time and in space. Such information is required to quantify the magnitude of the stratospheric source for ozone in the troposphere and to find out how the tropospheric content of ozone is changing both on an episodic, annual and long term basis. This question is of considerable importance both for climate, for the oxidation efficiency in the troposphere and for the ozone concentration close to the ground.

The sites in the TOR network were located over a 50 degree latitudinal range between 30° N and 80° N. Some sites were located close to the Atlantic ocean and were frequently exposed to background air unaffected by recent emissions. Other sites were located in regions close to source areas and thus provided data on the extent of boundary layer pollution. Some sites were located to be frequently downwind of important source regions, in order to obtain data on chemical transformation and transport of pollutants out of the source areas. Some sites were located at high elevation and thus provided data mostly from the free troposphere, while other sites always were within the atmospheric boundary layer. Table 8.3.1 gives the locations the ground based TOR network and of the vertical sounding stations. A comprehensive description of the TOR network can be found in Cvitaš and Kley (1994).

Table 8.3.1:   TOR sites with geographical coordinates, elevation and principal scientists

No.	Name of Site	Code	Latitude	Longitude	Altitude	Principal investigator
	In-situ measurements TOR sites		°N	°W(-)°E(+)	(m a.s.l.)	
	Northern Europe					
1	Ny Ålesund, Zeppelin Mountain	ZMT	78.9	11.9	474	Hov, Stordal
4	Åreskutan	ARE	63.4	13.1	1240	Oyola
3	Utö Island	UTO	59.8	21.4	7	Joffre, Laurila
19	Aspvreten	ASP	58.8	17.4	20	Oyola
2	Birkenes	BIR	58.4	8.3	190	Hov, Stordal
5	Rörvik	RVK	57.4	11.9	10	Lindskog
18	Lille Valby	LLV	55.7	12.1	15	Nielsen
	Western Europe					
9	Kollumerwaard	KOL	53.3	6.3	2	Beck
	Weybourne	WEY	53.0	1.1	10	Penkett, Clemitshaw
8	Mace Head	MHD	53.3	−9.9	10	Simmonds
10	Porspoder	PPD	48.5	−4.8	22	Toupance, Dutot
	Central Europe					
11	Schauinsland	SIL	47.9	7.8	1220	Volz Thomas
21	Wank	WNK	47.5	11.1	1776	Scheel
22	Zugspitze	ZUG	47.4	11.0	2937	Scheel
52	Jungfraujoch	JUN	46.5	8.0	3580	Delbouille, Zander
13	Sonnblick	SON	47.1	13.0	3106	Puxbaum, Radonsky
15	K-puszta	KPU	47.0	19.6	125	Haszpra
66	Krvavec	KRV	46.3	14.5	1720	Gomiscek
51	Puntijarka	PUN	45.8	16.0	980	Klasinc
	Zagreb-RB[1]	ZAG	45.8	16.0	180	Klasinc
	Southern Europe					
14	Pic du Midi	PDM	42.9	0.2	2877	Marenco
	Athens [1]	ATH	38.0	24.4	100	Varotsos
17	Izaña	IZA	28.3	−16.5	2370	Schmitt

Table 8.3.1 continued

No.	Name of Site	Code	Latitude	Longitude	Altitude	Principal investigator
	In-situ **measurements** TOR sites		°N	°W(-)°E(+)	(m a.s.l.)	
	Vertical soundings					
	Bjørnøya (Bear Island)		74.5	19		Hov, Stordal
	Hungriger Wolf		54	10		Bösenberg
23	Uccle		51	4		De Muer
24	Jülich		51	6		Kley, Smit
25	Observatoire de Haute Provence	OHP	44	6		Ancellet, Beekmann
	Athens		38	24		Varotsos

[1] Urban site with specific conditions, ozone levels not representative for a larger region.

b.   Instrumentation

The stations were equipped with high quality research type instrumentation. Station instrumentation is listed in Table 8.3.2, classified according to the priority given in TOR to the respective species. Since commercial instrumentation was not available in many cases, a number of new instruments with improved sensitivity and accuracy designed for automatic operation resulted from the sub project.

Table 8.3.2:   Details of instrumentation

Species	Priority	Technique
Ozone	1	UV absorption
NO, NO_2, NO_y [1]	1	Chemiluminescence with specific converters
CH_4, CO, CO_2	1–2	GC
C_2–C_9 HC	1	Automated GC or flask sampling - GC
> C_5 HC	1	As above
$J(NO_2)$, $J(O^1D)$	1	Filter Radiometer
Meteorology	1–2	Standard instrumentation
CFC tracers	2	GC (ECD)
Aldehydes and ketones	2	HPLC
PAN	2	GC (ECD), GC (LMA3)
H_2O_2	2	UV fluorescence in solution

[1] NO_y is the sum of NO_x and its oxidation products

Several approaches for automating VOC measurements were made and implemented at the TOR stations. A prototype for automatic sampling of VOCs was built to given specifications by Chrompack, the necessary parameters for optimal separation being determined in subproject work (Mowrer and Lindskog, 1991). The instrument was used at some TOR stations as well as in other subprojects. An intercomparison of VOC analysis, conducted in TOR demonstrated that a reasonable accuracy could be obtained for the more volatile species with careful work (Hahn, 1997).

For the nitrogen oxides (NO and NO_2), a sensitive and accurate chemiluminescence analyser with a photolytic converter has resulted from subproject work (ECO Physics Model CLD 770 Alppt and PLC 760). The instrument employed at several TOR stations and has been sold throughout the world (Volz-Thomas in Kley and Cvitaš, 1994).

Automated instruments for the measurements of peroxides and formaldehyde have been developed (Slemr et al., 1994) and are available from AEROLASER GmbH. A photo-electric sensor for the measurement of the photolysis rate of NO_2 (Junkermann et al., 1989) was improved for field use and in aircraft. It is now commercially available through Meteorologie Consult GmbH, as well as instrument for the photolysis rate of ozone, which is the primary process in formation of OH radicals. A sonde was also developed at the Forschungszentrum Jülich for the balloon-borne measurement of hydrogen peroxide (patent obtained, licence contract with UNISEARCH).

Table 8.3.3 shows a summary of the species that were measured at the TOR sites according to the data available in the TOR data base. The time periods that the sites were operational for a given component as well as specification of the instrumentation used are summarised in Cvitaš and Kley (1994) and in Hahn (1996).

c. Standardisation and calibration

Standardisation of the instrumentation and the measurement procedures was an essential part in quality assurance. Responsibilities for developing standards and calibration procedures were as follows:

ozone	Studsvik, Sweden (Pedro Oyola)
NO_x	FZ Jülich
HCs, CFCs	IFU (Garmisch-Partenkirchen)
PAN, H_2O_2	NILU (Lillestrøm) and UEA (Norwich)
ozone sondes	FZ Jülich

Table 8.3.3:   Summary of TOR measurements available in the TOR data base at RIVM (5.11.96)

No	Site name	O_3	NO	NO_2	NO_y	PAN	VOC	CH_4	CO	CFC	H_2O_2	meteo	O_3 profile	Meteo profile
1	Ny Ålesund (N)	×												
2	Birkenes (N)	×					×	×		×		×		
3	Utö (SF)	×		×		×	×					×		
4	Areskutan (S)	×										×		
5	Rörvik (S)	×	×	×		×	×					×		
7	West-Beckham (GB)	×		×			×							
8	Mace Head (EI)	×						×						
9	Kollumerwaard (NL)	×	×	×		×	×	×	×	×		×		
10	Porspoder (F)	×		×			×							
11	Schauinsland (D)	×	×	×	×		×				×	×		
12	Garmisch-P. (D)												×	×
13	Sonnblick (A)	×												
14	Pic du Midi (F)	×												
15	K-Puszta (H)						×							
16	RBI (CR)	×												
17	Izaña (E)	×				×	×					×		
18	Lille Valby (DK)	×	×		×									
19	Aspvreten (S)	×	×	×		×	×				×	×		
19	Aspvreten (S)	×	×	×		×	×				×	×		
20	Tour du Donon (F)	×	×	×			×							
21	Wank (D)	×										×		
22	Zugspitze (D)	×										×		
23	Ukkel (B)												×	×

Table 8.3.3:   continued

No	Site name	O_3	NO	NO_2	NO_y	PAN	VOC	CH_4	CO	CFC	H_2O_2	meteo	O_3 profile	Meteo profile
24	Jülich (D)												×	×
25	OHP (F)												×	×
27	Zeppelinfjellet (N)	×		×		×	×	×						
51	Puntijarka (Cr)	×												
66	Krvavec (Slo)	×												
70	Kloosterburen (NL)	×	×	×										
71	Moerdijk-c (NL						×					×		
213	Tänikon (CH)	×												
222	Waldhof (D)	×					×							
345	Kosetice (CZ)						×							
346	Rucava (LIT)						×							

Notes: Stations 1 (Ny Ålesund) and 27 (Zeppelinfjellet) are at the same location but at different altitudes (see Table 8.3.1). The mountains Wank and Zugspitze are both in the vicinity of Garmisch-Partenkirchen. The Dutch TOR station was moved from Kloosterburen to Moerdijk and then to Kollumerwaard.
The British site West Beckham was later moved to Weybourne

Ozone instruments were brought to Studsvik in Sweden to be checked against a reference instrument approximately every year, and standard procedures were agreed upon for other species to standardise *in-situ* or at a laboratory close to the measurement site. Considerable effort was spent to inter-compare and harmonise CO, CH_4 and non-methane hydrocarbon measurements within TOR. IFU, as the central laboratory, distributed a number of stabilised and pre-analysed aliquots of a large sample of ambient air among the participating TOR laboratories to serve as a reference gas mixture. Three different NIST certified standard gas mixtures were used as the common primary standards. Two intercomparison rounds were completed. The procedures and results are summarised by Hahn. Another European intercomparison preceded the one carried out within TOR (De Sager *et al.*, 1993), and most of the TOR laboratories also participated there.

The NO_x measurements were harmonised by using standardised calibration procedures based upon gas phase titration (GPT) and by comparing the individual NO calibration gases to a master standard at FZ Jülich. The latter was compared during the NARE 1993 intensive with the primary standard of the

NOAA Aeronomy Laboratory in Boulder, Colorado, and was found to be in excellent agreement (deviation < 3 %; see McKenna *et al.*, 1995). The calibration facility for the ozone and water vapour sounding equipment used in TOR in Jülich is described in detail in Kley and Cvitaš (1994, pp. 160–168) and in Smit *et al.* (1997). A formal international intercomparison of peroxy radical measurements (PRICE) was held in cooperation between TOR and the EU project OCTA (Volz-Thomas *et al.*, 1996).

The instrument developments within TOR have greatly improved the measurement capabilities in many other national and international projects, like EMEP and a range of EU funded projects, *e.g.* OCTA, FIELDVOC, TOASTE. The ozone sonde calibration facility has now become the World Centre for quality assurance of ozone and water vapour sondes under WMO (Global Atmosphere Watch GAW) (WMO Report, 1995).

d.   Data Centre

Since the objectives of the TOR project required evaluation of the measurements from all sites, it was judged to be essential to have a central data centre providing controlled exchange of data. Such a data centre should enable collection of all data from the project, regular correspondence with the owners of the data and support for its users. RIVM hosted the TOR data base.

The data archive contains data with a multiform character resulting from the diverse nature of the sites within the TOR network. This situation called for handling, storage and exchange of data designed to accommodate the non-uniform structure of the measurements. Ground-based data exchange is performed through the ISO 7168 format. Incoming new data files required extensive quality control on correct ISO formatting. It was a general principle to have TOR PIs retain responsibility for the quality assurance of the data. Clarifying ambiguities and suspicious values by TOR PIs was necessary, in particular, at the beginning of the project. Later on, the number of necessary contacts with the PIs decreased. RIVM offered to host the database until the end of 1997. Table 8.3.4 shows the access scheme to be used during this period.

Table 8.3.4: Access to the TOR database in the period 1996–1998

Period	Status of data in the release schedule
1.1.1996 to 30.6.1997	Data open to TOR and non-TOR PIs; however, co-authorship should be offered if a scientific paper originates from the data
From 1.7.1997	Public access with no co-authorship requirement; the data may be released on a CD-ROM or WWW

**The TOR data is now freely available and can be accessed at www.nilu.no/projects/tor/tor.html

e. Atmospheric transport and chemistry models

In the TOR project a hierarchy of models was developed and/or applied:

*   0-dimensional statistical models, box models, "heuristic" models;

*   Trajectory chemistry models based on idealised meteorology, such as the Harwell trajectory model;

*   Trajectory chemistry models based on real meteorology, *e.g.* different versions of the EMEP model and the Dutch MPA model;

*   2-dimensional climatological models like the different meridional global chemistry models (in use at the universities of Oslo and Bergen, and the zonal "channel" model in use in Oslo);

*   Climatological chemistry-transport models were developed at the University of Oslo (on the basis of the NASA GISS GCM);

*   LOTOS, a European scale lower troposphere 3-D chemistry-transport model, developed at TNO;

*   Mesoscale chemistry transport model with meteorology derived from mesoscale weather prediction models, *i.e.* the (EURAD model developed in EUMAC and the model developed in TOR at the University of Bergen in Norway).

The use of the models was seen as an integral part of the analysis of the scientific questions asked in each task. Coupled NWP-chemistry models can realistically be run only for limited periods of time (weeks), off-line 3-D model can be run for months and simpler models can be run for months or even years. These models have therefore been used in conjunction because they address different questions.

8.3.3 Results and conclusions

Four task groups were organised to review the scientific information available from the TOR project. The four task groups prepared six reports (Hov, 1997). The task groups represent a comprehensive overview of the TOR results but the material available is by no means exhausted and will be the subject for further research. A dedicated TOR issue of *J. Atmos. Chem.* has been published with nearly 20 papers (Kley *et al.*, 1997).

The findings of the TOR subproject in terms of trends in photo-oxidant concentrations, processes influencing the concentration of photo-oxidants, regional and continental distribution of photo-oxidants and photo-oxidants on a global scale are summarised below. The results were also discussed and put in perspective to findings elsewhere in EUROTRAC and in the scientific community in general in the report from the EUROTRAC Application Project (Borrell *et al.*, 1997).

a.   Trends in photo-oxidant concentrations and precursors

At the beginning of TOR, analyses of available records, *e.g.* those from Hohenpeissenberg (Attmannspacher *et al.*, 1984; Wege *et al.*, 1989) and Arkona (Feister and Warmbt, 1987), and the reanalysis of historic measurements (Volz and Kley, 1988; Kley *et al.*, 1988; Crutzen, 1988) had suggested that the ozone concentrations in the 1980s were higher than what they had been around the turn of the century and that photochemical formation was the likely cause for that increase. A clearer picture has emerged from research within or related to TOR of the magnitude and the nature of the increase in tropospheric ozone concentrations.

Analysis of historic records

A quantitative method to measure ozone was used continuously from 1876 until 1911 at the Observatoire de Montsouris, Paris (Volz and Kley, 1988). The 24-h average ozone concentration was around 10 ppb, about a factor of 3–4 smaller than is found today in many areas of Europe and North America. Analysis of several Schönbein records led to similar conclusions about the pre-industrial ozone concentration and would suggest that the tropospheric background was 10 ppb in both hemispheres (see Anfossi *et al.*, 1991; Sandroni *et al.*, 1992) and in the free troposphere over Europe (Marenco *et al.*, 1994). This agreement must be viewed with caution, however, because of the known problems associated with the Schönbein method. Kley *et al.* (1988) concluded from an extensive laboratory evaluation of the method that these data are only semi-quantitative in nature and should not be used for trend estimates. In particular, the close agreement between ozone concentrations at Pic du Midi at 3000m altitude and at Montsouris in the Paris basin is difficult to explain, because in the absence of local photochemical production daily average ozone concentrations at a surface site like Paris, which is heavily influenced by dry deposition at night, are expected to be lower than those in the free troposphere.

After 1910, only few and mostly sporadic measurements of tropospheric ozone were made using both optical and chemical techniques, which were reviewed by Crutzen (1988), Kley *et al.* (1988) and Staehelin *et al.* (1994). In Fig. 8.3.1, a comparison of historic measurements that were made using quantitative techniques is shown with measurements made in the late 1980s. On average, ozone concentrations in the troposphere over Europe (0–4 km) today are a factor of two higher than in the earlier period. Fig. 8.3.1 also shows that little can be inferred about a possible increase in tropospheric ozone before 1950, because of the variance between the different sites. In this context, it is interesting to note that the data from Montsouris (1876–1911; 40 m a.s.l.) and those from Arosa (1950–1956; 1860 m a.s.l.) do not show a single day with ozone concentrations above 40 ppb (Volz-Thomas, 1993; Otmans and Levy, 1994).

Trends in ozone concentrations over the past decades

The modern ozone measurements are mainly based on UV absorption, and were started in the 1970s at several remote coastal and high altitude sites.

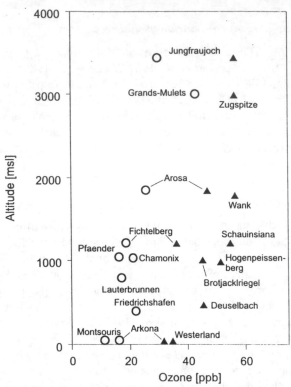

Fig. 8.3.1: Surface ozone concentrations observed in late summer at different locations in Europe (Staehlin *et al.*, 1994). The open circles summarise historic data collected before and during the 1950s, the triangles are from measurements made after 1988. The data have been plotted against the altitude of the different sites.

The records for Mauna Loa, Hawaii (Otmans and Levy, 1994) and the Zugspitze TOR station (Scheel *et al.*, 1993) are shown in Fig. 8.3.2. The trends observed at the various remote sites were evaluated in TOR and are presented in Fig. 8.3.3. All stations north of about 20° N exhibit a positive trend in ozone that is statistically significant, when taken over the whole period. On the other hand, a statistically significant negative trend of about –7 % per decade is observed at the South Pole.

For the most part, the trends increase from –7 % per decade at 90° S to +7 % per decade at 70° N. There are particularly large positive trends observed at the high elevation sites in southern Germany (1–2 % per year). These large trends perhaps reflect a regional influence above and beyond the smaller global trend (Volz-Thomas, 1993).

The trends observed in the northern hemisphere are largely due to the relatively rapid ozone increase that occurred in the 1970s. Over the last decade, no or only little ozone increase has occurred in the free troposphere. Indeed, ozone concentrations at some locations in the polluted boundary layer over Europe have even decreased over the last decade. In Delft in the Netherlands ground level ozone concentrations decreased in the 1970s as a result of increasing NO_x

concentrations, with $O_x = O_3 + NO_2$ slightly increasing (Guicherit, 1988; Low *et al.*, 1992). In the 1980s and 1990s ground level ozone concentrations as well as the oxidant (O_x) levels as measured in Kollumerwaard in the northern part of the Netherlands, have decreased somewhat.

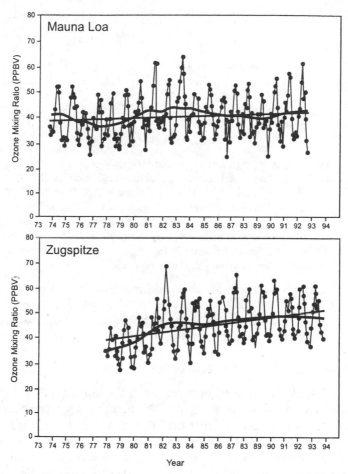

Fig. 8.3.2:   Trends in tropospheric ozone concentrations observed at the Zugspitze TOR station since 1978 (Scheel *et al.*, 1993; Sladkovic *et al.*, 1994) in comparison to the trend observed at Mauna Loa, Hawaii since 1973 (Otmans and Yevy, 1994).

Measurements of ozone at Mace Head in the period April 1987 to June 1992 show an average upward trend of 2.5 ppb per decade in the summer in polluted air which has passed over continental Europe, while in polluted air in winter there is a downward trend of 1.3 ppb per decade 1992 (Simmonds, 1993). A more reliable estimate of the nature and the possible causes for the observed changes is only possible with longer records than those available today.

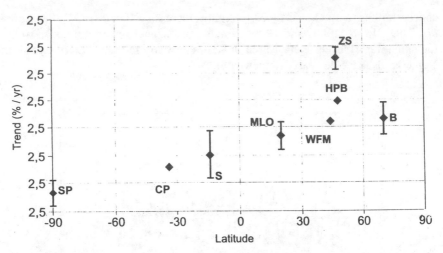

Fig. 8.3.3: Trends in tropospheric ozone concentrations observed at different latitudes. Only coastal and high altitude observatories are included. SP: South Pole, 90° S, 2800 m a.s.l., start 1975; CP: Cape Point, 34° S, 1982; S: Samoa, 14° S, 1975; MLO: Mauna Loa, 20° N, 3400 m a.s.l., 1973; WFM: Whiteface Mountain, 43° N, 1600 m a.s.l., 1973; ZS: the Zugspitze, 47°N, 3000 m a.s.l., 1978; W: the Wank, 47° N, 1800 m a.s.l., 1978; HPB: Hohenpeissenberg, 48° N, 1000 m a.s.l., 1971; Arkona, 53° N, 1956; B: Barrow, 70° N, 1973. The solid line is a linear fit through the data, excluding the sites in southern Germany (HP, W and Z) (based upon Volz-Thomas, 1993).

Free troposphere concentrations are obtained from ozone sondes or measurements at high altitude sites such as the South Pole, Mauna Loa and the Zugspitze. London and Liu (1992), Logan (1997) and Miller *et al.* (in preparation) have analysed the global ozone sonde records. All studies of the ozone sonde records show, on the average, increases in free tropospheric ozone at northern mid-latitudes of around 10 % per decade since 1970.

Although the North American records are not of the same length and quality as the European records, it seems likely that the trend in free tropospheric ozone over North America has been smaller than that observed over Europe. The new studies also show evidence that the upward trend over Europe is smaller since about 1980 than it was before. Logan (1997) argued that the measurements made at Hohenpeissenberg, Lindenberg and possibly other European stations might be influenced by SO_2 leading to a positive bias in the trends. In polluted areas, local titration of ozone by NO_x can also influence measurements of ozone, but these effects should not be important outside of the atmospheric boundary layer. De Muer and De Backer (1994) have corrected the Uccle data set allowing for the known instrumental effects, including SO_2 interference. The ozone trend in the upper troposphere was only slightly reduced (10–15 % per decade, 1969–1991) and remained statistically significant. However, below 5 km the trend was indeed reduced from around +20 % per decade to +10 % per decade and became statistically insignificant.

Indirect information on ozone trends

Several studies in TOR provide information about the ozone level to be expected for zero anthropogenic NO_x and VOC emissions. When the relationship between organic C_3-C_8 nitrates and O_x (= O_3 + NO_2) was studied in the Schauinsland measurements and extrapolated to zero concentration of the organic nitrates, an ozone concentration of 20–30 ppb was found (Flocke *et al.*, 1994). This value represents the sum of the contributions from stratospheric intrusions and photochemical ozone produced from the oxidation of CO, CH_4, and C_2-C_8 in the presence of NO_x, since alkyl nitrates are a necessary by-product during photochemical ozone formation in the oxidation of the higher alkanes. The extrapolation is in good agreement with the ozone concentration at the altitude of Schauinsland (1200 m a.s.l.) derived from the available historic ozone measurements in Fig. 8.3.1.

Penkett *et al.* (1994) on a specific day in December found a negative correlation between O_3 and CO with a slope of approximately 1:6 in the concentration ratio O_3:CO. The O_3 concentration was extrapolated to approximately 15 ppb at zero CO, and this number was interpreted as the ozone present in the surface troposphere due to stratospheric intrusions, a value in good agreement with the available historic data at sea level in Fig. 8.3.1.

Trends in precursor concentrations

The existing measurements do not span a sufficient time period for an analysis of the long-term trends in NO_x and hydrocarbon concentrations. The longest continuous record of individual hydrocarbons has been measured at Birkenes near the south coast of Norway since the summer of 1987 as a part of TOR. Solberg *et al.* (1994) have shown that there is a statistically significant upward trend in the concentrations of acetylene and propane and butane and also in the sum of C_2-C_5 hydrocarbons. On the other hand, there is a downward trend in the concentration of alkenes (ethene and propene). By comparing the observed changes in nonmethane hydrocarbon concentrations with changes in the large scale transport patterns, Solberg *et al.*, (1994) concluded that climatological variability is an important factor for observed changes in concentrations in addition to changes in emissions. A clear picture of how emission reductions affect the atmospheric concentrations of fairly short-lived species is only possible with the help of longer measurement series and at more locations.

An analysis from the subproject ALPTRAC of the concentration of nitrate and lead in an ice core from a high-altitude Alpine glacier (Puxbaum and Wagenbach, 1994) showed that the concentrations of both species increased strongly after 1940, while having remained almost constant in the period before. Nitrate is the final product of NO_x oxidation and is removed from the atmosphere by heterogeneous processes such as rain-out. Therefore, changes in the concentration of nitrate in the ice should reflect the changes in the concentrations of the precursor NO_x. The onset of the increase in nitrate concentrations coincides with the start of the increase in tropospheric ozone over Europe which was found to be around 1940 to 1950 (Kley *et al.*, 1988; Staehlin *et al.*, 1994).

Furthermore, the simultaneous increase in the lead and nitrate concentrations indicate that automobile exhaust is a major source for the nitrate at Monte Rosa and, hence, a major cause for the ozone increase.

Trends in other photo-oxidant concentrations

Long-term records of other photo-oxidants, such as hydrogen peroxide or peroxyacetyl nitrate, are sparse. For peroxyacetyl nitrate, a continuous record from the Dutch air quality network shows an increase of almost a factor of 3 in the 1970s (Guicheret, 1988) when O_x concentrations were slightly increasing and ozone concentrations decreasing (see above). The increase in PAN, which is formed from peroxy radicals and NO_2 whereas ozone is formed from the reaction of peroxy radicals with NO, suggests that the potential for the formation of photo-oxidants was still increasing in the 1970s in the heavily polluted areas of Europe. In the 1980s the concentration of PAN at Delft has stabilised, and the same is seen in the 1990s at the TOR site Kollumerwaard.

The atmospheric measurements of hydrogen peroxide made in the US and in Europe (many of the latter were made in the framework of TOR) are not of a sufficient length to allow a trend assessment, and the concentration of H_2O_2 is also highly variable in space and with time. Sigg and Neftel (1991) used a record derived from Greenland ice-cores to argue in favour of an increase in atmospheric hydrogen peroxide concentrations, but the integrity of such a reactive and light sensitive species in the firn layer, *e.g.* before the ice is formed, needs to be established.

b.   Climatology of ozone over Europe

The TOR sites were located to show the differences between the climatology of ozone and precursors within Europe compared to adjacent areas. The measurements show that there is excess ozone in the boundary layer over Europe in the summer, while there is a wintertime ozone reduction.

This is seen when the measurements at the Arctic site in Ny Ålesund on Spitzbergen at almost 79° N are compared with measurements taken 2000 km further south at the mainland station Birkenes not far from the Norwegian south coast, and with measurements at Mace Head on the coast of western Ireland and Porspoder at the west coast of France. The annual cycles of the daily hourly ozone maximum for four years (1988–1991) for the Ny Ålesund site and the Birkenes site are shown in Fig. 8.3.4. In Ny Ålesund, there is very little annual variability. There is a small May minimum and an overall decline in concentration in summer with an annual minimum in concentration in July. The scatter in the daily maximum ozone concentrations in May reflects the occurrence of surface ozone depletion in April–May at many Arctic coastal sites (Barrie *et al.*, 1988; Solberg *et al.*, 1996). At Birkenes there is a November minimum of less than 30 ppb, and a May–June maximum of 50 ppb. The European influence is seen at Birkenes but not in Ny Ålesund (Hov and Stordal, 1992). The amplitude in the annual variation in ozone is about 10 ppb at Mace Head (Simmonds, 1993).

Ozone (ppb) Zeppelinfjellet 1988-1991

Enkeltmålinger og 30 dagers glidende middel
Max = 49 1704 Min = 18 2105
Middel (s1, s2, s3, s4, år) : 34 37 31 35 34

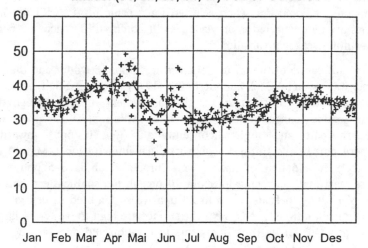

Jan Feb Mar Apr Mai Jun Jul Aug Sep Oct Nov Des

Ozone (ppb) Birkenes 1988-1991

Enkeltmålinger og 30 dagers glidende middel
Max = 61 805 Min = 21 1210
Middel (s1, s2, s3, s4, år) : 29 45 41 29 37

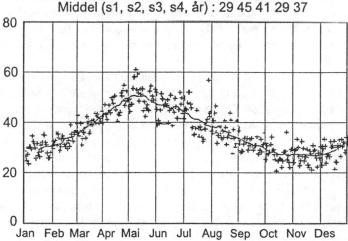

Jan Feb Mar Apr Mai Jun Jul Aug Sep Oct Nov Des

Fig. 8.3.4:   Daily maximum of hourly ozone for 1988-1991 at the Zeppelin Mountain in
Ny Ålesund (in ppb), upper part and at Birkenes, lower part. The full line is the
30-day running mean. The average concentrations for January, February and March
(JFM), AMJ, JJA, OND and the year are 34, 37, 31, 35 and 34 ppb, respectively, at the
Ny Ålesund site, at Birkenes the same averages are 29, 45, 41, 29 and 37 ppb,
respectively (Hov and Stordal, 1993).

The seasonal variation and mean concentrations are comparable at Mace Head and in Ny Ålesund when exception is taken for the cases with boundary layer ozone loss in the Arctic spring. The ozone concentration at Mace Head has a maximum in April for unpolluted air as identified by CFC11 concentrations below the average value and there is an annual minimum in July. The concentrations at Porspoder when selected for oceanic origin with the help of trajectories are quite similar as at Mace Head (monthly difference < 5 ppb) and exhibit a similar seasonal variation.

Further quantification of the ozone deficit over Europe in winter and the surplus in summer was given by Beck and Grennfelt (1994) who found that, based on the 1989 measurements, in summer there is a gradient in the average diurnal maximum ozone concentration with lower values (30–40 ppb) in the northwestern part and higher concentrations (60–70 ppb) towards the southeastern part of the European network. In winter (October–March) on the average there is a deficit of ozone over Europe which is 0–5 ppb near the northwest coast and increasing to about 10 ppb in the southeastern part of the network. Very large deficits of up to 20 ppb were identified in central Europe where the concentrations of NO_x near the surface are high in winter. A large part of this apparent deficit is thus likely due to titration and does not necessarily represent a deficit in O_x.

Beck and Grennfelt (1994) identified four sites (Mace Head in Ireland, Svanvik and Jergul both in northern Norway, and Strath Vaich in the UK) which together were denoted as reference sites, where the diurnal variation in ozone concentration was very small and the summertime and wintertime ozone concentration averages were very similar. According to these sites, the concentration of ozone in surface air at the western boundary of Europe in westerly air flows is about 31 or 32 ppb. On a seasonal basis, boundary layer processes may add 30–40 % in summer and subtract about 10 % in winter, while in photo-oxidant episodes hourly values of 100 ppb or more can be reached as a result of chemical formation within the boundary layer. The ozone concentration in the atmospheric boundary layer is determined by exchange of ozone between the boundary layer and the free troposphere, dry removal at the surface and *in-situ* chemical production and loss. During photochemical episodes in the atmospheric boundary layer, the ozone concentration there often exceeds the lower free tropospheric concentration so that there is a net flux of ozone upwards out of the boundary layer, but on average most of the time the main source of ozone in the boundary layer is a flux downwards from the free troposphere.

c.   Distribution of precursors and their role for the observed photo-oxidant distributions over Europe

The hydrocarbon measurements made in TOR were analysed by one task group for spatial and seasonal distribution. From a kinetic analysis of the summer/winter ratios of the alkanes, it was shown that the decrease of hydrocarbons in spring is not solely due to reaction with hydroxyl radicals because of the increasing UV radiation, but is to a large part determined by atmospheric transport processes. In order to make the data set useful for modelling studies, the data from each site were separated into polluted sectors, where the concentrations are governed by local influences, and sectors which are more representative of a larger region. The separation was made either by trajectory analysis or on the basis of local information such as the local wind field or NO_y concentrations. The huge data set collected in TOR can also be used to provide information on emission inventories. However, this requires the exact knowledge of the meteorological terms, *e.g.* wind field, height of the mixed layer and turbulence, on a larger scale and can thus only be done with the help of detailed model calculations. The study in TOR pointed to a number of potential deficiencies in European emission inventories.

It has been noted that there is a local, natural contribution to the concentrations of alkenes at some TOR measuring sites. This has been observed *e.g.* at Schauinsland (Klemp *et al.*, 1993) and at Birkenes (Solberg *et al.*, 1996). The measurements at Schauinsland have lead to the conclusion that biogenic emissions of olefins such as ethene, propene, 1-butene and 1-hexene may have a considerable impact on the total VOC reactivity in air that is not directly influenced by anthropogenic emissions (Klemp *et al.*, 1993). The contribution of the biogenic hydrocarbons to ozone formation has not yet been quantified.

d.   Process oriented studies

A number of studies were conducted in TOR using experimental data and models in order to gain insight into the specific processes that control the ozone budget at a given location.

Catalytic efficiency of NO_x

An important question in the quantitative understanding of the tropospheric ozone budget is that of the efficiency of NO_x as a catalyst in photochemical ozone formation, *i.e.* how many ozone molecules are formed for each molecule of NO_x emitted? The efficiency depends in a non-linear fashion on a large number of factors such as the VOC to NO_x ratio and their concentrations, the UV radiation flux, pressure, temperature and humidity and, last not least, the ozone concentration itself. It may therefore vary considerably with latitude, time of year, altitude and precursor concentrations. The question was investigated by Liu *et al.* (1987) and Hov (1989) with a chemical box model and it was found that the efficiency should be around 10 (molecules of ozone formed for each NO_x molecule emitted) in polluted air masses and was postulated to increase to around 100 for very clean conditions. The dependence on the initial ozone concentration was not investigated. Most important, Liu *et al.*, postulated that the efficiency could be somewhat higher in winter because of the longer lifetime of NO_x in their model.

Volz-Thomas *et al.* (1993) investigated the same question based upon experimental data from Schauinsland using the correlation between O_x (defined as: $[O_x] = [O_3] + [NO_2]$) and the oxidation products of NO_x (defined as: $[NO_z] = [NO_y] - [NO_x]$), with the latter serving as a proxy for the NO_x that has been destroyed before the air mass reaches the measuring site. The data in Fig. 8.3.5 suggest that in summer, approximately five O_3 molecules are formed per NO_x molecule which is oxidised in the polluted air masses that reach Schauinsland from the city of Freiburg. The negative relationship between O_x and NO_z which is sometimes observed in winter, on the other hand, suggests that under these conditions destruction of NO_x occurs on the expense of O_3 due to NO_3 related chemistry (Volz-Thomas et al., 1993).

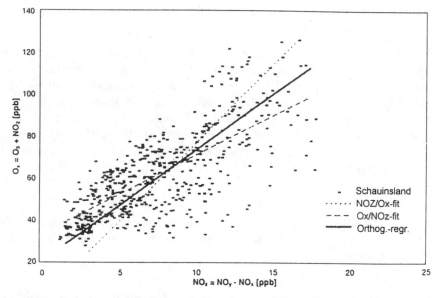

Fig. 8.3.5:   The observed relationship between the oxidant concentration ($O_x = O_3 + NO_2$) and oxidised NO_x ($NO_z = NO_y - NO_x$) at the TOR station Schauinsland for the wind direction from Freiburg (northwest) for summer (Volz-Thomas et al., 1993). Freiburg is located approximately 10 km away. In summer, approximately five O_3 molecules are formed per NO_x molecule oxidised to NO_z, while in winter there is a negative relationship observed between O_x and NO_z which supports that destruction of NO_x occurs on the expense of O_x. The sum of all species derived from NO_x is defined as NO_y, while NO_x is the active form of NO_y which serves to produce ozone through the oxidation of NO by peroxy radicals followed by photolysis of NO_2. The oxidation products of NO_x are defined as NO_x (= $NO_y - NO_x$), with HNO_3, NO_3^-, organic nitrates and N_2O_5 as the main components. The amount of NO_z is a measure of the fraction of NO_x that is no longer in the active form to produce ozone.

The experimental approach, which was followed independently by Trainer et al. (1993) and was adopted in several other studies, is not unambiguous, however: The products of photochemical NO_x oxidation are mainly PAN and HNO_3. While, under practical considerations, formation of HNO_3 is a permanent loss for NO_x in the ABL, PAN can decompose during transport and recycle the NO_x , thus leading to additional ozone formation. On the other hand, losses of HNO_3 by deposition during transport are much more effective than losses of O_x, thus leading to an overestimation of the catalytic efficiency of NO_x.

The experimental results from TOR are in good agreement with an analysis by Derwent and Davies (1994), who calculated on the basis of observed behaviour of ozone in the rural ozone network over the UK that about six ozone molecules were formed for each NO_x molecule oxidised in the summer, while in the winter there is close to a 1–1 relationship between O_3 and NO_x loss, indicating that the reaction $NO + O_3 \rightarrow NO_2 + O_2$ and then further oxidation of NO_2 to NO_z can account for most of the ozone chemistry in the winter months. The calculations suggested that the efficiency increased to about 20 ozone molecules produced per NO_x molecule oxidised for initial NO_x concentrations in the sub ppb range.

Correlation of O_3 and CO

The slope of the linear regression of O_3 versus CO can also be used to estimate the amount of ozone formed from the precursors emitted into the air masses together with CO, provided that the emission ratio between CO and the other precursors is known. At the Zugspitze, a positive correlation is observed in summer, whereas in winter the slope is often negative indicating a loss of O_3 (Scheel *et al.*, 1993; Sladkovic *et al.*, 1994). The results from Schauinsland demonstrate the importance of using O_x (the sum of O_3 and NO_2) instead of O_3 when such an analysis is made in polluted air: The usual positive correlation is observed during summer, with a slope around 0.3 O_3/CO. The observed slope is larger during periods of fair weather and increases with photochemical age of the air masses. This behaviour is expected since CO is consumed by chemistry after the initial CO production from short-lived VOC ceases, whereas ozone is still being produced or maintained by mixing with air from aloft. In winter, a strong negative correlation is found between O_3 and CO in polluted air masses at Schauinsland. However, a large fraction of the observed negative slope is due to the titration of O_3 to NO_2 by NO emissions, a process which does not constitute a permanent loss.

Radical yield from VOC

From simultaneous measurements of organic nitrates, ozone, NO_x and VOC made at Schauinsland in the plume of Freiburg, it was shown that the fraction of the smaller organic nitrates (C_3 and C_4) was larger than what is expected from laboratory data. The oxidation of VOC in these air masses leads, on average, to about four peroxy radicals for each hydrocarbon molecule oxidised by OH, if possible contributions from biogenic VOC are neglected. As a consequence, more ozone is produced in a shorter time than is predicted by photochemical models. Preliminary results from laboratory studies seem to confirm the hypothesis that the large fraction of small organic nitrates is a consequence of rapid decomposition of the peroxy radicals from the oxidation of larger hydrocarbons.

The findings are supported by a budget analysis based on the measured decay of anthropogenic hydrocarbons and NO_x between Freiburg and Schauinsland. This experiment was conducted in close collaboration with sub project TRACT. The average OH concentration over the transport time of about 3 h derived from the hydrocarbon decay was $5–8 \times 10^6$ molecules/cm^3 around noon time despite the presence of high NO_x concentrations (up to 70 ppb). The sink of OH by reaction with NO_2 could only be balanced if radical amplification was postulated in consistence with the results from organic nitrates (Kramp *et al.*, 1994). There remains the strong possibility that the additional peroxy and OH radicals come from the decomposition of biogenic VOC, *e.g.* terpenes, which would then require smaller amplification factors for the anthropogenic VOC (Kramp and Volz-Thomas, 1997). However, the decomposition of the biogenic VOC would then have to produce $< C_4$ alkylperoxy radicals and HO_2 at the rate required to explain the findings from the organic nitrates. In any case, the large OH concentrations lead to a faster removal of NO_x than is currently assumed in photochemical models, which means that the regime where ozone formation is controlled by the concentration of NO_x is quickly reached.

Local ozone formation rate

The local gross photochemical O_x formation rate (P_{O_3}) was determined at several TOR sites from the photostationary state of NO_x and O_3 (PSS). The results are summarised in chapter 4. At all sites, P_{O_3} was found to increase about linearly with the product of the NO_x concentration and the photolysis frequency of NO_2, the former being an indicator of local precursor concentration and the latter a measure of the UV radiation flux. Fairly large production rates of up to 50 ppb/h were found at Rörvik and Schauinsland when high NO_x levels occurred on sunny days during noon time. At cleaner sites, such as Izaña, NO_x concentrations exceed 2 ppb only occasionally due to local emissions and P_{O_3} usually remains below 10 ppb per hour. Of course, even these production rates must not be interpreted as being representative of the free troposphere around the Canary islands, but is largely driven by precursors from the island itself, which are advected to the site during daytime by thermal upslope. Free tropospheric NO_2 levels, as observed at night, are around or below 0.1 ppb, translating into a gross production rate of 5 ppb per day in these air masses (Schultz *et al.*, 1996).

When normalised to the primary controlling factors, *i.e.* UV radiation and precursor concentrations, similar ozone formation rates are found at all TOR sites, where appropriate data for PSS analysis were available. Quite similar results were derived from PSS for a rural site in North America (Trainer *et al.*, 1991).

The values obtained for P_{O3} from PSS were compared to those derived from the concentrations of NO and of peroxy radicals (HO_2 and its organic homologues, RO_2), that were made at the Schauinsland and Izaña TOR sites using the technique of matrix isolation and ESR spectroscopy. At Izaña, the formation rates derived from PSS and from the measured RO_2 and NO concentrations are in excellent agreement, whereas substantial disagreement exists at Schauinsland. There, the ozone formation rate as derived from the radical measurements and NO is about a factor 3 smaller than what is predicted from PSS. According to the radical measurements by MIESR, P_{O3} never exceeds 10 ppb/h. Further work is required in order to understand the causes for this disagreement.

Influence of clouds on the balance of ozone

In clouds there is a separation of HO_2 and NO because HO_2 is very soluble and enters the aqueous phase while NO is not water soluble. The reaction between HO_2 and NO is therefore not efficient in clouds. On the contrary, HO_2 in the aqueous phase leads to ozone destruction. The estimated effect of cloud chemistry is that the chemical O_3 production rate in the lower troposphere where most clouds are found, is reduced by 30–40 %, while O_3 destruction reactions are enhanced by up to a factor of 2 (Lelieveld and Crutzen, 1990). However, the resulting O_3 reduction in the lower troposphere also lowers the O_3 dry deposition flux. As a consequence, the global amount of O_3 in the troposphere may be reduced by 10–30 % due to cloud chemistry, when compared to a cloud-free atmosphere ((Jonson and Isaksen, 1993; Dentener *et al.*, 1993).

Convection versus chemistry

Calculations were carried out with a coupled 3-D numerical weather prediction and chemical transport model for Europe and the North Atlantic (Flatøy *et al.*, 1995). In central Europe the chemistry is calculated to be the dominant term on sunny summer days in the change in the concentration of ozone in the atmospheric boundary layer with a net generation of more than 10 ppb/h on some days, which is in reasonable agreement with the results derived from the measurements (see above). At locations like southern Scandinavia which are far downwind of the main source regions, photochemical episodes with high ozone concentrations can often be observed, but the local chemical generation of ozone there is usually low because the precursors are depleted, although relatively large production rates are sometimes observed at Rörvik based on PSS analysis. The removal terms are mainly horizontal advection, convection and dry deposition. Convection is the dominant loss term when it occurs (up to 7 ppb/h).

e. Contribution of exchange between the free troposphere and the boundary layer for average and peak levels of ozone and other photo-oxidants

Ozone and precursors have longer lifetimes in the free troposphere than in the atmospheric boundary layer. The vertical redistribution of ozone and its precursors between the boundary layer and higher altitudes thus has a strong influence on the ozone distribution in the troposphere. Upward flow in convective systems serves to bring ozone and precursors from the continental atmospheric boundary layer into the free troposphere, where their lifetimes are long enough to be transported over large distances. This upward flux is, however, balanced by downward mesoscale flow, where ozone and odd-nitrogen species are brought from the free troposphere into the planetary boundary layer where their lifetime is shorter Lelieveld and Crutzen, 1994). The exchange between the atmospheric boundary layer and free troposphere is an episodic process and is therefore difficult to include explicitly in global models. The importance of convection for the understanding of free tropospheric ozone has been demonstrated in TOR from the analysis of vertical soundings with a 3-D model (Flatøy *et al.*, 1995).

A model of the exchange processes between the atmospheric boundary layer and the free troposphere by convection was developed within TOR (Beck *et al.*, 1994). The calculations are based on ozone sonde measurements at Uccle in Belgium. The calculated flux is highly episodic, which is the nature of convective events, and maximises in summer. If the net upwards convective flux for Uccle is assumed to be representative for continental northwest Europe, it corresponds to an upward flux of 1.0 Gmol/d, or about 5 % of the downward flux of ozone from the stratosphere to the troposphere over the northern hemisphere. Convection is an important process for vertical transport over continents in the summer, but vertical transport is also significant in large scale rising or sinking cells, in frontal zones and over mountainous regions. The vertical transport of precursors, notably NO_x and possibly PAN, out of the boundary layer is another process which needs to be accounted for since this intensifies the *in-situ* tropospheric ozone chemistry there.

f.   Influence of export from North America on European ozone concentrations and free tropospheric ozone

The impact of North American emissions on the global ozone budget was estimated by Parrish *et al.* (1993), who observed a strong correlation between ozone and CO with a consistent slope $O_3/CO = 0.3$ at several island sites in eastern Canada. Jacob *et al.* (1993) used a 3-D model and the same measurements to estimate that pollution from North America contributes 30 Tg of O_3 to the northern hemisphere in summer of which 15 Tg is due to direct export and 15 Tg is due to export of NO_x leading to O_3 production in the free troposphere. This anthropogenic source of O_3 is about one third of the estimated cross-tropopause transport of O_3 in the northern hemisphere in summer. Since North America accounts for about 30 % of man-made NO_x emissions in the northern hemisphere, it can be concluded that in the northern hemisphere anthropogenic sources contribute an amount comparable to the flux from the stratosphere to tropospheric ozone.

Chemical processes are slow in the upper free troposphere and fast in the marine boundary layer, while the transport is fast in the upper free troposphere and slow in the atmospheric boundary layer. Evidence for the influence of export from North America on ozone levels in the free troposphere over Europe and the North Atlantic is not seen in the data from the surface sites Mace Head and Porspoder (see above). The low ozone concentrations observed at these sites in marine air gives a clear indication for the efficient removal of ozone (and NO_x) in the marine boundary layer.

Fig. 8.3.6 compares the seasonal cycles of ozone obtained at the high altitude TOR stations Izaña , Tenerife, in the southern North Atlantic and the Zugspitze in southern Germany with those at Mauna Loa in the Pacific. During winter, the ozone concentrations are around 40 ppb and are quite similar at all three stations. All stations show a similar increase until a maximum of 55 ppb is reached at Mauna Loa in April. While concentrations at Mauna Loa, which is furthest away from populated areas, decrease during summer, concentrations at Izaña and

Zugspitze continue to rise and remain at the high level until June and August, respectively.

The vertical soundings from Jülich also show much lower concentrations of ozone to be present in the free troposphere in winter (around 40–50 ppb) than in spring and summer (50–70 ppb; Smit *et al.*, 1993). Because of their location, both Jülich and the Zugspitze could already be influenced by European emissions.

At the Zugspitze, this conclusion is supported by the positive correlation between ozone and CO in summer and negative in winter. However, at Izaña high ozone concentrations persist until August in air masses that arrive from the northern part of the North Atlantic (Schmitt *et al.*, 1993). Since Izaña is located at the western boundary of Europe and because most of the air masses arrive from north westerly directions, the higher ozone concentrations in summer are at least partially the result of export of ozone from the North American continent into the free troposphere over the Atlantic and Europe.

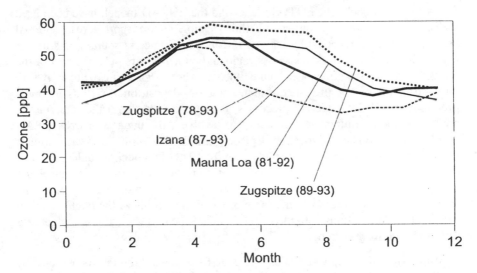

Fig. 8.3.6:   Average seasonal cycle of ozone observed at the TOR station Izaña and compared with those from Zugspitze on the European continent and Mauna Loa, Hawaii (Otmans and Levy, 1994; Sladkovic *et al.*, 1994; Schmitt, 1994). All three stations show an increase in early spring. Ozone in the Pacific at Mauna Loa decreases during summer, however, while relatively high concentrations persist until August at the Zugspitze. The concentrations at Izaña fall in between, which is a result of the combination of low concentrations that are advected from the southern parts of the North Atlantic or the Saharian Desert and high concentrations that are advected from the northern part of the hemisphere (Schmitt, 1994). The highest concentrations of up to 80 ppb occur frequently together with increases in methane and other trace gases that are indicative of anthropogenic pollution, such as CO, hydrocarbons, and PAN. This is most evident in spring, when temperatures are cold enough that PAN can survive long range transport to Tenerife. It is unclear at present how much ozone levels at Izaña are influenced by stratospheric intrusions, however.

The lower ozone concentrations observed in winter in the free troposphere over Europe strongly suggest that the catalytic efficiency of NO_x for ozone formation in the free troposphere is much lower during that time of the year. Possible reasons are destruction of NO_x and ozone through night time chemistry and cloud processes and a faster exchange between the free troposphere and the atmospheric boundary layer than was assumed in the early model calculations.

g.   Fluxes across regional boundaries and export from Europe

The export of ozone and precursors from Europe should be of similar magnitude as that from North America. However, it occurs mainly at the eastern boundary, where measurements are not available. The measurements at Mace Head, which have been analysed in TOR for export of ozone from Europe (Derwent *et al.*, 1994), cannot be considered representative because only a small fraction of ozone and its precursors from European sources leaves Europe in westerly directions. In addition, Mace Head is located at sea level and cannot provide information on transport in the free troposphere.

Two Eulerian models, the LOTOS model and the EURAD model, have been used in TOR task group 4 to investigate the question on transport across regional boundaries and to identify the regions in Europe that act as sources or sinks of ozone. Since horizontal fluxes across regional boundaries are only one part of the budget equation for atmospheric constituents, where also vertical fluxes due to large-scale motion, small-scale turbulence and dry deposition as well as chemical production and loss terms, processes due to clouds, and the amount of anthropogenic and biogenic emissions are involved, the budgets of ozone, PAN and NO_x were considered together. Episodic simulations for a one week summer-smog episode in 1990 were performed with the EURAD model. It could be shown that the transport of ozone from the free troposphere towards the atmospheric boundary layer is a major source of ozone for the near-surface layers of the troposphere. This source is in the same order of magnitude as the photochemical production. Calculations with the LOTOS model focused on seasonal budgets for different regions in Europe.

A comparison of the episodic calculations by EURAD and the results obtained by LOTOS on the basis of a two-month simulation leads to similar relative amounts of ozone transported out of Europe. For the altitude range 0–2 km, losses due to net transport are about 20 % of the ozone that is produced photochemically. For those regions in Europe (central Europe for EURAD, western Europe for LOTOS) with high precursor emissions, the net transport is about 40 % of the photochemical production of ozone. The calculations with the EURAD model show that large amounts of ozone are transported into the lower troposphere (altitudes below 2 km). The contribution of this term is about 60 %–70 % of the photochemical production. This may be typical for summer-smog conditions characterised by high-pressure systems controlling the dynamic features of the troposphere. Further investigations of the role of large-scale subsidence for ozone in the ABL are needed. The general role of large-scale motions for the ozone budget is not clear yet: at the end of the summer-smog episode calculated with

EURAD a front passes over central Europe leading to losses of ozone within the ABL due to upward motions. The role of clouds is also thought to play an important role for the vertical distribution of ozone and its precursors as shown by Flatøy *et al.* (1995).

h.   Influence of transport from the stratosphere on tropospheric ozone

The contribution of stratosphere-troposphere exchange (STE) to the tropospheric ozone budget, although much smaller than the terms of *in-situ* photochemical ozone production or destruction, is comparable to the net photochemical ozone source in the troposphere as recent model calculations indicate and is hence an important issue for the TOR project. Although STE has now been studied since four decades, uncertainties still subsist concerning the estimation of the average cross tropopause ozone flux and its seasonal variation and the importance of other processes besides tropopause folds is still uncertain. Two distinct algorithms to detect the presence of tropopause folds from ozone and meteorological radio soundings have been developed within one of the TOR task groups (KMI/CNRS and University of Cologne) and have been applied to several sites of the TOR network. A third algorithm, based on the connection of foldings to frontogenetic processes and using objective analysis, gives fold statistics on a global scale. Episodes of stratospheric air injection are also detected from ozone and ^{7}Be measurements at mountain sites. The latter two studies were contributed by the EUMAC project to the task group.

A principal result of the work is that folds occur much more frequently than previously assumed. Over Europe, more folds are observed over Uccle than the OHP, which is consistent with a more frequent occurrence of the polar jet at the more northerly site Uccle. A best estimate for the northern hemispheric ozone flux due to tropopause folds alone of 6.0 (3.7–8.5) $\times$ 10^{10} mol cm^{-2} s^{-1} is derived, comparing favourably with GCM simulations or tracer studies ($O_3/^{90}Sr$, $O_3/NO_y/N_2O$).

8.3.4 Final remarks

The TOR subproject has made significant contributions to the knowledge of tropospheric ozone. The subproject has had a significant influence on environmental policy, as documented in the EUROTRAC Application Project Report (Borrell *et al.*, 1996), both within the EU, UNECE, WMO and UNEP (IPCC).

a.   What was achieved?

The main achievement in TOR was the development of suitable instrumentation for high-quality measurements of ozone precursors (NO$_x$, NO$_y$, VOC), intermediates (carbonyl compounds, RO$_2$) and photolysis rates and the implementation of quality assurance procedures for these measurements, the establishment of a large high-quality network and the accumulation of a significant data base has a lot of the potential data still missing. A large success was the exchange of ideas that lead to a much better understanding of the complexity of the ozone problem, *i.e.* the complex role of transport and chemistry. New scientific strategies were developed in TOR on how to tackle the ozone problem from the experimental side. Examples are the concept of O_x for studying chemical ozone formation although used before by Dutch specialists, the O_x/NO_z relation for studying of the catalytic efficiency of NO_x, the use of alkyl nitrates as a measure of peroxy radicals and the role of individual hydrocarbons in ozone formation and the use of the H_2O/O_3 correlation for the ozone budget in the upper troposphere.

A hierarchy of models was developed and/or applied in TOR for the analysis of the data. Three dimensional transport models with coupled chemistry modules were developed in TOR and have been used to study important transport processes, such as convection, and for budget studies. Additional 3-D modelling related to TOR issues, in particular transboundary fluxes, was made in the subproject EUMAC, with which a close link was established and which used the TOR data for model evaluation.

The data gathered in TOR have been analysed by the individual PIs and in the six task group reports in order to provide information on the distribution of photo-oxidants and precursors in the boundary layer and in the free troposphere over Europe and in the air masses that are transported over the North Atlantic. At a few sites, *i.e.* where measurements had been started before the implementation of TOR, long enough records exist to allow a trend analysis. The high quality data sets collected at some TOR sites and the data collected in dedicated additional field experiments (some of the latter were conducted in close collaboration with other subprojects, *e.g.* TRACT and ACE, or in collaboration with EU projects, *e.g.* TOASTE, OCTA, BOA, HANSA, and national programmes) provide detailed mechanistic information on the chemical and physical processes that control the budget of ozone and its precursors in the polluted boundary layer and in the free troposphere. The vertical soundings were analysed for distribution and seasonal variation of ozone in the free troposphere over Europe and provide insight into processes such as the exchange between the boundary layer and the free troposphere and between the troposphere and the stratosphere.

b.　What was not achieved?

It was not possible to obtain rigorous answers on all the questions that were asked when the subproject was formulated. This reflects the usual problem in experimental sciences, namely that, after some insight has been gained, the problem identifies itself as being of much larger complexity than what has been anticipated when detailed information was not available. The particular problem with a secondary pollutant like ozone is the inseparable influence of transport and chemistry on the concentrations that are observed at a given location. The chemical life time of ozone varies from minutes in urban areas (due to titration by NO) and a few days in the clean boundary layer (due to photolysis followed by formation of OH radicals) to several months in the free troposphere (where photochemical transformation is rather slow because of the low water vapour concentration). The net chemical balance of O_3 depends in a non linear fashion on the concentrations of NO_x, VOCs, H_2O and O_3 itself and on the UV radiation flux and is thus closely coupled with the atmospheric life cycles of other trace gases. At continental surface sites, dry deposition plays an important role in the net balance of ozone, in addition to advection and vertical exchange. For these reasons it is impossible to obtain quantitative answers on the complete budget of ozone over Europe from measurements alone, even at a large number of sites with very sophisticated equipment. It is the combined influence of the chemistry along the transport path and the atmospheric transport processes that control the observed concentrations at a given time and location. Hence, answers about the chemical processes can only be given after the - in most cases dominant - influence of transport has been quantified. For a large region, such as Europe, this can only be achieved with the help of a fully coupled model of chemistry and transport or possibly a hierarchy of coupled models for the different regimes. The development of such models was, and indeed still is, the major task of atmospheric science within and outside of EUROTRAC. TOR has provided significant input into a better understanding of the chemistry and the role of transport for the budget of ozone over Europe. It has identified important processes that are now being implemented into numerical models and it has found ways how to investigate the local photochemical balance and how to obtain quantitative answers on important chemical aspects from the measurements without having to use numerical models for the interpretation. The TOR data base will continue to serve as a source for model validation and improvement as well as for scientific analyses related to photo-oxidants over Europe.

Another problem was associated with the nature of TOR being implemented as a EUROTRAC subproject. It has proven extremely difficult to implement the network of high-quality sites in sufficiently short time and at the same time develop all the data quality and management procedures and the heuristic models for data interpretation. Reasons for delays were the inherent bottom-up structure of the subproject and, in particular, the funding situation. Since the individual contributors had to seek funding from their own national governments, no control was given to the steering group and, hence, influence on the location and the measuring programme of individual site could only be sought through scientific argumentation in the TOR community and the start of the measurements was in many cases delayed because of national funding problems. While the scientific argumentation within TOR certainly stimulated the work in the project as a whole, it was not possible to enforce the implementation of all priority 1 measurements with the pre-specified high quality equipment at all TOR stations and in the short time that would have been required. The integration of scientific groups from European countries where regional air quality was not a big theme before EUROTRAC, was certainly a challenge that has lead to a great improvement of the situation of atmospheric science in Europe as a whole.

The participants in TOR also underestimated the amount of labour needed to set up a site and make such a large suite of research type measurements with high standards for data quality and complicated instruments at monitoring pace. As a result, most groups were completely exhausted by just producing the data and doing the quality control. Only very few groups had the resources to conduct the full suite of priority 1 measurements over several years and to exploit their data with statistical methods and heuristic models at the same time. The high scientific value of the observations made in the network and the necessary validation procedures resulted in a time lapse between the actual measurements, submission to the database and, finally, use of the data by other TOR and/or non-TOR participants. Model development was also much slower than anticipated, because of the growing necessity for more complexity and new approaches that went along with the growing understanding of the system. Therefore, only a limited amount of measurements have really been analysed with 3-D models so far.

8.3.5 Further information

The TOR contribution is shown in context in chapter 2 of this volume. A set of papers were also published in a special issue of *J. Atmos. Chem.* (Kley *et al.*, 1997), and a chapter on tropospheric ozone processes by A. Volz-Thomas and B. Ridley in the 1994 assessment of stratospheric ozone depletion (WMO Report, 1995).

A complete account of the subproject can be found in volume 6 of this series (Hov, 1997) which contains the following detailed reports from the TOR task groups and also contributions from the individual principal investigators, the titles of which are given in the next section.

An Overview of Tropospheric Ozone Research,
Øystein Hov, Dieter Kley, Andreas Volz-Thomas, Jeannette Beck,
Peringe Grennfelt and Stuart A. Penkett

Task Group 1   **Spatial and Temporal Variability of Tropospheric Ozone over Europe**

H.E. Scheel, G. Ancellet, H. Areskoug, J. Beck, J. Bösenberg, D. De Muer,
A.L. Dutot, A.H. Egelov, P. Esser, A. Etienne, Z. Ferenczi, H. Geiß,
G. Grabbe, K. Granby, B. Gomiscek, L. Haszpra, N. Kezele, L. Klasinc,
T. Laurila, A. Lindskog, J. Mowrer, T. Nielsen, P. Perros, M. Roemer,
R. Schmitt, P. Simmonds, R. Sladkovic, H. Smit, S. Solberg, G. Toupance,
C. Varotsos and L. de Waal

Task Group 2a   **The Emission and Distribution of Ozone Precursors over Europe**.

A. Lindskog, S. Solberg, M.l Roemer, D. Klemp, R. Sladkovic, H. Boudries,
A. Dutot, R. Burgess, H. Hakola, T. Laurila, R. Schmitt, H. Areskoug,
R. Romero, L. Haszpra, J. Mowrer, N. Schmidbauer and P. Esser

Task Group 2b   **Photochemical Ozone Production Rates at Different TOR Sites**

A. Volz-Thomas, D. Mihelcic, H.-W. Pätz, M. Schultz, B. Gomiszec,
A. Lindskog, J. Mowrer, P. Oyola, K. Hanson, R. Schmitt, T. Nielson,
A. Eggelov, F. Stordal and M. Vosbeck

Task Group 3a   **Exchange of Ozone Between the Atmospheric Boundary Layer and the Free Troposphere**

J.P. Beck, N. Asimakopoulos, V. Bazhanov, H.J. Bock, G. Chronopoulos,
D.De Muer, A. Ebel, F. Flatøy, H. Hass, P. van Haver, Ø. Hov, H.J. Jakobs,
E.J.J. Kirchner, H. Kunz, M. Memmesheimer, W.A.J. van Pul, P. Speth,
T. Trickl and C. Varotsos

Task Group 3b   **Stratosphere-Troposphere Exchange – Regional and Global Tropopause Folding Occurrence**

M. Beekmann, G. Ancellet, S. Blonsky, D. De Muer, A. Ebel, H. Elbern,
J. Hendricks, J. Kowol, C. Mancier, R. Sladkovic, H.G.J. Smit, P. Speth,
T. Trickl and Ph. Van Haver

Task Group 4   **Ozone and its Precursors in Europe: Photochemical Production and Transport across Regional Boundaries**

M. Memmesheimer, H.J. Bock, A. Ebel and M. Roemer

8.3.6 TOR: Steering group and principal investigators

Steering group

Dieter Kley (Coordinator)	Jülich
Jeannette Beck	Bilthoven
Peringe Grennfelt	Göteborg
Ivar S.A. Isaksen	Oslo
Stuart A. Penkett	Norwich

Principal investigators (*) and their contributions to the TOR final report (Hov, 1997) are listed below.

Ozone Monitoring and Measurements

G. Ancellet* and M. Beekmann,
   Ozone Observations in the Free Troposphere: Results of the TOR Station 25

A.L. Dutot, P. Colin, A. Etienne, H. Boudries, G. Toupance*, P. Perros and
M. Maillet,
   Tropospheric Ozone and Precursors at the Porspoder Station, France

L. Klasinc*, V. Butkovic, T. Cvitaš, N. Kezele, I. Lisac and J. Lovric,
   Ozone measurements in Zagreb and on Mount Medvednica in Croatia

C C. Varotsos*, P. Kalabokas and G. Chronopoulos,
   Characteristics of the Tropospheric Ozone Content above Athens, Greece

H. Puxbaum, B. Gomiscek* and K. Radunsky,
Measurements of Trace Constituents at the High Alpine Background Station
Sonnblick

J. Slemr* and W. Junkermann,
Measurements of Low Molecular Weight Carbonyl Compounds

T. Laurila*, H. Hakola, H. Lättilä and T. Koskinen,
Photochemical Observations at Utö -a Moderately Polluted Site with a
Pronounced Seasonal Cycle

T. Nielsen*, K. Granby, A.H. Egeløv, P. Hummelshøj and H. Skov,
Atmospheric Nitrogen Compounds, Photochemical Oxidants and Products

B.D. Belan*, V.V. Zuev, V.E. Zuev, V.E. Meleshkin, T.M. Rasskazchikova,
Atmospheric Monitoring at the TOR Station in Tomsk

Vertical Profiles and Transport of Ozone

Jens Bösenberg*, G. Grabbe, V. Matthias and T. Schaberl,
Distribution and Vertical Transport of Ozone in the Lower Troposphere
determined by LIDAR

D. De Muer*, Ph. Van Haver and H. De Backer,
Vertical Profiles of Ozone and Meteorological Parameters at Uccle, Belgium

T. Trickl*,
Vertical Soundings of Tropospheric Ozone with the IFU UV Lidar

R. Zander*, Ph. Demoulin, E. Mahieu, G. Roland, L. Delbouille and C. Servais,
Total Vertical Column Abundances of Atmospheric Gases Derived from IR
Remote Solar Observations made at the Jungfraujoch Station

H.G.J. Smit, D. Kley* and W. Sträter,
Vertical Distribution of Ozone and Water Vapour over Jülich, and The
Evaluation of the ECC Ozone Sondes under Quasi Flight Conditions

Modelling Studies

M. Memmesheimer*, A. Ebel, H.J. Bock, H. Elbern, H. Hass, H.J. Jakobs,
G. Piekorz,
EURAD in TOR: Simulation and Analysis of a Photosmog Episode

J.P. Beck*, E.C. Kirchner, W.A.J. van Pul, D. De Muer, P. Grennfelt*,
P.J.H. Builtjes*, M.G.M. Roemer, R. Bosman, P. Esser, M. Vosbeek and
W. Ruijgrok,
Continental Ozone Issues; Monitoring of Trace Gases, Data Analysis and
Modelling of Ozone over Europe

I.S.A. Isaksen*, J. E. Jonson and T. Berntsen,
Model Studies of Ozone on Regional Scales in the Troposphere.

M.G.M. Roemer*, R. Bosman, T. Thijsse, P.J.H. Builtjes*, J.P. Beck,
M. Vosbeek and P. Esser,
Budget of Ozone and Precursors over Europe

8.3.7 References

Anfossi, D., Sandroni, S., Viarengo, S.; 1991, Tropospheric ozone in the nineteenth century: The Moncalieri series. *J. Geophys. Res.* **96D**, 17349–17352.

Attmannspacher, W., Hartmannsgruber, R., Lang, P.; 1984, Langzeittendenzen des Ozons der Atmosphäre aufgrund der 1967 begonnenen Ozonmeßreihen am Meteorologischen Observatorium Hohenpeißenberg, *Meteorol. Rdsch.* **37**, 193–199.

Barrie, L.A., J.W. Bottenheim, R.C. Schnell, P.J. Crutzen, R.A. Rasmussen, 1988, Ozone destruction and photochemical reactions at polar sunrise in the lower Arctic atmosphere, *Nature* **334**, 138–140.

Beck, J.P., Grennfelt, P.; 1994, Estimate of ozone production and destruction over northwestern Europe. *Atmos. Environ.* **28**, 129–140.

Beck, J.P., Pul, W.A.J., De Muer, D., De Backer, H.; 1994, Exchange of ozone between the atmospheric boundary layer and the free troposphere. *EUROTRAC annual report 1993, TOR*, EUROTRAC ISS, Garmisch-Partenkirchen, pp. 44–51.

Borrell, P., Builtjes, P., Grennfelt, P., Hov, Ø., (eds); 1997, *Photo-oxidants, Acidification and Tools: Policy Applications of EUROTRAC Results*, Springer, Heidelberg, 216 p.

Crutzen, P.J.; 1988, Tropospheric ozone: An overview. in: I.S.A.Isaksen (ed),*Tropospheric Ozone*, Reidel, pp. 3–32.

Cvitaš, T., Kley, D.; 1994, *The TOR Network*, EUROTRAC ISS, Garmisch-Partenkirchen.

De Muer, D., De Backer, H.,Haver, P.van; 1994, Trend analysis of 25 years of regular ozone soundings at Uccle (Belgium), in: P.M. Borrell, P. Borrell, T. Cvitaš, W. Seiler (eds), *Proc. EUROTRAC Symp. '94*, SPB Academic Publishing bv, The Hague, pp. 330–334.

De Saeger, E., Tsani-Bazaca, E.; 1993, EC intercomparison of VOC measurements. in: P.M. Borrell, P. Borrell, T. Cvitaš, W. Seiler (eds), *Proc. EUROTRAC Symp. '92*, SPB Academic Publishing bv, The Hague, p. 157.

Dentener, F. J., Lelieveld, J., Crutzen, P.J.; 1993, Heterogeneous reactions in clouds: consequences for the global budget of O_3. *Proc. CEC/EUROTRAC Symp.* Varese, Italy.

Derwent, R.G., Davies, T.J.; 1994, Modelling the impact of NO_x or hydrocarbon control on photochemical ozone in Europe. *Atmos. Environ.* **28**, 2039–2052.

Derwent, R.G., Simmonds, P.G., Collins, W.J.; 1994, Ozone and carbon monoxide measurements at a remote maritime location, Mace Head, Ireland from 1990–1992, *Atmos. Environ.* **28**, 2623–2637.

Feister, U., Warmbt, W.; 1987, Long-term measurements of surface ozone in the German Democratic Republic, *J. Atmos. Chem.* **5**, 1–21.

Flatøy, F., Hov, Ø., Smit, H.; 1995, 3-D model studies of vertical exchange processes of ozone in the troposphere over Europe. *J. Geophys. Res.* **100**, 11465–11481.

Flocke, F., Volz-Thomas, A.,Kley, D.; 1994, The use of alkyl nitrate measurements for the characterization of the ozone balance at TOR-Station No. 11, Schauinsland, in: P.M. Borrell, P. Borrell, T. Cvitaš, W. Seiler (eds), *Proc. EUROTRAC Symp. '94*, SPB Academic Publishing bv, The Hague, pp. 243–247.

Guicherit, R.; 1988, Ozone on an urban and regional scale - with special reference to the situation in the Netherlands, in: I.S.A. Isaksen (ed), *Tropospheric Ozone*, D. Reidel Publ., pp. 49–62.

Hahn, J.; 1996, *EUROTRAC Data Handbook*, EUROTRAC ISS, Garmisch-Partenkirchen, 67 p.

Hahn, J.; 1997, Intercomparison and Harmonisation of CO, CH4, and non-methane Hydrocarbon Measurements within TOR, in: Hov, Ø. (ed), Tropospheric Ozone Research, Springer Verlag, Heidelberg, pp. 341–350.

Hov, Ø.; 1989, Changes in Tropospheric Ozone: A Simple Model Experiment, in: R.D.Bojkov, P.Fabian (eds), *Ozone in the Troposphere*, Deepak Publ.

Hov, Ø., Stordal, F.; 1993, Measurements of ozone and precursors at Ny Aalesund on Svalbard and Birkenes on the south coast of Norway, ozone profiles at Bjørnøya, and interpretation of measured concentrations, 1992. *EUROTRAC annual report 1992, TOR*, EUROTRAC ISS Garmisch-Partenkirchen, pp. 175–183.

Hov, Ø. (ed); 1977, *Tropospheric Ozone Research*, Springer Verlag, Heildelberg, 499 pp.

Jacob, D., Logan, J.A., Gardener, G.M., Yevich, R.M., Spivakowsky, C.M., Wofsy, S.C.; 1993, Factors regulating ozone over the United States and its export to the global atmosphere. *J. Geophys. Res.* **98**, 14817–14826.

Jonson, J. E., Isaksen, I.S.A.; 1993, Tropospheric ozone chemistry: The impact of cloud chemistry, *J. Atmos. Chem.* **16**, 99–122.

Junkermann, W., U. Platt, A. Volz-Thomas; 1989, A photoelectric detector for the measurement of photolysis frequencies of ozone and other atmospheric molecules, *J. Atmos. Chem.* **8**, 203–227

Klemp, D., Flocke, F., Kramp, F., Pätz, W., Volz-Thomas, A., Kley, D., 1993, Indications for biogenic sources of light olefins in the vicinity of Schauinsland/Black Forest (TOR station no. 11). in: J. Slanina, G Angeletti, S. Beilke (eds), *Air Pollution Research Report 47*, Commission of the European Communities, Brussels, pp. 271–281.

Kley, D., Isaksen, I.S.A., Penkett, S.A.; 1986, Joint European experiment to study tropospheric ozone and related species TOR (Tropospheric ozone research). *A research proposal submitted to EUROTRAC on behalf of the TOR participating scientists by the TOR steering group*.

Kley, D., Volz, A., Mülheims, F., 1988, Ozone measurements in historic perspective. in: I.S.A. Isaksen (ed), *Tropospheric Ozone*, Reidel, pp. 63–72.

Kley, D., J. Beck, P. Grennfelt, Ø. Hov, S.A. Penkett; 1997, *J. Atmos. Chem.* **28**, 1–359.

Kramp, F., A. Volz-Thomas; 1997, On the Budget of OH Radicals and Ozone in an Urban Plume From the Decay of C_5–C_8 Hydrocarbons and NO_x, *J. Atmos. Chem.* **28**, 263–282.

Kramp, F., Buers, H.J., Flocke, F., Klemp, D., Kley, D., Pätz, H.W., Schmitz, T., Volz-Thomas, A.; 1994, Determination of OH concentrations from the decay of C_5–C_8 hydrocarbons between Freiburg and Schauinsland: Implications on the budgets of olefins, a contribution to subproject TOR, in: P.M. Borrell, P. Borrell, T. Cvitaš, W. Seiler (eds), *Proc. EUROTRAC Symp. '94*, SPB Academic Publishing bv, The Hague, pp. 373–378.

Lelieveld, J., Crutzen, P.J.; 1990, Influences of cloud photochemical processes on tropospheric ozone, *Nature* **343**, 227–233.

Lelieveld, J., Crutzen, P.J.; 1994, Role of deep cloud convection in the ozone budget of the troposphere, *Science* **264**, 1759–1761.

Liu, S. C., M. Trainer, F. C. Fehsenfeld, D. D. Parrish, E. J. Williams, D. W. Fahey, G. Hubler, P. C. Murphy; 1987, Ozone production in the rural troposphere and the implications for regional and global ozone distributions, *J. Geophys. Res.* **92**, 4191–4207.

Logan, J.A.; 1997, Trends in the vertical distribution of ozone: An analysis of ozonesonde data. *J. Geophys. Res.*, in press.

London, J., S. Liu; 1992, Long-term tropospheric and lower stratospheric ozone variations from ozonesondes observations, *J. Atmos. Terr. Phys.* **5**, 599–625.

Low, P.S., P.S. Kelly, T.D. Davies; 1992, Variations in surface ozone trends over Europe, *Geophys. Res. Lett.* **19**, 1117–1120.

Marenco, A., Philippe, N., Hervé, G.; 1994, Ozone measurements at Pic du Midi observatory. *EUROTRAC annual report 1993, Part 9 TOR*, EUROTRAC ISS, Garmisch-Partenkirchen, pp. 121–130.

McKenna, D.S., *et al.*; 1995, Oxidising Capacity of the Tropospheric Atmosphere: *Final Report to the European Commission on contract EVSV-CT91-0042*, EU.

Miller, A.J., Tiao, G.C., Reinsel, G.C., Wuebbles, D., Bishop, L., Kerr, J., Nagatani, R.M., DeLuisi, J.J., Comparisons of observed ozone trends in the stratosphere through examination of Umkehr and balloon ozonesonde data, in preparation.

Mowrer, J., Lindskog, A.; 1991, Automatic unattended sampling and analysis of background levels of C_2–C_5 hydrocarbons, *Atmos. Environ.* **25A**, 1971–1991.

Oltmans, S. J., Levy, H.II; 1994, Surface ozone measurements from a global network, *Atmos Environ.* **28**, 9–24.

Parrish, D.D., Holloway, J.S., Trainer, M., Murphy, P.C., Forbes, G.L., Fehsenfeld, F.C.; 1993, Export of North American ozone pollution to the North Atlantic Ocean, *Science* **259**, 1436–1439.

Penkett, S.A., Bandy, B.J., Burgess, R.A., Clemitshaw, K.C., Cardenas, L., Carpenter, L., 1994, Measurements of NO_x, NO_y, CO an ozone at Weybourne and Mace Head. *EUROTRAC annual report 1993, TOR*, EUROTRAC ISS, Garmisch-Partenkirchen, pp. 275–284.

Puxbaum, H., Wagenbach, D.; 1994, *EUROTRAC annual report 1993, ALPTRAC,* EUROTRAC ISS, Garmisch-Partenkirchen.

Sandroni, S., Anfossi, D., Viarengo, S.; 1992, Surface ozone levels at the end of the ninteenth Century in South America, *J. Geophys. Res.* **97**, 2535–2540.

Scheel, H.E., Sladkovic, R., Seiler, W., 1993, Ground-Based measurements of ozone and related precursors at 47°N, 11°E, in: P.M. Borrell, P. Borrell, T. Cvitaš, W. Seiler (eds), *Proc. EUROTRAC Symp. '92*, SPB Academic Publishing bv, The Hague, pp.104–108.

Schmitt, R., Ozone in the free troposphere over the North Atlantic: Production and long-range transport, *EUROTRAC Annual Report 1993, Part 9,* TOR, EUROTRAC ISS, Garmisch-Partenkirchen 1994, pp. 162–165.

Schmitt, R., P. Matuska, P. Carretero, L. Hanson, K. Thomas, Ozone in der freien Troposphäre: Produktion und großräumiger Transport, *Abschlußbericht zum Vorhaben 07 EU 764*, Bundes minister für Forschung und Technologie, Meteorologie Consult GmbH, Glashütten 1993, 71 p.

Schultz, M., D. Mihelcic, A. Volz-Thomas, R. Schmitt, Die Bedeutung von Stickoxiden für die Ozonbilanz in Reinluftgebieten - Untersuchung der Photochemie in Reinluft anhand von Spurengasmessungen auf Teneriffa, *Berichte des Forschungszentrum Jülich*, JÜL–3170, 1996.

Sigg, A., Neftel, A., *Nature* **351** (1991) 557.

Simmonds, P.G., 1993, Tropospheric ozone research and global atmospheric gases experiment, Mace Head, Ireland. *EUROTRAC Annual Report 1992, Part 9, TOR*, EUROTRAC ISS, Garmisch-Partenkirchen 1993, pp. 234–242.

Sladkovic, R., Scheel, H.E., Seiler, W., Ozone climatology of the mountain sites, Wank and Zugspitze, Transport and Transformation of Pollutants in the Troposphere, in: P.M. Borrell, P. Borrell, T. Cvitaš, W. Seiler (eds), *Proc. EUROTRAC Symp. '94*, SPB Academic Publishing bv, The Hague 1994, pp. 253–258.

Slemr, J., Dietrich, J. Sheumann, B., Koemp, P., Kern, M., Junkermann, W., Werle, P., Intercomparison of Formaldehyde measuring techniques, in: P.M. Borrell, P. Borrell, T. Cvitaš, W. Seiler (eds), *Proc. EUROTRAC Symp. '94*, SPB Academic Publishing bv, The Hague 1994, pp. 963–966.

Smit, H.G.J., D. Kley, H. Loup, W. Sträter, Distribution of Ozone and Water Vapor Obtained from Soundings over Jülich: Transport versus Chemistry, in: P.M. Borrell, P. Borrell, T. Cvitaš, W. Seiler (eds), *Proc. EUROTRAC Symp. '92*, SPB Academic Publishing bv, The Hague 1993, pp. 145–148.

Smit, H.G.J., W. Sträter, M. Helten, D. Kley, D. Ciupa, H.J. Claude, U. Köhler, B. Hoegger, G. Levrat, B. Johnson, S.J. Oltmans, J.B. Kerr, D.W. Tarasick, J. Davies, M. Shitamichi, S.K. Srivastav, C. Vialle, G. Velghe; 1997, JOSIE: The 1996 WMO International intercomparison of ozonesondes under quasi flight conditions in the environmental simulation chamber at Jülich, in: R. Bojkov, G. Visconti (eds), *Proc. XVIII Quadrennial Ozone Symp.* L'Aquila, Italy, in press.

Solberg, S., Schmidbauer, N., Semb, A., Stordal, F., Hov, Ø., Boundary layer ozone depletion in the Norwegian Arctic, *J. Atmos. Chem.* **23** (1996) 301–322.

Solberg, S., Stordal, F., Schmidbauer, N., Hov, Ø., 1Non-methane hydrocarbons (NMHC) at Birkenes in South Norway, 1988-1993. NILU Report 47/93, 1994.

Staehelin, J., Thudium, J., Buehler, R., Volz-Thomas, A., Graber, W., Trends in surface ozone concentrations at Arosa (Switzerland), *Atmos. Environ.* **28** (1994) 75–87.

Trainer, M., et al., Correlation of Ozone with NO_y in Photochemically Aged Air, *J. Geophys. Res.* **98** (1993) 2917–2925.

Trainer, M., M.P. Buhr, C.M. Curran, F.C. Fehsenfeld, E.Y. Hsie, S.C. Liu, R.B. Norton, D.D. Parrish, E.J. Williams, B.W. Gandrud, B.A. Ridley, J.D. Shetter, E.J. Allwine, H.H. Westberg, Observations and modelling of the reactive nitrogen photochemistry at a rural site, *J. Geophys. Res.* **96** (1991) 3045–3063.

Volz, A., Kley, D., Evaluation of the Montsouris series of ozone measurements made in the nineteenth century. *Nature* **332** (1988) 240–242.

Volz-Thomas, A., Flocke, F., Garthe, H.J., Geiß, H., Gilge, S., Heil, T., Kley, D., Klemp, D., Kramp, F., Mihelcic, D., Pätz, H.W., Schultz, M., Su, Y., in: P.M. Borrell, P. Borrell, T. Cvitaš, W. Seiler (eds), *Proc. EUROTRAC Symp. '92*, SPB Academic Publishing bv, The Hague 1993, pp. 98–103.

Volz-Thomas, A., Gilge, S., Heitlinger M., Mihelcic, D., Müsgen, P., Pätz, H.W., Schultz, M., Borrell, P., Borrell, P.M., Slemr, J., Behmann, T., Burrows, J.P., Weißenmayer, M., Arnold, T., Klüpfel, T., Perner, D., Cantrell, C.A., Shetter, R., Carpenter, L.J., Clemitshaw, K.C., Penkett, S.A., Peroxy Radical Intercomparison Exercise: A Joint TOR/OCTA Experiment at Schauinsland 1994, in: P.M. Borrell, P. Borrell, K. Kelly, T. Cvitaš, W. Seiler (eds), *Proc. EUROTRAC Symp. '96*, Computational Mechanics Publications, Southampton 1996, pp. 621–626.

Volz-Thomas, A., Trends in photo-oxidant concentrations, In: Photo-Oxidants: Precursors and Products, in: P.M. Borrell, P. Borrell, T. Cvitaš, W. Seiler (eds), *Proc. EUROTRAC Symp. '92*, SPB Academic Publishing bv, The Hague 1993, pp. 59–64.

Wege, K., Claude, H., Hartmannsgruber, R., 1989, Several results from the 20 years of ozone observations at Hohenpeissenberg, in: R.D. Bojkov, P. Fabian, A (eds), *Ozone in the Atmosphere*, Deepak publ.1989, pp. 109–112.

WMO report No. 104, *Report of the fourth WMO meeting of experts on the quality assurance/science activity centers (QA/SACs) of the global atmosphere watch.* Jointly held with the first meeting of the coordinating committees of IGAC-GLONET and IGAC-ACE at Garmisch-Partenkirchen, Germany, 13–17 March 1995, WMO TD.No. 689, 1995.

8.4 TRACT: Transport of Air Pollutants over Complex Terrain

Franz Fiedler[1] and Peter Borrell[2]

[1]Universität Karlsruhe, Forschungszentrum Karlsruhe,
Inst. für Meteorologie und Klimaforschung, D-76128 Karlsruhe, Germany
[2]P&PMB Consultants, Ehrwalder Straße 9, D-82467 Garmisch-Partenkirchen,
Germany

8.4.1 Introduction

Understanding the mass balance of the many chemical constituents either emitted to the atmosphere from anthropogenic or biogenic sources or formed within the atmosphere belongs to the most challenging problems in atmospheric sciences. Depending of the size of the volume for which a system with open boundaries is defined, different processes come into play determining the local distribution and temporal variation of the concentrations within the system.

Most of the European areas with large cities, dense population and heavy industrialisation have hilly and mountainous terrain shape. Therefore the fate of pollutants emitted to the atmosphere cannot be described and followed by simplified methods without taking into account the many effects of the mountainous terrain.

8.4.2 Aims of the subproject TRACT

It was the principal aim of the EUROTRAC Subproject TRACT (Transport of Air Pollutants Over Complex Terrain) to address the spectrum of processes which are relevant to describe the transport, the turbulent diffusion and in part also some chemical aspects within the lower atmosphere over mountainous terrain.

In the past detailed studies of transport and turbulent diffusion of air pollutants over complex terrain have mainly been made over smaller distances in the order of tens of kilometres. In almost all studies, simple configurations of the land surfaces were chosen to meet the conditions of horizontal homogeneity which underlie the theories of the model concepts used. In other studies, transport and diffusion processes could only be taken into account by using mean-flow trajectories. When the transport and the mixing of pollutants is considered over longer distances the influence of the horizontal inhomogeneities of the earth surface plays an important role since at each barrier the air is deflected, both vertically and horizontally. At the same time the level of turbulence and convective activity in the lower atmosphere changes dramatically. However the parameterisation assumptions presently used in numerical models incorporate these effects very poorly. They must be improved by detailed and accurate observations of the processes governing atmospheric transport over complex terrain. But without detailed analyses of the phenomena involved little progress in the parameterisation can be expected.

It was clear from the beginning of the project that an observational data base of a much broader scope would be needed to perform a detailed analysis of the processes involved.

First it would be necessary to take into account all the features which are connected to the daily cycle of the lower atmosphere, especially within the atmospheric boundary layer. This makes it necessary to observe the atmosphere in an area through which the air moves within a single day. Over this time scale neither the individual sources nor the source types of the chemical species can be considered to be constant. The chemical species also undergo complex transformations within such scales of time and space. Therefore, a realistic field study over typical distances of 300 km × 300 km cannot be undertaken without a reasonable balance between observing the meteorological conditions over the complex terrain and taking into account the chemical transformation of the species during the transport.

Particular attention was paid to the following points.

*   The effects of orography on atmospheric transport and exchange processes within the planetary boundary layer over complex terrain. In particular, turbulent dispersal, channelling and mountain induced wind systems were investigated in different scales ranging from industrialised regions to the transalpine scale (TRANSALP sub-group).

*   Handover of air pollutants from the atmospheric boundary layer to the free troposphere. This process is especially important for the long-range transport of air constituents and may contribute significantly to the transalpine exchange. Secondary flow systems along mountain sides are a particularly important mechanism for the handover processes.

*   To establish mass balances of pollutants (especially SO_2, NO and NO_2) in mesoscale regions including the respective emissions and deposition.

In the project plan the work was focussed from the beginning on preparing a large field campaign to obtain the observational database necessary for the detailed understanding of the processes involved.

Within TRACT, special emphasis was given to the interaction of field observations and mesoscale modelling. Therefore the observations were prepared in such a way that consistent data sets for the needs of model initialisation would be provided. The data would also provide a basis for model evaluation.

The subproject organised four field campaigns; three, which formed the project, TRANSALP, were tracer experiments to examine the channelling of flows through the high Alps. The other, the TRACT campaign itself was focussed on the meteorological and chemical problems associated with complex terrain, and held in the region comprising south-western Germany, south-eastern France and northwestern Switzerland. The next section deals with the TRANSALP campaigns. The remainder, which constitutes the majority of this report, is devoted to the TRACT campaign.

8.4.3 TRANSALP

The Alps constitute an appreciable topographical and meteorological break between northern and central Europe and the Mediterranean. The peaks, glaciers and upper valleys are sensitive ecosystems while many of the lower valleys are sources of appreciable pollution. An understanding of the air flow into the region and through and between the valleys is necessary to any measures intended to reduce the impact of pollutants in the Alps.

The project TRANSALP (Transport of tracer gases across the Alps) was a sub-group of TRACT set up to study transport of air masses through very complex terrain by means of tracer release experiments. Three possible processes for transalpine transport of trace substances were considered:

*   injection from the planetary boundary layer into the free troposphere;

*   channelling of air through the north to south directed valleys; for example the Reuss, Gotthard and Leventina valleys;

*   coupling of the air masses between adjacent valleys when the boundary layer height exceeds that of the intervening mountains.

The project consisted of field campaigns backed by modelling activities (Anfossi *et al.*, 1996; Ambrosetti *et al.*, 1998).

a.   TRANSALP Experimental campaigns

There were three field campaigns.

1989   To investigate the split in a northerly flow between the Leventina and Blenio valleys in Ticino in southern Switzerland

1990   To investigate the transport in a northerly flow from the Leventina valley across the Gotthard pass between southern Switzerland and Italy and the Swiss plateau

1991   To investigate the transport in a southerly flow between the Swiss plateau and southern Switzerland and Italy

In each case a chemically inert perfluorocarbon tracer, perfluoro-methylcyclohexane (C_7H_{14}) or perfluorodimethylcyclohexane (C_8H_{16}), was released, and its dispersion and transport tracked with at sampler stations situated in the valleys downwind from the release site. In the 1990 experiment there were some 40 sampling sites; in 1991 more than 60. Fig. 8.4.1 shows the distribution in 1990.

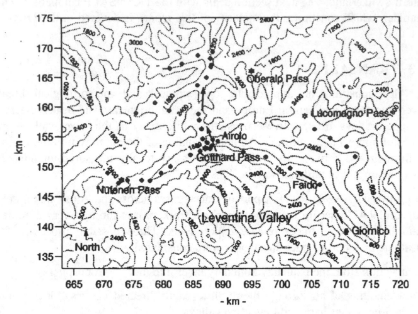

Fig. 8.4.1:   The area of the 1990 TRANSALP campaign, showing the tracer release points (squares) and sampling locations.

The samplers for the tracer were supplemented in 1990 and 1991 by a motor glider (METAIR) which flew pre-defined tracks along the valleys of interest to investigate the vertical distribution of the tracer. A variety of meteorological measurements were made, many with the help of the Swiss weather service (Ambrosetti *et al.*, 1998).

b.   *TRANSALP experimental results*

The 1989 TRANSALP campaign

The main tracer release in 1990 campaign was made during a general anti-cyclonic situation so that the flow is principally determined by daily south to north valley breezes, the "Inverna", that blow up from the Po valley to the central Alps. The release took place in the Leventina valley about 5 km south of the bifurcation between the Leventina and Bleniovalleys. The tracer was found in both valleys, and comparison between equivalent samplers indicated that about 20 % of the tracer found its way into the Blenio. More detailed analysis showed that the centre line of the plume to be along the south west side of the Levantino.

The results are consistent with a clear channelling of the flow, the plume being shifted towards the south western side due to differential heating of the valley sides. The plume enlarges more quickly than would be expected from experiments in flat terrain, presumably due to the effects of thermal eddies in the confined valley (Ambrosetti *et al.*, 1998).

The 1990 TRANSALP campaign

The 1990 release was again made during an anti-cyclonic situation under "Ivernone" wind conditions in which the normal southerly "Iverna" valley breezes are supplemented by strong southerly flows aloft. The release was made in the Leventina valley about 25 km south of the Gotthard pass. As Fig. 8.4.1 shows, the sampling stations were placed so as to detect flows across the Gotthard, Nufenen, Oberalp and Lucomagno passes. The arrival of the tracer was seen at most stations but the concentrations were smaller than hoped for because of wind speeds were lower than forecast. The motor glider detected the tracer in the surrounding valleys and very high tracer concentrations above the top of the Gotthard pass.

It is clear from the results that channelling of the tracer occurred across the Gotthard, Nufenen and Lucomagno passes. As tracer was also found high above the valley bottom as well as much to the north of the Gotthard pass, some transfer to the free troposphere also occurred. Some evidence for night time flow reversal was also obtained (Ambrosetti *et al.*, 1998).

The 1991 TRANSALP campaign

The weather conditions in 1991 were chosen to ensure that there would be a flow from north to south. The tracer was released on the eastern shore of Lake Lucerne about 45 km north of the Gotthard pass. Samplers were placed in valleys to the south of the release site and also in the valleys that run east and west. Three motor gliders were also employed to take samples aloft. The results obtained at some of the sampling sites across the Gotthard are shown in Fig. 8.4.2 (Nodop *et al.*, 1992).

From the results it was possible to identify two transport routes under north-south flow conditions. The speed of transport was nearly identical with the average valley wind speed. There is also transport in the west to east direction which indicates the importance of local thermal wind fields in determining the dispersion. The results at Airolo in Fig. 8.4.2 show an interesting effect: the tracer arrives in the valley about an hour before it does at the two more elevated sites suggesting that the flow is channelled to the bottom and that the valley "fills up" with tracer. The valley drains to the East. Flights in the South of the area on the following day show a general distribution of tracer up to 2500 m a.s.l. within the area confirming the trans-alpine transfer of pollutants (Nodop *et al.*, 1992, Ambrosetti *et al.*, 1998).

c.   **Modelling activities in TRANSALP**

AS the experiments were carried out in a limited area with non-reactive tracers, the models were used to produce the wind fields and to simulate the dispersion of the tracer. The diagnostic wind field model, CONDOR, was used to drive the Lagrangian dispersion model, ARCO, and the wind field model, MINERVE, was used with the dispersion model, SPRAY. The combinations produced satisfactory agreement both with each other and, considering the complexity of the terrain, with the experiments (Desiato et. al., 1998; Anfossi *et al.*, 1998). Some wind field simulations were also done with the RAMS model (Martilli and Graziani, 1998).

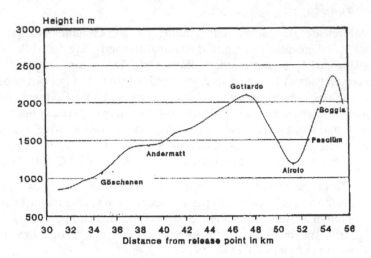

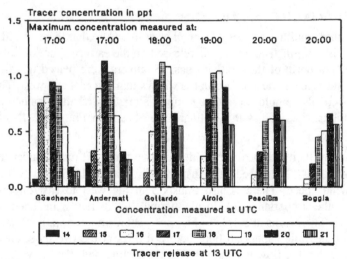

Fig. 8.4.2:   Tracer plume evolution across the St. Gotthard pass, as seen at 6 ground-level sites.

d.   Conclusions from TRANSALP

The campaigns provide direct evidence for transport of air pollutants through alpine valleys and also over the ridges. The initial dispersion is by channelling along the valley bottoms but there is evidence for transfer to the free troposphere as well. However as the processes are totally dependent on the overall and local weather situations for the time of day the results cannot be regarded as typical. Much more work will be required to estimate for example budgets for the transfer between alpine valleys or for transport into the Alps from the surrounding regions.

8.4.4 The TRACT field experiment

As a typical region for European conditions, the area of south-west Germany with some parts of eastern France and northern parts of Switzerland was chosen. At the western side the Vosges mountains form a natural boundary. The foothills of the Alps formed the southern boundary of the experimental area. Within this region, the mountains range up to heights of about 1500 m. It is intersected by the broad valley of the Rhine running from north to south and by the Danube valley running from south-west to north-east. There are also many smaller valleys as well. The width of the valleys allows different meteorological phenomena like channelling and flow partitioning to develop together with some other smaller scale processes which are important to the dispersion of air pollutants in mountainous terrain. These processes include mountain and valley winds and the initiation of convection at mountain tops which contribute to vertical mixing, a process often known as "mountain venting".

The area chosen had been used for previous studies such as the TULLA experiment (Fiedler *et al.*, 1991) and so a reasonably good data base for the meteorological conditions and anthropogenic emissions from industries and traffic was already available. Thus the TRACT measurements started with a better understanding of the area than would have been the case for a completely new area.

The selected area is shown in Fig. 8.4.3. There are always two contradictory influences when choosing an experimental site. First the area should be large enough to give the different air pollutants, stemming from either single intense sources or from industrial regions or from road traffic, time to mix completely and fill the whole boundary layer and to start chemical reaction. One would like to choose as large an area as possible, to provide enough time for the mixing and chemical reactions.

However with the inhomogeneity of the atmospheric conditions, created by the terrain shape, land use, the scattered locations of the emission sources by industry and the roads, it is necessary to make as many accurate observations of the atmospheric system as possible. Therefore, if the area is chosen is too large, the available instruments and manpower will not be sufficient to provide the required information base.

It is noticeable that in many field campaigns the paucity of measurements is then replaced by a broad spectrum of assumptions or speculation.

a.   The observation systems

Three types of observation system were used in the TRACT experiment:

1.   Continuous surface observations of meteorological and chemical variables, partly with stations in an operative meteorological network and partly with additional temporary stations;

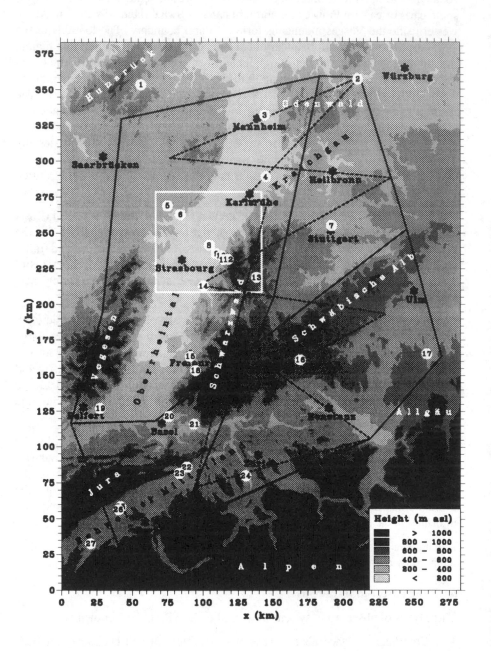

Fig. 8.4.3: The TRACT area comprises Baden-Württemberg in Germany, the northern part of Switzerland and the eastern part of France. The nested area, the locations of the vertical sounding stations and some of the flight tracks are shown.

2.   A reasonably dense network of vertical sounding stations, indicated in
     Fig. 8.4.3, for meteorological variables such as wind speed, wind direction,
     temperature, moisture and pressure. Measurements were made at frequent
     intervals in the intensive measurement periods;

3.   Ten aircraft mainly equipped with chemical instruments measuring
     concentrations of different species. Five of these flew along prescribed
     tracks, indicated by the black lines in Fig. 8.4.3; the remainder were
     employed in the nested area.

The field measurements were made between the 10th and 23rd September 1992;
the time schedule is shown in Fig. 8.4.4.

Before the direct field measurements were made a period (7th–9th September)
was devoted to quality control and quality assessment (section 8.4.4c). It was the
first time that a large effort had been devoted to instrument calibration and intercomparison.

Ground based measurements from the operational and temporary stations were
made for the whole measurement period. They provided the background
information for both meteorological and chemical variables.

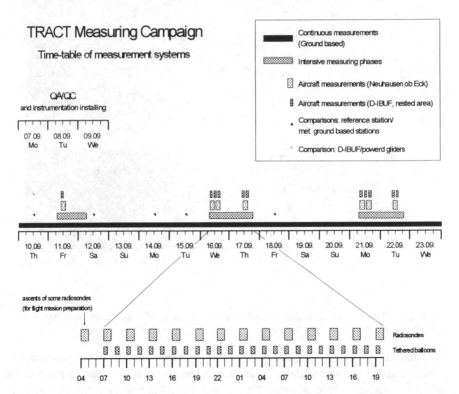

Fig. 8.4.4: Time table of the TRACT experiment (7th to 23rd September 1992) indicating
continuous measurements and embedded intensive measuring phases

Three intensive measuring phases were included within the field phase starting on 10th (24 h), 16th (36 h) and 21st (36 h) September. During these intensive periods a network of 11 radiosonde stations and other vertical soundings were activated to measure the atmospheric conditions at three hourly intervals (see Fig. 8.4.3). A list of stations is given in Larsen *et al.* (2000) and Zimmermann (1995).

In order to study the interactions of the atmosphere and the orography in more detail, an additional nested area was installed at a cross section of the Rhine valley (indicated by the square around Strasbourg in Fig. 8.4.3). Along this cross section line additional measurements with tethered balloons were made at half hourly intervals. The nested area, which contained 16 stations, is shown in greater detail in Fig. 8.4.5. A list of stations can be found in Larsen *et al.* (2000) and Zimmermann (1995).

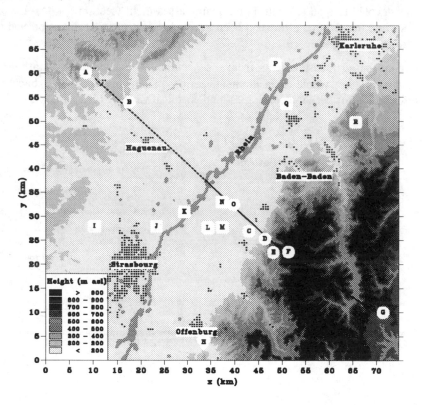

Fig 8.4.5:   The nested area across the Rhine valley north of Strasbourg with the northern area of the Black Forest. The locations of the measuring stations are shown.

Ten aircraft, which constituted the most expensive and most sophisticated measurement systems available to the campaign, were active during parts of the intensive periods. Five aircraft flew along the flight tracks within the main experimental area (see black lines in Fig. 8.4.3). One aircraft and four motor gliders operated within the nested area.

In order to minimise the sampling time and so to obtain as near a simultaneous measurement as possible, the main area was sub-divided into two sub-areas. From the aircraft measurements, the inflow of the air into and outflow of air from the area, together with the conditions inside the experimental area was determined. However it was also necessary to observe the top of the boundary layer and the interaction between the free atmosphere and the underlying boundary layer. Therefore, depending on the weather conditions, the aircraft flew along the tracks of the western and the eastern box, starting at Neuhausen ob Eck (Fig. 8.4.3 station 16). During synoptic flow conditions from the north-west or the south-east, the area was sub-divided into northern and southern boxes.

The vertical structure of the concentration distribution was obtained by one aircraft flying a curtain pattern between the lowest level ($\sim$ 150 m above ground) to a height a little above the boundary layer. A second aircraft measured the conditions in approximately the centre of the boundary layer. A further aircraft sampled the concentrations just above the boundary layer flying cruising along the track indicated in Fig. 8.4.3 by the dashed line.

A comprehensive description of the co-ordinated measurements during the TRACT experiment is given in the Field Phase Report of TRACT (Zimmermann, 1995). Full descriptions of the preparatory phase is given in the general TRACT Operational Plan (Fiedler, 1992), the TRACT Aircraft Operational Plan (Wachs *et al.*, 1992) and a Quality Assurance Project Plan for Aircraft Measurement Systems (Mohnen *et al.*, 1992).

The TRACT operation centre was located nearly in the middle of the experimental area at the Neuhausen ob Eck air field (station 16 in Fig. 8.4.3), from where all aircraft for the main investigation area operated. For the nested area along the cross-section through the Rhine valley a local operation centre was established at the airport at Offenburg. (Fig. 8.4.5, station H)

b.   Data bases

In such an extensive field activity as TRACT a variety of observational data are collected. In addition to having to compile and archive the data, it is necessary to inter-compare the data sets from different groups for completeness and accuracy. So five different data sub-centres were created to treat the various groups of data:

background data (from operational and temporal ground based networks)

vertical sounding data from radiosondes and tethered balloon systems

aircraft data (mainly chemical species)

emission data

turbulent flux and energy balance data

After compilation and processing all the data were collected together into the main TRACT data base. The complete data base is located at the Institute of Meteorology and Climate Research of the University of Karlsruhe. It is accessible via internet and may also be obtained on a CD-ROM (Zimmermann, 1999).

The data formed the starting point of the individual analyses of the studies from groups participating in TRACT. The data has also been used in various other subsequent air pollution studies, because it forms a useful comprehensive data source for evaluations of combined atmospheric and chemical model studies.

c.   Quality assurance and control

The results and conclusions from most field experiments are based on the comparison of measurements from a large number of ground stations and airborne platforms. It is essential, therefore, that the measurements meet the highest expectations with respect to their accuracy, precision, completeness, comparability and representativeness; experience shows that the necessary data quality can only be obtained through the implementation of a rigorous quality assurance programme. The project description (Fiedler, 1989) stated the quality assurance principles of the programme, somewhat before these became so fashionable, and the programme was fully implemented in the campaign.

Quality control measures were implemented specifically for (Zimmermann, 1995):

aircraft measurements (main area aircraft);

aircraft measurements (nested area aircraft);

radiosondes;

ground-based meteorological measurements; and

ground-based chemical measurements.

Aircraft measurements (main area aircraft)

A detailed quality assurance plan, setting data quality objectives for the principal quantities to be measured, was drawn up for the five aircraft operating in the main area (Kanter *et al.*, 1996) Table 8.4.1.

Table 8.4.1: Data quality objectives for the main area aircraft campaign.

Quantity	Data quality objective
temperature	± 0.5 K
dew point	± 1.0 K
O_3	± 20 % or 2 ppb (whichever is greater)
SO_2	± 30 % or 2 ppb (whichever is greater)
NO	± 30 % or 1 ppb (whichever is greater)
NO_2	± 20 % or 1 ppb (whichever is greater)
NOy	± 30 % or 2 ppb (whichever is greater)
H_2O_2	± 35 % or 0.2 ppb (whichever is greater)

A ground-based mobile calibration laboratory was established (Kanter *et al.*, 1996) to provide known flows of the calibrated gases to the aircraft. At the beginning of the campaign the analytical equipment on each aircraft was calibrated with samples from the mobile laboratory, an onerous task which for the five machines took nearly two days of round the clock working (Mohnen *et al.*, 1992).

An intercomparison was then carried out with formation flights of all five aircraft at three different altitudes to check the measurement system under actual flight conditions. At the end of each flight mission the instruments on pairs of aircraft were again inter-compared with "fly-bys".

Two aircraft were equipped to take grab samples of non-methane hydrocarbons. These were subsequently analysed by gas chromatography using for calibration samples obtained from the National Centre for Atmospheric Research (NCAR) in Colorado.

The ozone and NO_y measurements met the data quality objectives without restriction. The values found for sulphur dioxide and nitric oxide were generally close to or below the detection limits, and there were some difficulties with hydrogen peroxide. The results showed some surprising deviations in the "standard" measurements of altitude, temperature and relative humidity. (Kanter *et al.*, 1996)

Aircraft measurements (nested area aircraft)

The four motor gliders and the single normal aircraft operating in the nested area measured ozone together with a number of other standard parameters. An intercomparison was made with a two formation flight by the motor gliders. The normal aircraft could not fly sufficiently slowly and so flew the same track as the motor gliders several times during the flight period. The ozone measurements met the data quality objectives set for the aircraft in the main area. Once again some deviations were found in the "standard" parameters. (Koßmann *et al.*, 1994; Zimmermann, 1995).

Radiosondes

Radiosondes were used at five sites to obtain vertical profiles of the temperature, pressure, relative humidity and the horizontal wind speed and direction. Two types of sonde were used and a single intercomparison between the two types showed that the parameters were within the manufacturers' specifications (Zimmermann, 1995).

Ground-based meteorological measurements

An appreciation of the meteorology was essential to TRACT and so an effort was made to compare the instruments used at the six principal stations. A mobile station was set up to measure:

total radiation,	relative humidity,	solar radiation,
wind speed,	surface radiation temperature,	wind direction,
air temperature,	air pressure, and	wet bulb temperature.

It was then driven to the various sites, at each of which a three-hour intercomparison was made. The results, although pinpointing one or two problems, were satisfactory (Kalthoff and Kolle, 1993, Zimmermann, 1995)

Ground-based chemical measurements within the nested area

For the temporary stations within the nested area, a calibration laboratory was set up at Ehrlenhof by the Association pour la surveillence et l'étude de la pollution atmosphérique en Alsace. The various instruments were brought initially to the laboratory for comparison with the accepted standards and then were brought again after the campaign was finished. The results obtained for ozone and sulphur dioxide were excellent (with a few percent) but problems were experienced with the nitrogen oxides which, however, still appear to be within the data quality objectives for the aircraft in the main area (Heinrich *et al.*, 1994, Zimmermann, 1995).

Conclusions on quality assurance

Quality assurance is always a carefully defined compromise between the optimism of the investigators concerning their own measurements and the reality of making measurements under difficult field conditions. It allows one to set realistic goals and, when the results are analysed, to see which differences are truly significant and which lie within the noise. For TRACT, effective measures were taken which ensured that the data obtained was "adequate for the purpose for which it was intended" (Mohnen, 1996).

8.4.5 Emission inventory

In regional studies of air quality, the emissions of pollutants from anthropogenic sources are of vital importance, and a field study of the size of TRACT would be incomplete without a detailed analysis of the emissions within the area of investigation as a function of time and space. The TRACT area is a densely populated area and includes industrial centres such as Stuttgart, Mannheim-Ludwigshafen, Karlsruhe, Strasbourg, Freiburg and Basel. These areas are connected by roads forming important traffic routes running north to south along the Rhine valley or from west to east across the northern mountains of the Black Forest.

The emission inventory was constructed so as to provide a data set consistent with the chemical modules incorporated into atmospheric models both in temporal and spatial resolution and also from the variety of the chemical species included (Obermeier *et al.*, 1996; Berner *et al.*, 1996).

Therefore the emission inventory concentrated mainly on those species which are precursors for ozone and other photo-oxidants in the boundary layer, since the experiment was designed to be conducted under fair weather conditions. Nitrogen oxides, carbon monoxide, sulfur dioxide and non-methane hydrocarbons were determined at hourly intervals for the whole region on a 1 km by 1 km grid. Statistics on road traffic counts, housing density, numbers of inhabitants and industrial activity were used to calculate the emissions which were then interpolated to obtain the gridded data. Large point sources were identified with a special survey and analysed individually.

In all a detailed and reasonably reliable emission data set was obtained for the whole period from 9th to 22nd September, 1992 in hourly time steps with a spatial resolution of 1 km by 1 km.

In this way, the concept of an integrated study was developed which included, as a function of time and space, the calculation of the emissions inside the investigation area, the analysis of meteorological transport and mixing conditions and the development of the concentration fields during the day.

8.4.6 Modelling activities

Understanding observed phenomenological distributions of air pollutants on a regional scale is, in the current view, not possible without the use of well-developed models. It is recognised that the behaviour of the concentrations of different species is the result of the combined activity of atmospheric transport on the mean wind from the source to the receptor, of turbulent diffusion and of chemical transformation during transport. Additionally, solar radiation gives rise to photochemical reactions, and the source distributions and deposition play important roles. The conditions are still more complex if the area is bounded by complex terrain.

The measurements in TRACT were organised in such a way that all data bases required for the use of atmospheric models including the chemical transformation were available (Fiedler, 1993). As well as the atmospheric variables are needed for initialisation of the model, terrain and land use data were available for the TRACT area at high resolution.

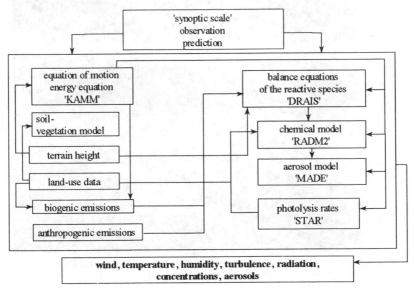

Fig 8.4.6:   The scheme of the atmospheric and chemical model system KAMM used for the TRACT project.

As the horizontal resolution must be as high as possible in order to avoid smearing effects of the emissions in grid boxes that are too large close to intensive point and line sources, a non-hydrostatic model system KAMM (Karlsruhe Atmospheric Mesoscale Model; Adrian and Fiedler, 1991; Fiedler *et al.*, 1999) was combined with the chemical code RADM2, modified for European conditions (Vogel *et al.*, 1999). The complete model system with its different sub-modules is shown in Fig. 8.4.6. Particular emphasis is given to the harmonised parameterisation of turbulence with respect to turbulent transport, momentum, heat and moisture and the turbulent transport of chemical species. The same inter-dependence is present for the interaction at the surface between the atmospheric variables and the evaporation and transpiration of vegetation and the deposition and biogenic emissions of various chemical species. Model simulations are virtually the only means of apportioning the relative contribution of the different processes to the observed concentration fields.

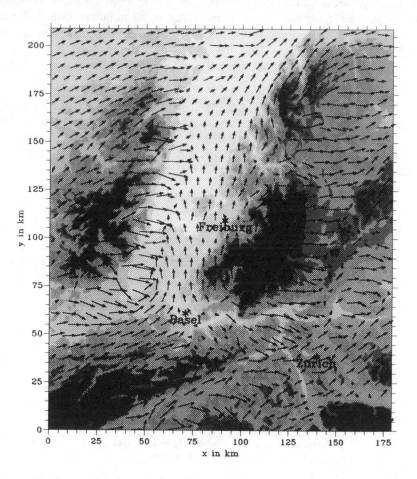

Fig. 8.4.7: Simulated wind field at 15 m above ground level for an ideal westerly synoptic flow.

Sensitivity studies of various sorts have been made in order to clarify the way in which complex terrain influences the transport, diffusion and even the chemical transformation processes. In Fig. 8.4.7 an interesting example is given for the general flow structure within the experimental area of TRACT and particularly in the Rhine valley. In the figure the flow at 15 m above ground (independent of the terrain height) is shown for an ideal westerly synoptic flow. It is seen that the topography deforms the flow field at lower heights dramatically. In the south-west there is an inflow through the Burgundy Gate. There the flow splits with the main branch flowing to the north about 150 km along the Rhine valley. A second branch passes along the southern end of the Black Forest and then flows to the north-east seeking a way between the Black Forest and the Swabian mountains.

There is a similar picture for the case of the TRACT intensive measuring period on 16th and 17th September, 1992. In Fig. 8.4.8 the modelled wind field for the 16th September at 12:00 CET is shown. In general the same structure is visible in comparison to the ideal case presented in Fig. 8.4.7, but details are much more complicated. This model result has been obtained by using the large scale synoptic conditions to drive the mesoscale model by splitting up the meteorological conditions into a basic state (synoptic large scale field) and deviations from the basic state which is the result of the mesoscale model.

The north-westerly flow around Metz, Nancy and Saarbrücken is consistent with the synoptic situation. At the Burgundy Gate, west of Basel, the inflow south of the Vosges appears again. At the southern edge of the Black Forest a strong flow develops following partly the Danube valley towards Ulm and partly crossing the Swabian mountains towards Stuttgart. How complex the flow is, which develops in mountainous terrain, is visible from the stagnating air between the Vosges and Black Forest, west of Freiburg and also in the northern Rhine valley around Mannheim. The arrows in the white insets show the observed wind measurements at the same time. Although many of the ground observations are influenced by buildings, trees, hedges or forest and so are not representative for a comparison of the calculated wind field within a grid box, it is astonishing how well the model shows up the details of the wind flow in the TRACT area.

Another interesting feature is the converging flow region on the lee side of the Black Forest, west of Stuttgart, where it is expected from the wind flow alone that the emissions from the Karlsruhe area and the Stuttgart area must overlap. It can also be concluded from the meteorological conditions that in areas with identical emission rates but differing in wind speed and direction, the observed concentrations and the ongoing chemical transformations will develop quite differently. Therefore the concentration pattern shows a large variability.

Due to pressure disturbances connected to mountain ranges smaller scale secondary flow systems often develop both in front of and, in a more pronounced way, on the lee side of mountains. These larger flow features are much more likely to produce vertical transport than the general smaller scale turbulence.

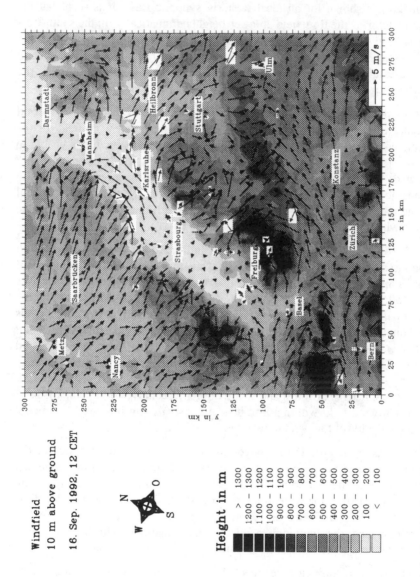

Windfield

10 m above ground

16. Sep. 1992, 12 CET

Height in m

> 1300
1200 – 1300
1100 – 1200
1000 – 1100
900 – 1000
800 – 900
700 – 800
600 – 700
500 – 600
400 – 500
300 – 400
200 – 300
100 – 200
< 100

Fig. 8.4.8:   Simulated wind field at 10 m a.g.l. at 12:00 CET for the intensive measuring phase of 16th–17th September.

Lakes, such as Lake Constance, act as disturbances in the surface, and combined with mountains around, are a strong influence on secondary flows. A model simulation for this area is shown in Fig. 8.4.9 where the potential temperature and the wind field is presented. These secondary flows are intensified, when the model resolution is improved.

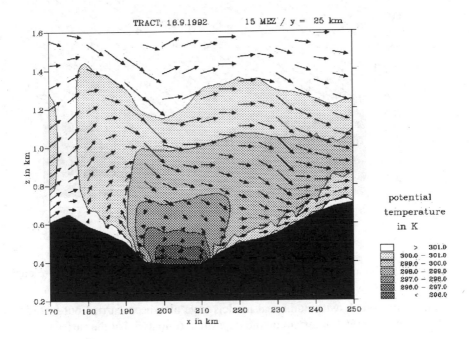

Fig. 8.4.9:   Simulated potential temperature and wind field for a vertical cross section at Lake Constance with an intensive secondary flow system.

Fig. 8.4.10 shows an example from a model run with a horizontal resolution of 250 m. At the bottom of smaller valleys, a counter flow develops as part of a standing eddy with a horizontal rotation axes. In complex terrain it is through these eddies that the emitted material is removed from lower levels of the atmosphere and handed over to higher levels. Mountain effects are not included in the usual parameterisations employed in numerical models which are theoretically only then valid for horizontally homogeneous terrain. It is clear that models which have a coarser grid size cannot resolve these features, and must create errors in the transfer due to vertical mixing. The hand-over due to these secondary eddies, connected to mountains, must be considered as an important mechanism for effective and rapid exchange, inside the boundary layer, of air close to ground with air at higher levels.

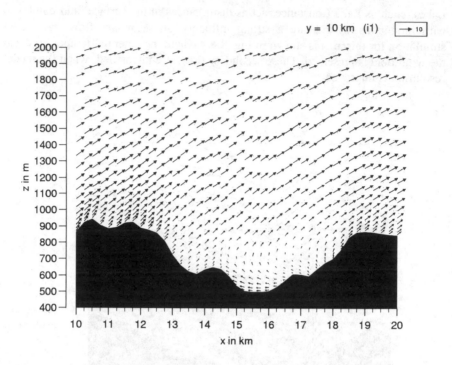

Fig. 8.4.10:   Simulated vertical cross section of the wind over the Murgtal with a resolution of 250 m.

Sensitivity studies with numerical models have also been carried out in order to determine the effect of terrain on the deposition of material at the surface (Uhrner, 1997). The results indicated that, during night hours, the deposition values calculated with coarser model resolution, without the secondary flow systems involved, may be a factor of two too low.

Similar studies have been undertaken to study the influence on biogenic emissions of temperature variations connected to mountainous terrain (Vogel *et al.*, 1995). For summer conditions with high temperatures, comparison of the emission rates were made for a particular terrain type and the corresponding mountainous terrain, keeping other conditions constant.

Since the surface temperatures at different heights in the mountains vary dramatically, compared with the temperatures over flat terrain, biogenic emission rates are larger when real topography is taken into account. As a consequence the contribution of biogenic emissions to ozone production is larger with real topography than with a flat terrain. Similar studies indicate that deposition rates are also a function of land use (Vogel *et al.*, 1991).

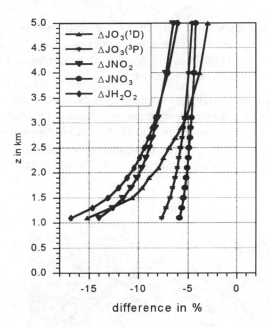

Fig. 8.4.11:   Simulated vertical profiles of the relative difference of the photolysis rate $\Delta J = (J_{Schauinsland} - J_{Offenburg})/J_{Offenburg}$ (17th September, 1992, 12:00 CET).

An unexpected effect of topography is the change of the photolysis rate in mountainous terrain (Ruggaber *et al.*, 1995; Vogel *et al.*, 1997). Fig. 8.4.11 shows that the photolysis rates at the mountain top are reduced by up to 15 % compared to the photolysis rates at the same height over flat terrain.

8.4.7 Observations and results in TRACT

In complex terrain the atmospheric conditions are variable so that observations should be made with a high spatial and temporal resolution. It was fortunate that the TRACT field campaign overlapped with REKLIP, the Regional Climate Project of the Rhine valley (Fiedler, 1995). REKLIP operated a surface network of 36 well-equipped micrometeorological stations for about five years. These were used together with the operational networks of the weather services, the environmental agencies and private operators, and also the TRACT temporary stations to provide a reasonably dense measurement system. Thus a much clearer insight in the wind flow conditions was obtained than would have been possible in earlier times.

Fig. 8.4.12 summarises, in the form of wind roses, the day-time surface wind observations for the whole TRACT area for the whole period. The general structure seen in the observations is similar to that found from the model results. It can be seen that the mountains have an influence up to 150 km downstream. In particular, the flow around the Black Forest is more pronounced than that around the northern and southern side of the Vosges.

It is known from climatological data that the area around Basel is often not foggy when the neighbouring areas are completely covered with fog. The divergence of flow in the horizontal directions leads to a downward compensating flow trough in which low level clouds and fog are dispersed.

It can also be seen that the Kraichgau between Heilbronn in the east and Karlsruhe and Mannheim in the west is an important pathway for the transport of pollutants. Another path leads from Basel towards Ulm in the southern area along the valley of the Danube. During the day, when strong convection strongly links the flow close to the ground with that of the higher altitudes. The influence of the terrain is

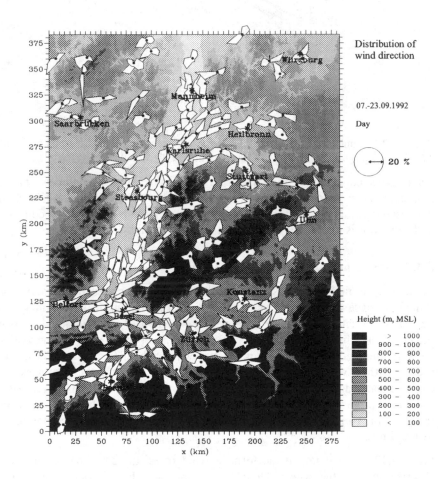

Fig. 8.4.12: Frequency distribution of the wind direction from ground station for the time period of the TRACT experiment during the day.

seen even at heights above the summits of the mountains. The flow in the lower levels follows the valley axes and convection then forces the flow at higher altitudes into the same direction.

The vertical structure of the atmospheric boundary layer is of vital importance to the concentrations of the air pollutants. Under neutral conditions the height of the boundary layer is determined by the strength of the geostrophic wind speed (and of the Coriolis force). Under non-neutral conditions the temperature structure plays a dominant role.

One important problem that is not yet fully understood is the height development of the boundary layer over complex terrain during the course of the day. Under stable conditions the height of the boundary layer usually follows the terrain height provided the mountain slopes are not too steep and the summits do not break through the boundary layer.

Much emphasis was given to resolving the boundary layer structure during the intensive measuring phases by launching radiosondes at 11 stations simultaneously, by using tethered balloons and also by flying some of the aircraft in a curtain-like pattern. An example of the vertical structure of the boundary layer is given in Fig. 8.4.13 for the morning hours and Fig. 8.4.14 for the conditions at noon. Vertical profiles of potential temperature, specific humidity and of the horizontal wind velocity are plotted along cross-sections through the named stations.

In the morning a shallow stable layer, of approximately 200 m in height, follows the terrain; the bulk of the boundary layer with less stable conditions is influenced by the residual layer from the convection of the previous day. The pronounced boundary between the friction layer and the free troposphere is marked by a strong inversion at around 1200 m a.s.l. The temperature structure is strongly correlated with specific humidity, the increase in potential temperature within the capping inversion being strongly correlated with a decrease in the specific humidity. It may be concluded that all non- or weakly-reactive chemical species would show a similar vertical structure. Within the boundary layer there are always marked changes of wind speed and direction. During the day, when a well-mixed boundary layer has developed potential temperature and specific humidity are nearly constant with height. The transition to the free troposphere is indicated by a strong inversion layer with the more stable conditions in the free troposphere above it. From the results it can be clearly seen that the boundary layer height decreases from north to south. At Wertheim it reaches 1200 m whereas at Freiburg and Lörrach it is only 900 m. The complexity of conditions in the real atmosphere is illustrated by this example: in the mesoscale there are interactions between synoptic conditions, the vertical structure of the boundary layer and the influence of the terrain.

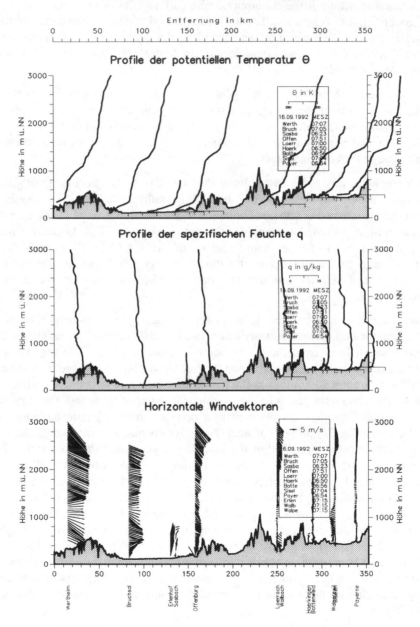

Fig. 8.4.13:   Measured vertical profiles of potential temperature, specific humidity and horizontal wind vector along the north-south cross section from Wertheim to Payerne on 16th September, 1992 at 7:00 CEST.

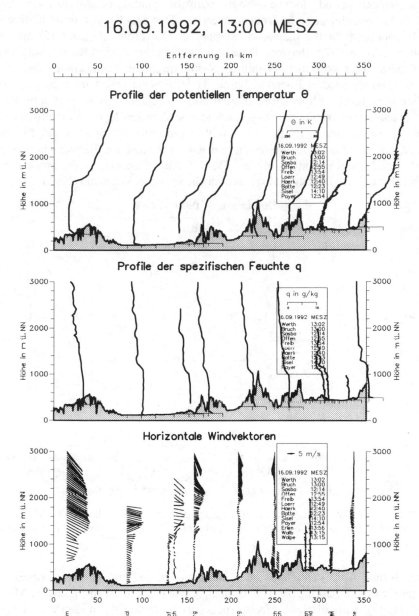

Fig. 8.4.14: Measured vertical profiles of potential temperature, specific humidity and horizontal wind vector along the north-south cross section from Wertheim to Payerne on 16th September at 13:00 CEST. The location of the measurement station is always at the left end of the temperature scale.

The 16th and 17th September during the intensive measuring phase turned out to be the "golden period" for the analysis, with the synoptic conditions remaining quite stable and almost no horizontal advection. This may be seen from the three-hourly soundings at the radiosonde stations at Oberbronn (Fig. 8.4.15) and at Musbach (Fig. 8.4.16). Between these ascents, between 7:00 in the morning and 16:00 in the afternoon, there is almost no shift in potential temperature, specific humidity, wind speed and wind direction. The differences are largely confined to the boundary layer with its strong diurnal cycle. The boundary layer is converted from a stable layer in the morning hours to a well-mixed layer in the afternoon. However the maximum height of the well-mixed zone differs by about 150 m, being about 1000 m at Oberbronn but only 850 m at Musbach. Further analyses of the meteorological conditions during TRACT and of accompanying spatial and temporal variations of chemical species may be found in Koßmann (1998).

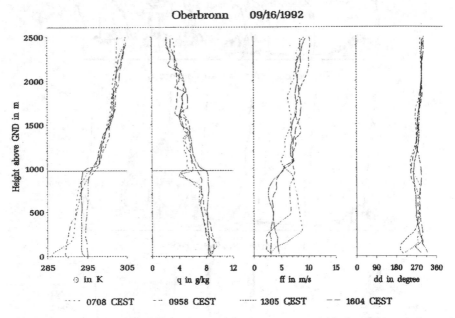

Fig. 8.4.15: Temporal evolution of potential temperature, specific humidity, wind speed and wind direction on 16th September at Oberbronn, 274 m a.s.l. The solid line indicates the boundary layer height at 13:00 CEST.

The influence of the changing vertical mixing of ozone, and its consequences for the daily variation, during TRACT campaign, has been analysed by Löffler-Mang et al. (1997), particularly for frontal passages or during the presence of valley wind systems. It was found that many dynamic and thermodynamic atmospheric processes are involved in the observed temporal changes of the concentrations. So the effects of these processes must first be separated before chemical production or removal may be identified in the observations.

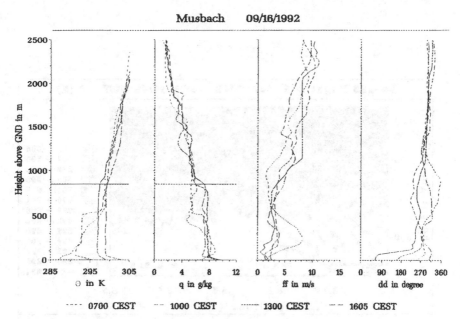

Fig. 8.4.16:   Temporal evolution of potential temperature, specific humidity, wind speed and wind direction on 16th September at Musbach, 694 m a.s.l. The solid line indicates the boundary layer height at 13:00 CEST.

It is this strong interaction between the dynamics and the chemistry which makes an integrative observation programme with an equally emphasis on the meteorological and chemical measurements essential to a proper understanding.

How important the spatial resolution is for understanding the transport and diffusion processes over complex terrain, is illustrated by the flight simulation in Fig. 8.4.17. The diagrams show profiles of the potential temperature (upper) and the specific humidity (lower). The boundary layer height is indicated by the dashed line. It can be clearly seen that, over the mountain plateau there is a well-mixed layer. With easterly winds the layer is transferred over the Rhine valley where it shifts outside the atmospheric boundary layer. This is a frequently occurring event and an important process for the efficient handover of material from the boundary layer to the free troposphere.

Fig. 8.4.18 shows the turbulence field for the same situation, and indicates another important aspect. Over the mountains the turbulence level is high and there is strong mixing associated with it in all directions. But as soon as the air moves over the Rhine valley the turbulence level is much reduced. Thus material transferred outside the boundary layer may then be transported much longer distances without strong vertical mixing and with little deposition to the surface. In this case, the boundary layer structure clearly follows the topographic structure, a phenomenon which is much more easily detected from the intensity of the turbulence than from the structure of mean quantities.

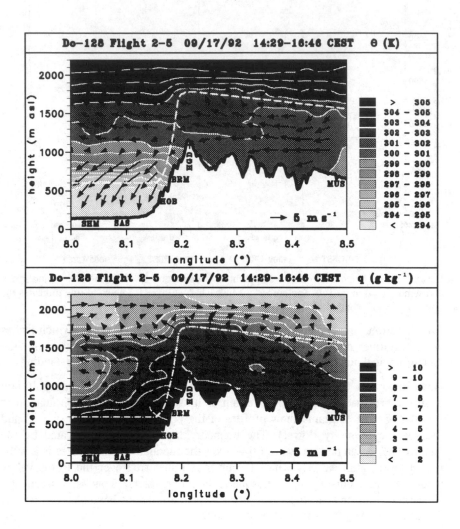

Fig. 8.4.17:   Interpolated cross sections of potential temperature, specific humidity and the horizontal wind vector from the Rhine river to Musbach in the Black Forest during the afternoon of 17th September 1992. The white dashed line indicates the top of the convective boundary layer as derived from aircraft measurements and vertical sounding systems. The horizontal range of the cross sections is about 36 km.

The handover process of polluted air from the boundary layer to the free atmosphere is also strongly influenced by secondary flow systems originating from pressure and temperature disturbances connected to the terrain structure.

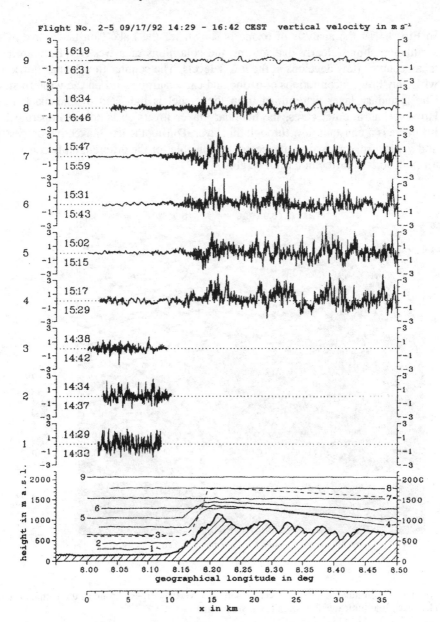

Fig. 8.4.18:   Flight pattern and vertical wind component along the cross section from the Rhine river to Musbach in the Black Forest during the afternoon of 17th September. The frequency of recording was 25 Hz. Solid lines above the orography (shaded) indicate aircraft flight legs. The starting point of each flight leg is marked by the leg number in the graph of the flight pattern. The dashed line above the orography represents the top of the convective boundary layer as derived from aircraft measurements and vertical sounding systems. Additionally, the leg numbers and the times of beginning and end of the legs are shown on the left side of the w-component graphs.

In Fig. 8.4.19 the analysis of ozone is shown for the Lake Constance area. The simulation shows clearly that, during the conditions of a westerly flow aloft, a weak counter flow develops at the lower levels. The counter flow brings down air with only low concentrations of ozone and carries air upward on the western side. The boundary layer determined from the model is indicated by the black line. However, as in other cases, the boundary layer inversion is not an impermeable lid: there is a constant flow through this layer. During the day, advective processes and vertical mixing by turbulence are responsible for the different daily courses of air pollutants at different terrain heights.

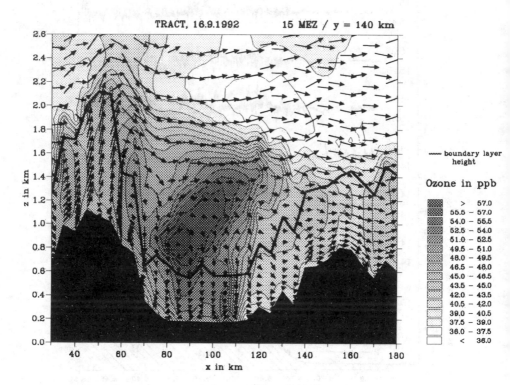

Fig. 8.4.19: Simulated ozone concentration for a vertical cross section in the Rhine valley. The black line indicates the boundary layer height.

In Figs. 8.4.20 and 8.4.21 the time variation of ground level concentrations of O_3, NO_2 and of NO and CO are plotted for the intensive measuring period, 16th–17th September, at four different locations. Schwarzwald-Süd and Freudenstadt are located within hills of the Black Forest whereas Rastatt and Kehl-Hafen are within the polluted region of the Rhine valley. The station with the highest elevation, Schwarzwald-Süd, shows high ozone concentrations for the whole day with only a weak variation during the whole period. In contrast, at the other stations which are located closer to the sources of NO and CO, there are sharp reductions in ozone during the night.

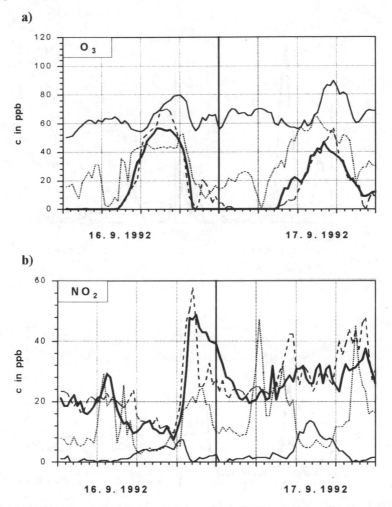

Fig. 8.4.20: Diurnal variation of measured ground level concentrations of a) O_3 and b) NO_2 at the stations Rastatt ——, Kehl-Hafen ---, Freudenstadt ·····, and Schwarzwald-Süd —— during the intensive measuring phase (16th–17th September, 1992).

The principal reason for the constant ozone concentration is the absence of deposition or the replacement of ozone by downward transport from higher altitudes of that lost by a weak deposition flux. The weaker thermal stability at higher levels is also an important influence: it permits a stronger coupling of the layers at different elevations and produces a more or less constant level of the concentrations during the whole day. It is only in the late afternoon that the vertical growth of the boundary layer within the Rhine valley reaches the Schwarzwald-Süd. Then an increase in the ozone level is observed for a few hours.

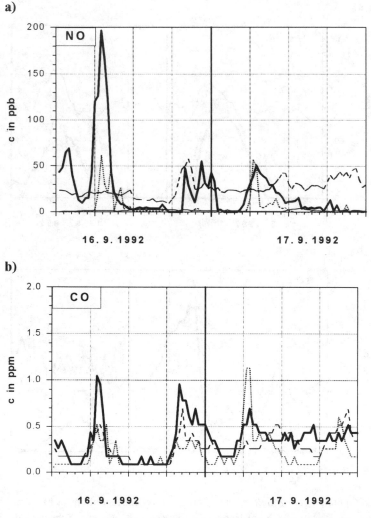

Fig. 8.4.21:   Diurnal cycles of measured ground level concentrations of a) NO and b) CO at the stations Rastatt ——, Kehl-Hafen ---, Freudenstadt ⋯⋯ and Schwarzwald-Süd —— during the intensive measuring phase (16th–17th September, 1992).

An important question concerning the processes which determine the decrease of ozone in the lower layers of the boundary layer was addressed by Güsten *et al.* (1998) with measurements in the nested area (see section 8.4.4a). From frequent captive balloon ascents during the late afternoon and night, the decrease of the ozone level was observed and it was possible to quantify the deposition of ozone at the ground using a profiling method. It appears that there is a discrepancy between the values of the ozone deposition determined directly from the deposition velocity and those determined from the isochron method. The deposition velocity gives values which are only between 20 to 50 % of those obtained by profiling. As well as deposition, chemical reactions of ozone with NO contribute to the depletion of ozone, which is confined to the shallow layer of at most 200 m above the ground.

This depletion processes for ozone during the daily cycle, obtained in a modelling study, are depicted in Figs. 8.4.22 and 8.4.23 for the nested-area cross section through the Rhine valley with the Hornisgrinde mountain of the northern Black Forest to the east. In the morning hours for the shallow layer up to 200 m, all ozone has been removed by deposition or by chemical reaction. In the upper levels higher ozone values still remain in the well mixed region left over from the previous day.

The mountain summits are surrounded by air with higher concentrations, so that the ozone values stay high throughout the whole day as shown in Fig. 8.4.20. In the early morning hours when vertical mixing is initiated again, the lower levels are filled up quickly by the downward transport of air with higher ozone concentrations. In addition, photochemical production of ozone in the whole domain of the boundary layer leads to a further increase of the ozone concentrations.

The increase of ozone near the ground, from almost zero to the maximum values seen in the afternoon (Fig. 8.4.21) is about 60 to 70 % due to the vertical mixing and about 30 to 40 % due to photochemical production within the boundary layers. However, in the proximity of strong sources of the precursors, especially NO, there is a net removal of ozone close to the ground throughout the day due to chemical reaction. Thus the increase of ozone in the morning hours observed within cities and close to roads with heavy traffic is simply due to the vertical mixing which more than compensates the photochemical loss.

Within the nested area there was sufficient information concerning the patchy ground structure to enable studies to be undertaken of the effective deposition and effective fluxes of other physical quantities such as momentum, heat and moisture. Hasager and Jensen (1999) developed a physical model for the calculation of representative fluxes of momentum within areas which form the grid size of present day mesoscale or weather forecast models.

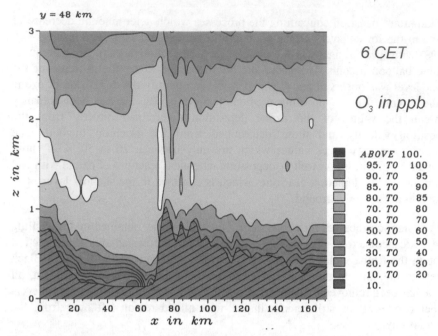

Fig. 8.4.22:   Vertical cross section for the calculated ozone distribution over the upper Rhine valley (Vosges Mountains in the west, Black Forest in the east) at 6:00 CET with the depleted shallow layer close to the ground.

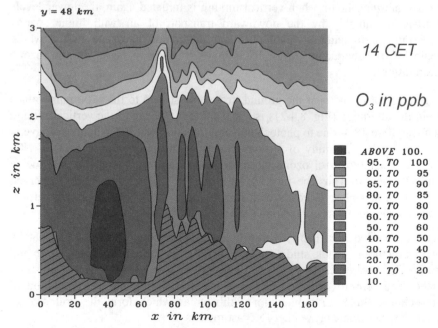

Fig. 8.4.23:   Vertical cross section for the calculated ozone distribution over the upper Rhine valley (Vosges Mountains in the west, Black Forest in the east) at 14:00 CET at the time of maximum ozone concentration.

As well as taking the vertical soundings of meteorological variables, the aircraft were able to provide some insight into the vertical structure of some of the chemical components. Standard species, such as ozone, could be measured continuously particularly in a curtain flight patterns. Volatile organic compounds (VOC) were sampled at prescribed locations by collecting air samples in small cans. A list of the VOC species analysed is shown in Fig. 8.4.24, the measurements in this case being made at a height of 500 m above the terrain. The figure also shows the rate constants for reaction with OH. As expected there is a rough anti-correlation between the measured VOC concentrations and the OH rate constants. From the measurements it was also found, that species with high OH rate constants showed a stronger decrease with height than those with low OH rate constants. For some of these species empirical formulae for the profiles could be derived. An overview of the findings is presented by Koßmann et al. (1997).

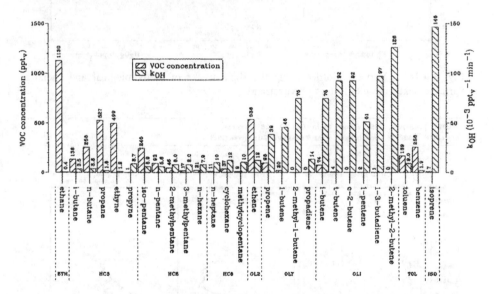

Fig. 8.4.24:   Mean VOC concentrations of the 14 air samples with NO_x load ≤ 2 ppb collected below 500 m a.g.l. Additionally, the OH rate constants k_{OH} and the corresponding VOC classes of the individual species in the chemical gas-phase mechanism RADM2 are shown.

Figs. 8.4.25, 8.4.26 and 8.4.27 show the vertical structure of meteorological variables and chemical species obtained from three successive ascending and descending curtain flights in the Swiss part of the TRACT experimental area between the Zürich and Zuger lakes. The left hand box shows the potential temperature, specific and relative humidity together with wind speed and direction. In the middle box, the vertical profiles of NO, NO_2 and NO_x (left three curves)and of O_3 and O_x (right two curves) are shown. Finally, the right hand box shows vertical profiles for SO_2 and H_2O_2.

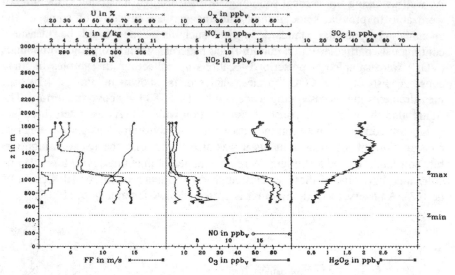

Fig. 8.4.25:   Vertical profiles along the flight tracks of meteorological and chemical variables taken close to Rüthi in Switzerland.

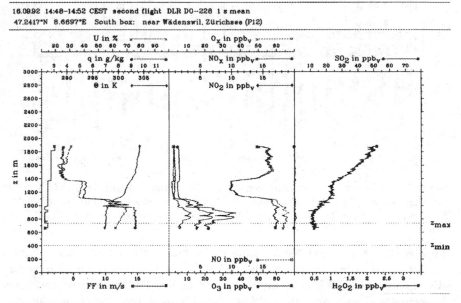

Fig. 8.4.26: Vertical profiles along the flight tracks of meteorological and chemical variables taken close to Lake Zürich.

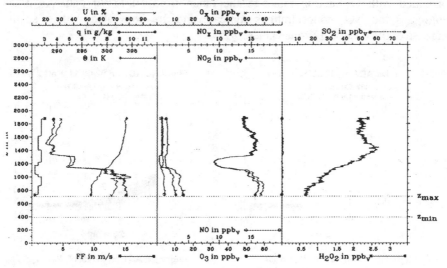

6.09.92 14:52–14:55 CEST second flight DLR DO–228 1 s mean
47.2164°N 8.4605°E South box: near Steinhausen, Zuger See (P13)

Fig. 8.4.27:   Vertical profiles along the flight tracks of meteorological and chemical variables taken close to the Zuger lake.

In all three figures the potential temperature increase corresponds to a decrease in specific humidity, and a rich vertical structure is visible. The nitrogen components have a similar structure to the specific humidity, except in Fig. 8.4.26, where high concentrations were found in the middle of the boundary layer.

Ozone shows a marked vertical structure in all three figures. underneath the strong capping inversion in the boundary layer, the ozone concentrations are much smaller, due presumably to chemical reaction since the specific humidity stays almost constant within this layer. Although the concentrations of NO and of NO_2 are rather low in this height range, all three vertical profiles are similar; NO is somewhat higher than NO_2 indicating that there is a plume from a strong source within this area.

In contrast to the other species, the concentrations of H_2O_2 increase with height. However, as profiles from the descending flights, P11 and P13, and the ascending flight, P12, show, it is clear that the measuring instrument had too slow a response time for the speeds chosen for aircraft ascent and descent. Finally, it should be mentioned that sulfur dioxide concentrations were very low during the whole measuring campaign.

These field measurements provided data not only for intensive analysis of the observed phenomena but also for model evaluation, and the observed vertical structure of the different chemical species is a sensitive and clear indicator of the quality of the models being used.

Furthermore the correlations between the concentrations of the various chemical components is noteworthy, as is shown for the concentrations of O_3 and of NO_x in Fig. 8.4.28, which shows results from flights made during the intensive measuring period, 11th September, 1992. At the beginning very low NO_x values are correlated with O_3 mixing ratios in the range from 50 to 70 ppb. As time progresses the NO_x concentrations increase and are correlated with a small decrease of the observed ozone values.

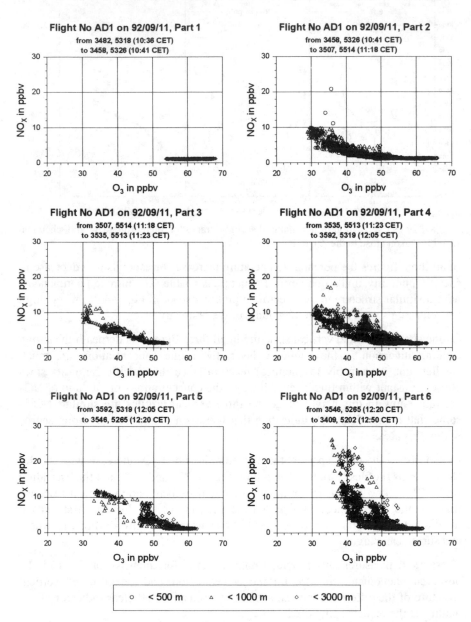

Fig. 8.4.28:   Relation of measured NO_x ($NO + NO_2$) and O_3 during aircraft ascents and descents at various times of the flight on 11th September, 1992.

8.4.8 Conclusions and reservations

a.   TRACT: a successful campaign.

The aim of the extensive field campaign of the EUROTRAC subproject TRACT was to study the processes of transport, turbulent mixing and chemical transformation of air pollutants during episodic conditions, particularly those that are involved in the daily cycles. The season chosen guaranteed a reasonable probability of fair weather conditions which meant that not only transport and diffusion processes could be studied but also the interaction during the day with gas phase chemical reactions.

Although, as is natural for field studies of this size, the synoptic conditions were not completely ideal for the whole observation period, a well-balanced study was achieved. The meteorological conditions during days on which strong thermal stability within the atmospheric boundary layer at night moved over to well mixed conditions connected to a rapidly growing boundary layer during the day were fully documented in a three-dimensional way. The TRACT study may be considered as one of the largest field studies of the boundary layer structure over complex terrain so far undertaken.

The relatively large number of vertical soundings from radiosondes and also from tethered balloon sondes with a three-hourly time step during the intensive measuring phases has provided an interesting insight into the vertical and horizontal structure of the boundary layer over mountainous terrain. A particular advantage was that a large number of ground based stations both from operational networks and from temporary stations were available in the area and these allowed very detailed studies to be made of the conditions close to the ground. Although aircraft stay in the air only for a few hours, which is a short time compared to the total length of an intensive measuring period, they provide valuable information about the three-dimensional distributions of the chemical species and meteorological variables. In particular, the availability of a large number of different measurements over the whole area gives us confidence about the specific details found in the various smaller areas, because those details appear in a consistent manner in many of the measured variables.

A reasonably clear picture has now emerged from TRACT about how the topography influences the horizontal flow system in the mesoscale domain of 300 km × 300 km. It was possible to demonstrate that the flow is influenced by the mountains more than 150 km downstream of the sources which has considerable influence for the mixing of emissions of different chemical species produced in a specific area.

Since the topography produces a variety of secondary flow conditions, quantitative results were derived from the observations for the exchange of air both in the vertical and horizontal directions. There are some examples where the exchange of polluted air in the vertical, due to secondary flow conditions, is more effective than that due to the usual small scale turbulence. Furthermore, it was shown how the turbulence level is distributed over some of the mountainous regions with a strong increase of turbulent intensity over the mountain barriers. Connected with the channelling and splitting of the horizontal flow through the mountains and the turbulence level connected to it, the different height development of the boundary layer could be quantitatively studied.

As well as an understanding of the physics of the boundary layer structure, valuable insight was gained into the three-dimensional distribution of a number of chemical species, particularly the nitrogen oxides and ozone and some hydrocarbons. The distribution is a function of the source locations and of the three-dimensional flow system.

Special studies were carried out which demonstrated the influence of a mountainous terrain on the three dimensional spatial distribution of the chemical species as a function of time. These studies include the influence on the general flow structure, on the deposition at the ground and on the photolysis rate, as well as the influence of the temperature level at different elevations on the chemical transformations These special studies show, that large errors may occur account if the dynamics of the relief on the mesoscale are not taken into account.

In summary, TRACT has provided a comprehensive and consistent data set. All the data needed to initialise complex mesoscale chemical-transport models, have been derived from the field study and are as complete as possible. The data range from larger scale synoptic data to the land use and emission data of the most important species. The TRACT data have already be used in subsequent model evaluation studies.

b.   Field campaigns: some reservations

While the successful results obtained from this complex field study have been emphasised, it is worthwhile mentioning some of the difficulties associated with large field campaigns in general and TRACT in particular.

The complexity of field measurements is often underestimated, particularly when chemical reactions and competing physical processes determine the final distribution of the different chemical species. Frequently the meteorological conditions are poorly determined and the effects of advection and turbulent phenomena on the measurements are no sufficiently appreciated. Too often, a poor compromise is made between the making enough measurements to determine the participating processes and making those which from the available resources are available. This often restricts the quantification of the different processes involved.

Also it is necessary to devise the appropriate measuring strategy for the particular aim of the field study in question, be it to determine a mass balance, to do a process study, to estimate quantities for a parameterisation or just to study the phenomena under particular conditions. In most of the field campaigns a mixture of aims is addressed by the different groups taking part but the overall strategy must include the measurements appropriate to the aims addressed.

Although much valuable data were collected and analysed by the large number of groups participating in TRACT, the accuracy of the results, say the mass balances could have perhaps been improved by examining the sampling requirements in a more rigorous way. For example the fluxes across the outer boundaries of the TRACT area were provided by the aircraft and the radio soundings during the hours of aircraft operation. But these times when aircraft are available were short compared to the time of a whole daily cycle. Even better perhaps would have been to make more measurements of the chemical species within the three-dimensional space of the research area.

Finally, atmospheric scientists should be always aware of the real scientific goal: that is to do only those measurements which are responsive to the effects being studied and to take into account the complexity of the system without making unjustified compromises. It is perhaps the willingness to make such compromises in field measurements that has condemned atmospheric science to progress more slowly than some other disciplines.

8.4.9   Acknowledgements

We would like to thank the many colleagues who took part in the TRACT and TRANSALP field campaigns, who submitted their data to the data bases in good time and who have been evaluating and publishing their results since. We are particularly indebted to the many colleagues, from the national networks and from the REKLIP campaign, for making their data available and so give us a more comprehensive data set with which to work. We must acknowledge the generous help from the many national funding agencies which supported the principal investigators involved. Among them are the following:

Bundesministerium für Bildung und Forschung (BMBF, Germany)

European Commission DG-XII, Brussels

Amt für Wehrgeophysik, Traben-Trabach

ASPA, Strasbourg

Bundesministerium für Verteidigung

Bundeswehr, Standort Neuhausen ob Eck

Deutscher Wetterdienst, Offenbach/Stuttgart/Hohenpeißenberg/Mannheim

Fischer Flug Inc., Offenburg

Meteo France, Illkirch

Schweizerische Meteorologische Anstalt, Zürich/Payerne

8.4.10   Further information

A complete account of the subproject can be found in volume 9 of this series (Larsen *et al.*, 2000) which includes some contributions from the individual principal investigators, the titles of which are given in the next section. Some papers on TRACT and TRANSALP were also published in a special issue of *Atmospheric Environment* (Desiato *et al.*, 1998, and subsequent papers).

8.4.11 TRACT: Steering group and principal investigators

Steering group

Franz Fiedler(Coordinator)	Karlsruhe
Domenico Anfossi	Torino
Rainer Friedrich	Stuttgart
Fritz Gassmann	Villigen
Franco Girardi	Ispra
Niels-Otto Jensen	Roskilde
Anne Jochum	Oberpfaffenhoffen
Paul Lightmann	Imperial College, London
Volker Mohnen	Garmisch-Partenkirchen, Albany
Bruno Neininger	Zürich
Robert Rosset	Toulouse
Eberhard Schaller	Garmisch-Partenkirchen
Alwig Stingele	Ispra
Heinz Wanner	Bern

Principal investigators

Here is a list of principal investigators (*) and their co-workers together with the title of their contribution to the TRACT final report (Larsen *et al.*, 2000).

J. Seier, P. Berner, R. Friedrich*, C. John and A. Obermeier
Generation of an Emission Data Base for TRACT

V. Mohnen*, H.-J. Kanter and F. Slemr
TRACT Data Quality Assessment

D. Anfossi*, E. Ferrero, D. Sacchetti, S. Trini Castelli, S. Finardi, A. Marzorati, S. Bistacchi, G. Bocchiola, G. Brusasca, P. Marcacci, G. Morselli, U. Pellegrini and G. Tinarelli
The TRACT Experiment: Contribution of Italian Participants to Measurements and Model Simulations

A.M. Jochum*, C. Strodl, H. Willeke, N. Entstrasser and H. Schlager
Transport Processes over Complex Terrain

G. Schayes* and P. Hakizimfura
Balloon Air Turbulence Measurements at Hegeney

N.O. Jensen*, C.B. Hasager, P. Hummelshøj, K. Pilegaard and R. Barthelmie
Surface Flux Variability in Relation to the Mesoscale

R. Vögtlin, M. Koßmann, H.-J. Binder, F. Fiedler*, N. Kalthoff, U. Corsmeier and H. Zimmermann
Investigation of Mixing Processes in the Lower Troposphere over Orographically Structured Terrain

A. Ebel*, H. Hass, H.J. Jakobs, M. Memmesheimer, J. Tippke and H. Feldmann
Simulation of the Episode of the TRACT Measurement Campaign with the
EURAD Model

K. Kuntze, M. Löffler-Mang, K. Nester*, G. Adrian and F. Fiedler*
Modelling of transport, mixing and deposition of air pollutants over
inhomogeneous terrain including chemical reactions

8.4.11   References

Adrian, G., F. Fiedler; 1991, Simulation of unstationary wind and temperature fields over complex terrain and comparison with observations. *Beitr. Phys. Atmos.* **64**, 27–48.

Anfossi, D., P. Gaglione, G. Graziani, K. Nodop, A. Stingele A. Marzorati, G. Graziani; 1996, The TRANSALP tracer campaigns: lessons learned and possible project evolution, in P.M. Borrell, P. Borrell, T. Cvitaš and W. Seiler (eds), *Proc. EUROTRAC Symp. '96*, Computational Mechanics Publications, Southampton, Vol. 1, pp. 751–756.

Anfossi, D., F. Desiato, G. Tinarelli, G. Brusasca, E. Ferrero and D. Sacchetti, 1998, TRANSALP 1989 experimental campaign - II. Simulation of a tracer experiment with Lagrangian particle models, *Atmos. Environ.* **32**, 1157–1166

Ambrosetti, P., D. Anfossi, S. Cieslik, G. Graziani, R. Lamprecht, A. Marzorati, K. Nodop, S. Sandroni, A. Stingele, H. Zimmermann; 1998, Mesoscale transport of atmospheric trace constituents across the central Alps: TRANSALP tracer experiments, *Atmos. Environ.* **32**, 1257–1272.

Berner, P., R. Friedrich, C. John, A.Obermeier; 1996, Generation of an emission data base for TRACT, in P.M. Borrell, P. Borrell, T. Cvitaš, K. Kelly, W. Seiler (eds), *Proc. EUROTRAC Symp 96*. Computational Mechanics Publ., Southhampton, Vol. 1, pp. 791–799.

Desiato, F., S. Finardi, G. Brusasca, M.G. Morselli; 1998, TRANSALP 1989 experimental campaign - I. Simulation of 3-D flow with diagnostic wind field models, *Atmos. Environ.* **32**, 1141–1156.

Fiedler, F.; 1989, *TRACT Project Description*, EUROTRAC ISS, Garmisch-Partenkirchen.

Fiedler, F., G. Adrian, M. Bär, J. Franck, K. Höschele, W. Hübschmann, K. Nester, T. Pfeiffer, P. Thomas, B. Vogel, S. Vogt, O. Walk; 1991, Transport und Umwandlung von Luftschadstoffen im Lande Baden-Württemberg und aus Anrainerstaaten (TULLA), KfK-PEF 88.

Fiedler, F.; 1992, TRACT Operational Plan. Inst. Meteorol. Klimaforsch., Universität Karlsruhe/Forschungszentrum Karlsruhe, 66pp.

Fiedler, F.; 1995, Klimaatlas Oberrhein Mitte-Süd 1995, Herausgeber: Oberrheinische Universitäten (Basel, Freiburg, Strasbourg, Karlsruhe),

Fiedler, F., I. Bischoff-Gauß, N. Kalthoff, G. Adrian; 1999, Modeling of the Transport and Diffusion of a Tracer in the Freiburg-Schauinsland Area, *J. Geophys. Res.*, in press.

Güsten, H., G. Heinrich, D. Sprung; 1998: Nocturnal Depletion of Ozone in the Upper Rhine Valley, *Atmos. Environ.* **32**, 1195–1202.

Hasager, C. B., N. O. Jensen; 1999, Surface-Flux aggregation in heterogeneous terrain, *Quart. J. Roy. Meteorol. Soc.* **125**, 2075–2102.

Heinrich, G., J. Weppner, H. Güsten, G. Clauss, H. Antz, A. Target; 1994, Quality assurance for trace gas sensors used in the nested area of TRACT, *EUROTRAC Annual Report 1993, part 10, TRACT*, EUROTRAC ISS, Garmisch-Partenkirchen, pp. 103–119.

Kalthoff, N, O. Kolle; 1993, Intercomparison of meteorological surface measurements during the TRACT campaign in summer 1992, *EUROTRAC Annual Report 1992, part 2, TRACT,* EUROTRAC ISS, Garmisch-Partenkirchen, pp. 104–119.

Kanter, H-J., F. Slemr, and V. Mohnen, 1996, Airborne chemical and meteorological measurements made during the 1992 TRACT experiment: quality control and assessment, *J. Air and Waste Management Association* **46**, 710–724

Koßmann, M., U. Corsmeier, R. Vögtlin, A.M. Jochum, C. Strodl, H. Willeke, B. Neininger, W. Fuchs, W. Graber; 1994, Aircraft intercomparison in the nested area during the TRACT campaign, *EUROTRAC Annual Report 1993, part 10, TRACT,* EUROTRAC ISS, Garmisch-Partenkirchen, pp. 94–102

Koßmann, M., H. Vogel, B. Vogel, R. Vögtlin, U. Corsmeier, F. Fiedler, O. Klemm, H. Schlager; 1997, The composition and Vertical Distribution of Volatile Organic Compounds in Southwest Germany, Eastern France and Northern Switzerland during the TRACT Campaign in September 1992, *Physics and Chemistry of the Earth* **21**, 429–433.

Koßmann, M., R. Vögtlin, U. Corsmeier, B. Vogel, F. Fiedler, H-J Binder, N. Kalthoff, F. Beyrich; 1998, Aspects of the convective boundary layer structure over complex terrain, *Atmos. Environ.* **32**, 1323–1348

Löffler-Mang, M., M. Koßmann, R. Vögtlin, F. Fiedler; 1997: Valley Wind Systems and their Influence on Nocturnal Ozone Concentrations, *Beitr. Phys. Atmos.* **70**, 1–14.

Martilli, A. and G. Graziani; 1998, Mesoscale circulation across the Alps: preliminary simulation of the TRANSALP 1990 simulations, *Atmos. Environ.* **32**, 1241–1256.

Mohnen, V., U. Corsmeier, F. Fiedler; 1992, Quality Assurance Project Plan: Aircraft Measurement Systems, Inst. Meteorol. Klimaforsch., Universität Karlsruhe/Forschungszentrum Karlsruhe 184pp.

Mohnen, V., 1996 Quality Assurance in Atmospheric Measurements and Assessments, in: P.M. Borrell, P. Borrell, T. Cvitaš, K. Kelly, W. Seiler (eds), *Proc. EUROTRAC Symp. '96,* Computer Mechanics Publications, Southampton, vol. 2, pp. 417–423.

Nodop, K., P. Ambrosetti, D. Anfossi, S. Cieslik, P. Gaglione, R. Lamprecht, A. Marzorati, A. Stingele, H. Zimmermann; 1992, Tracer experiment to study air flow across the Alps, in: P.M. Borrell, P. Borrell, T. Cvitaš and W. Seiler (eds), *Proc. EUROTRAC Symp. '92,* SPB Academic Publishing bv, The Hague, pp. 203–207.

Obermeier, A., J. Seier, C. John, P. Berner, R. Friedrich; 1996, TRACT: Erstellung einer Emissionsdatenbasis für TRACT, Forschungsbericht des Instituts für Energiewirtschaft und Rationelle Energieanwendung, Band 27, Stuttgart, ISSN 0938-1228.

Ruggaber, A., R. Dlugi, R. Forkel, W. Seidl, H. Hass, T. Nakajima, B. Vogel, M. Hammer, 1995, Modelling of radiation quantities and photolysis frequencies in the troposphere, in: A. Ebel, N. Moussiopoulos (eds), *Air Pollution III, Volume 4: Observation and Simulation of Air Pollution,* Computational Mechanics Publications, Southampton, pp. 111–118.

Uhrner, U.; 1997, Physikalische Ursachen für regionale Unterschiede in der nächtlichen Ozonverteilung, Diplomarbeit, Inst. Meteorol. Klimaforsch., Universität Karlsruhe/Forschungszentrum Karlsruhe.

Vogel, H., Bär, M., Fiedler, F.; 1991: Influence of different land-use types on dry deposition, concentration and mass balances: a case study, in: P.M. Borrell, P. Borrell, T. Cvitaš, W. Seiler (eds), *Proc. of EUROTRAC Symp. 92.,* SPB Academic Publishing bv, The Hague, pp. 559–562.

Vogel, B., F. Fiedler, H. Vogel; 1995, Influence of topography and biogenic volatile organic compounds emission in the state of Baden-Wuerttemberg on ozone concentrations during episodes of high air temperatures, *J. Geophys. Res.* **100**, 22907–22928.

Vogel, B., M. Hammer, N. Riemer, H. Vogel, A. Ruggaber; 1997, A method to determine the three-dimensional distribution of photolysis frequencies in the regional scale, in: P.M. Borrell, P. Borrell, K. Kelly, T. Cvitaš, W. Seiler (eds), *Proc. EUROTRAC Symp. '96*, Computational Mechanics Publications, Southampton, pp. 745–749.

Vogel, B., Riemer, N., Vogel, H., Fiedler, F.; 1999, Findings on NOy as an Indicator for Ozone Sensitivity based on Different Numerical Simulations, J. Geophys. Res. **104**, 3605–3620.

Wachs, P., U. Corsmeier, F. Fiedler; 1992, Subproject TRACT: Aircraft Operational Plan, Inst. Meteorol. Klimaforsch., Universität Karlsruhe/Forschungszentrum Karlsruhe.

Zimmermann, H.;, 1995, *Field phase report of the TRACT field measurement campaign*, EUROTRAC ISS, Garmisch-Partenkirchen, pp 196.

Zimmermann, H., 1999, contact *zdb@imk.fzk.de* to access the TRACT data.

Chapter 9

Biosphere-Atmosphere Exchange

The biosphere is the ultimate sink for atmospheric pollutants themselves or for the products of their chemical transformation in the atmosphere. Thus a knowledge of the deposition characteristics of trace substances is essential for the correct modelling of concentrations and fluxes. The biosphere is however not just a passive sink: it also a source of chemically active trace substances that can fuel the photo-oxidant production. In the south eastern United States for example the quantity of biogenic emissions in summer hinder the use of ozone reduction strategies based on the reduction of man-made emissions of volatile organic compounds. Within EUROTRAC there were two largely field subprojects devoted to studying the interface between the atmosphere and the biosphere.

9.1 **ASE** Air - Sea Exchange

9.2 **BIATEX** Biosphere-Atmosphere Exchange of Pollutants

BIATEX studied the deposition of selected trace substances to the biosphere; ASE looked at the interaction of the atmosphere with the sea.

9.1 ASE: An Overview of the Air-Sea Exchange Subproject

Søren E.Larsen (Coordinator)

Risø National Laboratory, Roskilde, Denmark

Substantial progress has been made in recent years in understanding the role of atmospheric transport of pollution for the marine ecosystems, as well as understanding the importance of the ocean as a source or sink for many of the important atmospheric trace gases influencing its composition and chemistry. The EUROTRAC subproject ASE has participated in and contributed significantly to this development. It has further contributed with results on a number of specific processes of importance for resolving the larger issues as well as with development of relevant measuring systems and computer programmes.

The individual studies have ranged far across professional boundaries, as is necessary for an interdisciplinary project such as ASE. The individual studies reported here cover subjects such as marine biology together with ocean and atmospheric transport and chemistry, as well as interfacial processes. The width of the instrumental development is equally large as many different instrumental principles are employed within modern environmental science.

9.1.1   Introduction

Due to its huge surface and large biological activity, the global ocean is a significant and important source as well as sink for many substances of importance for atmospheric composition, chemistry and radiation balances. Although the coastal seas cover only a small fraction of the total area of the ocean, it is disproportionately more important for the balance of many atmospheric trace constituents, because the biological activity in the coastal ocean is very large, compared to the open sea. Simultaneously the coastal ocean matters relatively more to humanity, because much human activity takes place in the coastal region, and therefore the coastal seas act as the main depository for anthropogenic pollution. They also simultaneously act as the immediate upstream boundary for the atmospheric flow across the coastal lands.

Conforming to this dual role of the ocean, most of the activities within ASE have been focused either on estimating the role of atmospheric deposition on the pollution of the coastal ocean or on the role of the ocean as a source/sink for trace constituents of importance in atmospheric chemistry.

9.1.2   Aims of the subproject ASE

The aim of ASE has been to study the air-sea exchange of material relevant to atmospheric chemistry in general, and specifically to the transport and transformation of pollutants across sea surfaces. Substances of interest have included both natural and man-made compounds, with the exchange being considered between the gas phase, particulate particles or from rain. The programme has concentrated on estimating emissions, atmospheric transformations and deposition of sulfur and nitrogen species, low molecular weight natural organo-halogen and hydrocarbon gases in the marine atmosphere, and on quantification of atmospheric reactivity of particulate material as well as emission from and deposition to the sea surface. The research has involved direct field measurements, as well as laboratory and field studies of the mechanisms controlling emission, deposition and atmospheric transformation processes. Regions studied have included the inner Danish waters, the North Sea, the Mediterranean and the eastern North Atlantic.

9.1.3   Principal scientific activities and results

a.   Structure of ASE and of this report

The ASE subproject was originally subdivided into five topic areas:

1   the contribution of biology and photochemistry of surface waters to sea-to-air trace gas fluxes;

2   trace gas emission measurements from surface sea water;

3   the role of atmospheric transformations of trace gases and aerosols in marine deposition processes;

4   the magnitude of the air-sea fluxes of natural and anthropogenic substances to European regional seas;

5   factors determining particle dynamics near air-sea surfaces.

However, as is apparent, there is an underlying unity to the five topics, since as their primary concern all have some aspects of mass transfer across the sea surface and its role in atmospheric chemistry and oceanic chemistry and biology. Therefore, in somewhat broader categories, the studies of the subproject centred around studies of processes of importance for deposition to the sea of atmospheric substances, and on studies of processes involved in exchange of biogases between the ocean and the atmosphere. This grouping was to some extent forced upon the subproject by the interests of the funding communities, and therefore on the form and focus of those ASE projects that were already organised when the project was originally formulated, and also on the joint and individual activities undertaken by the ASE groups during the lifetime of the project. With few exceptions these categories matched the working focus of the participating groups, in the sense that the first group involved studies of atmospheric processes, while the second group made most of their measurements in the liquid phase. This part therefore is organised into two further chapters according to these broad categories. Each chapter is introduced by several overview papers summarising the major results of the subproject, followed by contributions from individual groups. Some of the major joint activities conducted within ASE are also described.

b.   Results and activities of aerosols and deposition to marine systems

An important aspect of the ASE activities has been the effort to establish the magnitude and importance of atmospheric deposition of pollutants to European coastal waters. Some ASE projects have focused on the load of heavy metals from the atmosphere to the marine ecosystems, and others on the deposition of nutrients in the form of nitrogen compounds which give rise to many of the eutrophication events that have occurred in the enclosed northern European seas.

The research within ASE has to a large extent been responsible for the recognition of the importance of the atmospheric deposition to the pollution of the marine ecology over the North Sea and the inner Danish waters. Atmospheric transport has been found to account for, typically, 30 % or more of the total load for the substances studied here. Since the atmospheric contribution is generally more bio-available and is delivered to the surface where the largest biological activity takes place, the ecological significance of the atmospheric contribution is larger than indicated by its mass ratio. Simultaneously with the research within ASE, researchers in other regions of the world have found similar results for the importance of the atmospheric deposition to coastal seas in the vicinity of large industrialised regions.

Aspects of these results are reported by many of the groups, and are placed in a general framework in the review by T.D. Jickells and L.J. Spokes of the magnitude and significance of atmospheric inputs of metals and nutrients to marine ecosystems (Jickells and Spokes, 2000).

Many detailed studies have been carried out within this subject, both to ensure that necessary data was obtained, and also to clarify relevant questions concerning the transport involved, the chemical transformation and wet and dry deposition processes. These activities and results are described in the contributions from the individual groups, together with a couple of larger joint ASE activities.

One group of have studied the production, dynamics and role of marine aerosols, this is summarised by De Leeuw *et al.* (2000). The production of marine aerosols through various forms of droplet formation is important, both for the processes in the liquid phase controlling the exchange of gases across sea surfaces, and for the atmospheric processes involving multi-phase chemical transformations, as well as for deposition of aerosols and water-soluble gases to the ocean surface. Within ASE however, only atmospheric aspects of the formation of spray droplets were studied, these projects are described in detail in Larsen *et al.* (2000).

An important aspect of the research within ASE has been the joint experimental work over the North Sea, which has involved inter-calibration activities between the different groups at Mace Head in Ireland, and the 1991 North Sea Experiment, involving two ships and an effort to conduct a pseudo-Lagrangian experiment between these two ships. A smaller measuring campaign on the German research platform, Nordsee, was conducted just before the platform was dismantled. The main activity here, the NOSE experiment is described in overview paper by Schulz (2000).

c.   Results on biological processes and the air-sea exchange of biogenic gases

The ASE subproject has aimed to increase our knowledge about the role of the open ocean, as well as coastal waters, as sources and sinks for a number of gases of biological origin and atmospheric importance, and those with a biological component in their production/destruction cycle. Also here ASE groups have participated in, as well as contributed significantly to, the generally increase of knowledge about the source/sink cycles of many of these gases.

DMS (dimethylsulfide) is probably one of the biogenic gases that has been most extensively studied recently, both for its contribution to the global sulfur budget and for its possible significant contribution to the production of cloud condensation nuclei, through oxidation in the atmosphere. Here use of isotope signatures (Liss *et al.*) have indicated that about 30 % of the non-sea-salt sulfate in aerosols was derived from oxidation of DMS, but that this ratio must be considered highly variable. The pattern of DMS production in the surface waters has been mapped across the eastern Atlantic margin, showing that a production maximum in the spring follows the shelf break, where the biological production is highest. The marine cycle of DMS has been studied extensively, as has also the cycle of other marine trace constituents such as selenium. Another important contributor to the budget of atmospheric condensation nuclei is carbonyl sulfide, COS, for which the air-sea exchange budget has been determined in the Mediterranean and the eastern Atlantic.

Important natural atmospheric oxidising gases in the marine atmosphere are organo-halogens and non-methane hydrocarbons. The air-sea exchange for both, but especially the last species, have been extensively mapped and the ocean found to be a significant source for atmospheric constituents of these gases. A review presentation on the current knowledge on the ocean's role in the source/sink cycles for a number of the more important biogenic gases is described by Rapsomanikis (2000).

9.1.4   Positive and negative aspects of ASE

Within its lifetime, the ASE project has had the privilege to participate in, and contribute to, the strong development of knowledge about the exchange processes occurring across the ocean-atmosphere interface, which has taken place during this period, both with respect to scientific insight into the exchange processes, and with respect to knowledge about the absolute and relative importance of these exchange processes for a number of environmental ocean and atmospheric issues on global as well as local scales. This development has occurred simultaneously and to a large extent fuelled by the growing awareness in society about such environmental issues. The general awareness that we do not know enough about the global environment has led to questions on air-water pollution for the coastal countries of Europe.

ASE has provided a framework for multi-disciplinary research groups to study the environmental issues as they became apparent. In this context, the main handicap of ASE has been that it was never able to secure any long-term funding aiming specifically at the ASE objectives and it proved difficult to maintain a well defined ASE profile. In spite of this, groups participating in the subproject, found a very effective forum for the formulation of new concepts.

9.1.5   Acknowledgements

Support of the individual ASE contributions is acknowledged in the reports of the principal investigators. Here we wish to acknowledge the repeated support of EC-Environment, through the EIPO IGAC office, to the annual ASE workshops, hereby contributing significantly to the scientific productivity and continuing development of ideas and concepts within the subproject.

9.1.6   Further Information

The ASE contribution is shown in context in chapter 4 of this volume. A complete account of the subproject can be found in volume 9 of this series (Larsen *et al.*, 2000). The volume contains reports from the various principal investigators together with reviews of particular aspects of the field (see the list in section 9.1.7).

9.1.7   ASE: Steering group and principal investigators

Steering group

Søren E. Larsen (Coordinator)	Roskilde
Willy Baeyens	Brussels
Sauveur Belviso	Gif-sur-Yvette
P. Buat Menard	Bordeaux
Olivier F.X. Donard	Talence
J.L. Colin	Creteuil
Roy M. Harrison	Birmingham
Gerrit de Leeuw	The Hague
Peter S. Liss	Norwich
Spyridon Rapsomanikis	Mainz
Michael Schulz	Hamburg

Principal Investigators (*) and their contributions to the final report (Larsen *et al.*, 2000)are listed below.

Aerosols and Deposition to Marine Systems

T.D. Jickells* and L.J. Spokes; A Review of the Magnitude and Significance of Atmospheric Inputs of Metals and Nutrients to Marine Ecosystems.

M. Schulz*; The North Sea Experiment 1991 (NOSE): A Lagrangian-type experiment

    a. *M. Schulz*, T. Stahlschmidt, W. Maenhau*, F. Francois, J. Injuk, H. van Maldern and R. van Grieken*;* Trace element aerosol concentration and aerosol size distribution

    b. *M. Schulz*, W. Maenhaut*, F. Francois, R.M. Harrison*, M.I. Msibi, A.-M. Kitto, S. Yamulki, L.-L. Sørensen and W. Asman*;* Sulfur compounds

    c. *R.M. Harrison*, M.I. Msibi, A.-M. Kitto, S. Yamulki, L.-L. Sørensen, W. Asman*, O. Hertel, M. Schulz*, H. Bange, S. Rapsomanikis* and O. Krueger;* Nitrogen compounds

d. *W. Baeyens*, M. Leermakers, R.. Ebinghaus, H.H. Kock and J. Kuballa*;
    Mercury compounds

e. *J.H. Injuk, L. De Bock , H. van Maldern and, R. van Grieken**;
    Individual particle composition

f. *B. Quack, J. Kuss and J.C. Duincker**;
    Light halogenated organic compounds

g. *T. Jickells*, L. Spokes, M. Schulz*, A. Rebers, M. Leermakers
    and W. Baeyens*; Wet deposition of major ions and trace metals

h. *M. Schulz*, M. Schrader, S.E. Larsen*, F.A. Hansen and O. Krüger*;
    Evaluation of meteorological conditions during NOSE

G. de Leeuw*, S.E. Larsen* and P.G. Mestayer*;
    Factors Determining Particle Dynamics over the Air-Sea Interface

P.G. Mestayer* and B. Tranchant;
    Aerosol Dynamics Modelling in the Marine Atmospheric Surface Layer

G. de Leeuw* and A.M.J. van Eijk;
    Dynamics of Aerosols in the Marine Atmospheric Surface Layer

S.E. Larsen, L.-L. S.Geernaert, F.Aa. Hansen, P. Hummelshøj, N.O. Jensen and
    J.B. Edson;* Fluxes in the marine atmospheric boundary layer

J.L. Colin, R. Losno, N. Le Bris, M. Tisserant, I. Vassali, S. Madec, S. Cholbi,
    F. Vimeux, G. Bergametti, H. Cachier, C. Liousse, J. Ducret, , B. Lim,
    M.H. Pertuisot and P. Buat Menard**;
    Aerosol and Rain Chemistry in the Marine Environment

T.D. Jickells*, L.J. Spokes, M.M. Kane, A.R. Rendell and R.R. Yaaqub;
    Atmospheric Inputs to Marine Systems

M. Schulz*, T. Stahlschmidt, A. Rebers and W. Dannecker;
    Transformation of Polluted Air Masses when Transported over Sea with
    respect to Deposition Processes

Roy M. Harrison*, C.J. Ottley, M.I. Msibi and A.-M. Kitto;
    Chemical Transformations in the North Sea Atmosphere

W. Maenhaut* and F. François;
    Elemental Composition and Sources of Atmospheric Aerosols above and
    around the North Sea

R. Van Grieken*, J. Injuk, L. De Bock and H. Van Malderen;
    Study of Individual Particle Types and Heavy Metal Deposition for North
    Sea Aerosols using Micro and Trace Analysis Techniques

W. Baeyens*, M. Leermakers, C. Meuleman and Q. Xianren;
    Air-Sea Exchange Fluxes of Mercury and Atmospheric Deposition to the
    North Sea

W.A.H. Asman*, R. Berkowicz, O. Hertel, E.H. Runge, L.-L. Sørensen,
K. Granby, P.K. Jensen, J. Christensen,H. Nielsen, B. Jensen,
S.-E. Gryning, A.M. Sempreviva, N.A. Kilde, H. Madsen, P. Allerup,
Søren Overgaard, J. Jørgensen, F. Vejen and K. Hedegaard;
Deposition of Nitrogen Compounds to the Danish Coastal Waters

Biological Processes and Air-Sea Exchange of Biogases

S. Rapsomanikis*;
Oceans as Sources or Sinks of Biologically Produced Gases Relevant to
Climatic Change

D. Amouroux, C. Pecheyran, M.M. De Souza Sierra and O.F.X. Donard*;
Influence of Photochemical and Biological Processes on the Selenium
Cycle at the Ocean-Atmosphere Interface

P S Liss*, W J Broadgate, A D Hatton, R H Little, G Malin, N C McArdle,
P D Nightingale and S M Turner;
Biological Production of Trace Gases in Surface Seawater and their
Emission to the Atmosphere

S. Belviso, M. Corn, U. Christaki, J.C. Marty, C. Cailliau
and P. Buat-Ménard*;
Origin and Importance of Particulate Dimethylsulphoniopropionate
(DMSPp) in the Marine Cycle of Dimethylsulphide (DMS)

B.C. Nguyen*, J.-P. Putaud and N. Mihalopoulos;
Seasonal and Latitudinal Variations of Dimethylsulfide Emissions from
the North-East Atlantic Ocean

G. Uher, V.S. Ulshöfer, O.R. Flöck, S. Rapsomanikis* and M.O. Andreae;
The Biogeochemical Cycling of Carbonyl Sulfide and Dimethyl Sulfide at
the European Continental Margin

B. Bonsang*;
Air Sea Exchanges of Non-methane Hydrocarbons

J. Rudolph*, M. Ratte, Ch. Plass-Dülmer, R. Koppmann;
The Oceanic Source of Light Nonmethane Hydrocarbons

9.1.8   References

de Leeuw G., S.E. Larsen, P.G. Mestayer; 2000, Factors Determining Particle Dynamics
over the Air-Sea Interface, in: S. Larsen, F. Fiedler, P. Borrell (eds), 2000; *Exchange
and Transport of Air Pollutants over Complex Terrain and the Sea* Springer Verlag,
Heidelberg (in preparation)

Jickells, T.D., L.J. Spokes; 2000, A Review of the Magnitude and Significance of
Atmospheric Inputs of Metals and Nutrients to Marine Ecosystems, in: S. Larsen, F.
Fiedler, P. Borrell (eds), *Exchange and Transport of Air Pollutants over Complex
Terrain and the Sea,* Springer Verlag, Heidelberg (in preparation)

Larsen, S., F. Fiedler, P. Borrell (eds); 2000; *Exchange and Transport of Air Pollutants over Complex Terrain and the Sea* Springer Verlag, Heidelberg; Vol. 9 of the EUROTRAC final report (in preparation).

Rapsomanikis; S.; 2000, Oceans as Sources or Sinks of Biologically Produced Gases Relevant to Climatic Change, in: S. Larsen, F. Fiedler, P. Borrell (eds), *Exchange and Transport of Air Pollutants over Complex Terrain and the Sea* Springer Verlag, Heidelberg (in preparation)

Schulz, M.; 2000, The North Sea Experiment 1991 (NOSE): A Lagrangian-type experiment in: S. Larsen, F. Fiedler, P. Borrell (eds), *Exchange and Transport of Air Pollutants over Complex Terrain and the Sea* Springer Verlag, Heidelberg (in preparation)

9.2 BIATEX: Assessment and Achievements

Sjaak Slanina (Coordinator)

ECN, Petten, The Netherlands

9.2.1 Introduction

An understanding of the exchange of pollutants and trace substances between the atmosphere and the biosphere is essential to a full appreciation of the transport and chemical transformation of pollutants in the troposphere, and to the implementation of any policy for pollutant abatement. The biosphere is the ultimate sink for all pollutants or their oxidation products, and so the rates of deposition play a critical role in determining pollutant lifetimes in the atmosphere, their transport range and their concentration.

The biosphere is also a source of many compounds which are precursors for photo-oxidants, and so the contribution of biogenic emissions must be taken into account when determining the anthropogenic contribution to pollutant concentrations and deposition.

When BIATEX (Biosphere–ATmosphere EXchange of pollutants) was established as a subproject of EUROTRAC, the general notion prevailed that the absence of data on dry deposition and biogenic emissions would be an important barrier to obtaining insights into transport, export and import of pollution in Europe, as well as estimates on acid deposition and oxidant formation.

The project description reflects this state of affairs and was primarily oriented towards these problems.

A number of factors influenced the development of BIATEX, as will be explained in the next sections.

* At the time it was still unclear that eutrophication, exposure of ecosystems to too large amounts of nutrients such as phosphate and nitrate, was becoming a grave environmental problem in many parts of Europe, as important as acidification or exposure to increasing oxidant concentrations.

* In the initial phase of BIATEX, the critical load concept as a device for formulating environmental policy in Europe was developed. This critical load concept created a need to provide more accurate estimates of deposition loads of pollutants in Europe, in particular of the contribution of dry deposition, the most uncertain component.

* An environmental policy to limit oxidant formation in Europe became more important. A requirement for the policy development are assessments of biogenic, as well as anthropogenic, emissions of reactive organic

compounds. The cheapest environmental policy is based on limiting volatile organic compounds, but this policy is doomed if to great a quantity of volatile organic compounds are emitted by the biosphere.

*   The progress of environmental measures made it necessary to provide better estimates of pollution loads. Uncertainties in pollution loads of several hundred percent can be quite acceptable if very large exceedances (transgressions) of critical loads or critical concentrations are encountered. But environmental policy in many countries in Europe has now progressed to the point that quantitative decisions must be taken, for example on. whether emissions must be reduced by 60 or 90 %. The increase in costs is exponential for reductions over 50 %, so hard evidence is necessary to support large expenditures for abatement.

Many of the research groups engaged in BIATEX were already involved the application of results to the environment, and they, at least had some experience in supporting national environmental policy. However, no mechanism existed for transferring scientific data in a form suitable for the development of environmental policy on a European scale.

So the BIATEX community was not only confronted with a number of scientific problems regarding measurement and modelling of exchange fluxes, but also with the necessity of adapting BIATEX to the changing circumstances, and to developing suitable instrumentation to contribute to environmental policy development.

The very open and fruitful co-operation within BIATEX has been a great asset which has enabled us to reach many of the ambitious gaols described in the original BIATEX proposal. Frequent contacts within EUROTRAC, the SSC, the ISS and contacts with coordinators and investigators from other EUROTRAC subprojects helped in obtaining a good overall picture of priorities and possibilities.

Collaboration with national and international agencies, especially EMEP, should also be mentioned, which has been essential in developing the application of BIATEX results to a description of the exchange of pollutants on a European scale.

BIATEX was also supported by the European Communities at a later stage. This support was not only important financially for some activities within BIATEX but also paved the way to an integration of European research in this area.

9.2.2 Promises versus final products of BIATEX

The collaboration of some 30 European groups in research devoted to assessment of the exchange of pollutants between atmosphere and biosphere has been successful and has promoted coherence between European researchers.

This last point is not trivial: The development of an European Environmental Policy is dependent on a European consensus and hence on a coherent scientific data base for Europe. The contribution of EUROTRAC to European collaboration in environmental research has been very important in general and BIATEX is certainly not an exception in this.

The next question in a final report of such a large programme as BIATEX must, by necessity, be the evaluation of the results in comparison to the original objectives, as laid out in the description of BIATEX during the starting phase of the subproject.

The final report of BIATEX should provide answers to the following questions.

*   Has BIATEX answered the questions posed as the original objectives at the beginning of the subproject?

*   If these original objectives have not been met, what are the reasons for not delivering the promised results?

*   Which important gaps exist in the present state of knowledge?

The original description of BIATEX formulated the following objectives (the brackets indicate omitted sentences or parts of sentences in the original text) :

A   (...) to quantify fluxes of pollutants between atmosphere and biosphere, to study the mechanisms which are responsible for the observed exchange, and to provide **regional fluxes on seasonal and annual scales**. (...) Priorities are: nitrogen compounds, volatile organic compounds and sulfur compounds.

B   (...) data on spatial and temporal variations of some biogenic emission rates of some important trace constituents, (...) partial insight into mechanisms controlling emission fluxes and first order estimates of **regional fluxes on seasonal and annual scales**. (...)Priorities are: volatile organic compounds, sulfur compounds, nitrogen compounds.

C   (...) secondary products will be developed: Method development for eddy-correlation and gradient methods. Quality assurance and quality control for the whole measurement systems for exchange measurements. Development of mechanistic studies to generalise observations. Model development up to 3-D models describing deposition, and simpler models for biogenic emission. Assessment of results of individual and joint experiments.

Summarising the conclusions of the 1995 BIATEX workshop in Madrid the following evaluation of the achievements in these areas can be made:

A   Fluxes of pollutants between the atmosphere and the biosphere have been quantified, the mechanisms which are responsible for the observed exchange are now partially characterised, and **regional fluxes on seasonal and annual scales** have been provided.

The priorities are, according to the BIATEX results: nitrogen compounds, volatile organic compounds and sulfur compounds.

B   Data on spatial and temporal variations of biogenic emission rates of some important trace constituents have been obtained. Partial insight into mechanisms controlling emission fluxes has been provided and first order estimates of **regional fluxes on seasonal and annual scales** for biogenic emitted volatile organic compounds will be obtained before the end of 1995.

Priorities within BIATEX are: volatile organic compounds, nitrogen compounds and sulfur compounds.

C   Method development for eddy correlation and gradient methods has taken place, important progress has been achieved for reduced nitrogen compounds, some oxidants, VOCs and reduced sulfur compounds.

Quality assurance and quality control for the integral measurement systems for exchange measurements has not been performed in a rigorous way but first order qualifications for the precision and accuracy of a number of procedures are available.

Development of mechanistic studies to generalise observations has been partially successful *e.g.* in regarding the influence of a water layer on deposition, and systematic approaches on isoprene emissions, but have not been developed for other aspects.

Model development has been less successful than expected:
No 3-D models for deposition have been developed, but models are now available based on resistance analogy, which can provide better estimates of dry deposition fluxes.

No approaches for inhomogeneous situations have been developed.
A simple model for biogenic emission is currently under development and will be finished before the end of 1995.

The assessment of results of individual and joint experiments has been performed in a qualitative way however the integration of individual results in communal assessments is still a weak point.

To summarise the confrontation between promises and results, BIATEX has achieved or will achieve in the near future a number of the main objectives as promised in 1989, such as regional and seasonal deposition fluxes and emission fluxes of biogenic emitted organic compounds. Estimates of biogenic emissions of other compounds, for example NO or reduced sulfur, are not yet possible. Progress has been made in terms of additional information, but this information is still insufficient for estimates of emission fluxes on a European scale.

At the onset of BIATEX, far reaching plans were developed for model development in the area of deposition and biogenic emissions. The need for advanced models in this area has been clearly demonstrated, but the gap proved to be so large between the possibility of a systematic meteorological approach, and the quality of data for exchange measurements and available data sets of important parameters, that different approaches had to be chosen. In fact it was decided to employ resistance modelling to provide new estimates for the deposition of important compounds in Europe.

The ambitious goal to use biological mechanistic studies as a basis for extrapolation of exchange data in time and space, and under different biological and meteorological conditions has only partially been met. However, good co-operation between different disciplines, biology, biochemistry and atmospheric chemistry plus micrometeorology, has emerged within BIATEX. This co-operation will form the basis for future developments in this area.

International co-operation has been strengthened by EUROTRAC and certainly within BIATEX. One of the important developments in BIATEX is the increased coherence in the scientific community engaged in assessments of exchange between atmosphere and biosphere.

9.2.3 Principal results obtained in BIATEX

The reviews of the different areas of BIATEX which are published in this volume, provide a more detailed account of the results of BIATEX in the context of the general developments in the area of measurement and assessments of exchange of compounds between atmosphere and biosphere. But a number of principal achievements will be mentioned here.

1. The methodology to derive better deposition estimates for Europe, has been established. A model, based on a resistance analogy approach, has been derived by RIVM (Slanina, 1994). This method provides estimates of the resistance for turbulent transport in the atmosphere (R_a), based on parameters commonly available such as wind speed, estimates of heat fluxes *etc.* Detailed land use descriptions are used to derive the roughness height and the resistance for deposition due to the stagnant layer near the surface (R_b). Estimates for the surface resistance (R_c) are provided for Europe, based on the detailed land use descriptions mentioned earlier and on better surface resistance parameterisation, which have been developed by BIATEX participants. The deposition is calculated as passing through a layer at an altitude of 50 meters, to minimise effects of non-homogeneous terrain. Depositions of SO_2 and NH_3, calculated by means of this improved approach are presented in Figs. 9.2.1 and 9.2.2. The surface resistance and hence deposition fluxes at constant concentration vary over a much wider range as assumed in older models. The influence of surface wetness and variation in turbulence leads to much high depositions in countries like Netherlands, Denmark, Belgium and northern Germany compared to much dryer areas in Spain or Italy. The first results of deposition measurements in Portugal confirm the important role of water layers for the deposition of these compounds (Sutton *et al.*, 1997).

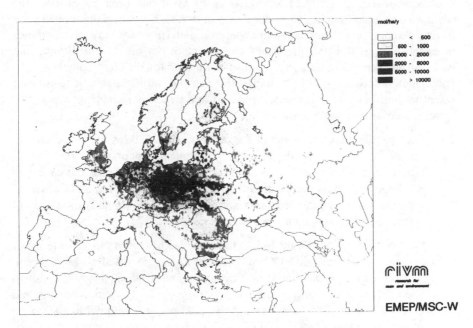

Fig. 9.2.1:   Deposition of acidifying compounds over Europe (mol ha^{-1} year^{-1}). The sum of wet and dry deposition is based on new estimates of dry deposition velocities.

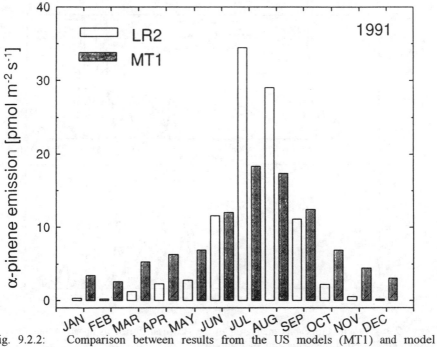

Fig. 9.2.2:   Comparison between results from the US models (MT1) and model calculations (LR2) performed at IFU and based on observations in the Bavarian forest over a period from 1987 to 1991.

Collaboration between BIATEX models and EMEP has been established, the parameterisation of deposition as developed in BIATEX is currently implemented in EMEP modelling. The change in deposition patterns is not only of importance in establishing, with better precision, exceedance of critical loads in Europe, but will also influence export and import of pollutants for different countries. The collaboration between BIATEX and EMEP provides an ideal vehicle to transfer scientific findings and conclusions into a format which is suitable as a basis for European environmental policy.

2.   The development of generalised descriptions of biogenic emissions in Europe has started. With the help of financial support from the BMFT, Germany, a good start has been made in this area. During the course of BIATEX joint experiments organised in the Bavarian forest, and individual experiments have provided more and better data on biogenic emissions of volatile organic compounds (Kesselmeier *et al.*, 1997a).

In 1994 and 1995, Günter Helas, has put together efforts to provide better descriptions of biogenic VOC emissions in Europe. An analysis of European research in this area showed that extrapolation of US data would result in large errors and that emission descriptions must be parameterised and validated with European data. An example of studies to derive a parameterisation for isoprene emissions under European conditions is given in Fig. 9.2.3.

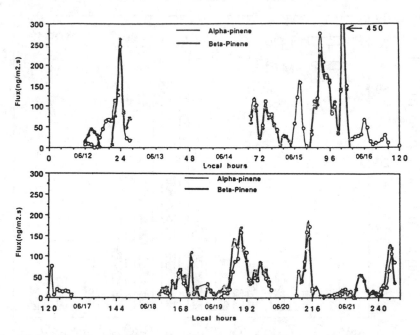

Fig. 9.2.3:   Pinene emissions measured in the Bavarian forest using an automated GC system developed at the University of Toulouse.

The best possible approaches for estimates of biogenic VOC emissions in Europe were discussed at the workshop and the approach was implemented in close collaboration with the GENEMIS subproject and EMEP. The first estimates of biogenic VOC emissions for Europe were made available at the end of 1995.

3.   The uncertainties in the surface resistance for deposition of gases like SO_2, NH_3, NO_2 and NO and hence in the contribution of dry deposition processes to the loads of these pollutants in Europe has been reduced considerably due to the results obtained in BIATEX. But a warning is necessary: When BIATEX started, the uncertainties in the deposition loads of different compounds were, depending on compound and location, very large, even up to 200 to 300 %. These uncertainties have been reduced considerably but remain of the order of 50 to 100 % depending on compound and area. Abatement policies in Europe will need a further decrease in these uncertainties in order to obtain good cost/benefit analyses for steadily increasing expenditure in this area. This will require a considerable effort in fundamental mechanistic research, in monitoring of exchange fluxes and in the development of models. (Sutton *et al.*, 1997; Erisman *et al.*, 1997)

4.   The first results on dry deposition of particles indicate that the deposition velocity of particles is higher than expected. Research on deposition of particles was incorporated at a late stage in BIATEX, according to recommendations of the SSC. Measurements of deposition velocities for aerosols has been carried out at different locations using different techniques. The University of Toulouse has applied eddy correlation techniques in combination with particle counting, the same approach has been used by Risø. ECN and TNO approached the problem by means of gradient measurements of sulfate in aerosol and ECN has studied the distribution of lead isotopes as these isotopes are deposited on particles and by way of the particles on vegetation. All these studies have indicated that the deposition velocity of particles over low vegetation as well as forests is much larger than formerly assumed on the basis of wind tunnel experiments, deposition modelling and the use of surrogate surfaces. Deposition velocities of 1 to 2 cm/sec have been observed over forests(Gallagher *et al.*, 1997).

These findings (Fig. 9.2.4) are not only important for the assessment of exceedance of critical loads for different compounds in Europe, but play also an important role in estimates of life time and transport of aerosols and hence in the radiative forcing caused by small particles (Gallagher *et al.*, 1997).

5.   In the cadre of BIATEX, the first reliable estimates of exchange fluxes of reduced sulfur compounds, other than SO_2, have been made (Fig. 9.2.5). Exchange of COS, CS_2, DMS, H_2S and other sulfur compounds have been studied. After the initial phase of BIATEX the decision was taken that these compounds would not have the highest priority in the second part of

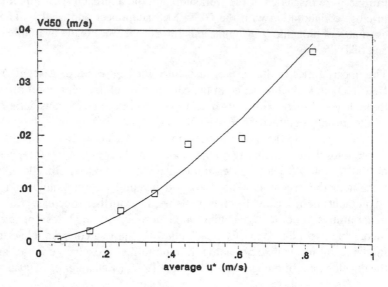

Fig. 9.2.4: Deposition velocity of sulfate aerosol measured in the Speulder forest, as a function of wind friction velocity, u_*.

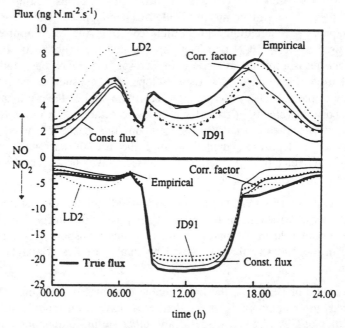

Fig. 9.2.5: Influence of photochemical reactions on NO and NO_2 fluxes. Constant flux indicates the exchange fluxes without correction for reactions, true flux is the value after optimal correction for chemical reactions.

BIATEX, but recent developments, for example, results of reduced sulfur emissions in Europe, have shown that the contribution of these compounds cannot be ignored in the future and that research on the exchange of these compounds is necessary to provide information on the sulfur balance in Europe in the near future (Kesselmeier *et al.*, 1997b).

6.   The influence of atmospheric reactions on the exchange between atmosphere and biosphere of different compounds has been shown to be quite important. The influence of the decomposition of ammonium nitrate on the deposition fluxes of nitric acid is a good example, another is the interaction of photo-oxidants. The reactions of NO_x and ozone have been characterised and modelled. It was known for some time already that especially measurements of the exchange fluxes of NO and NO_2 were influenced by the rapid reaction between NO, NO_2 and ozone under influence of solar radiation. In several experiments, *e.g.* at Speulderbos in the Netherlands and in locations in the UK, the influence of these reactions were investigated. The conclusion is that the shifts in equilibrium do not influence ozone and NO_2 fluxes very much, but modify NO fluxes considerably, even to the extent that deposition fluxes are measured while in reality emission of NO takes place. Integration of experimental work and modelling has been carried out successfully to solve this problem (Fowler *et al.*, 1997).

7.   Research on exchange of radiatively active gases has increased within BIATEX. Basically the same techniques as used for compounds important for acidification problems, eutrophication and increased oxidant levels are used to measure the exchange of radiative active gases. As this kind of research has expanded considerably in the last years, more and more BIATEX participants have performed exchange measurements of compounds like CO_2 and methane. The activities cover a wide range, *e.g.* methane emissions from wetlands in Scotland have been reported (Slanina, 1994), but biological mechanistic research has been carried out as well (Fig. 9.2.6) to characterise the formation and oxidation of methane in soils.

8.   At the start of BIATEX all work on exchange measurements was mainly carried out in northern Europe. European descriptions of the exchange of gases and aerosols between biosphere and atmosphere were all based on extrapolation of northern European data even though everybody realised that this extrapolation was questionable. During the second half of BIATEX the participation of southern European groups has been strengthened considerably. Important contributions from Italy (ISPRA and others), Portugal, Spain and Turkey have laid a proper basis to remedy this situation. By the end of 1995 it should be possible to verify a number of extrapolations made in assessments of depositions and biogenic emissions with data obtained in southern Europe.

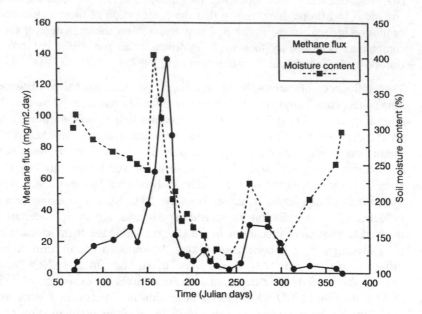

Fig. 9.2.6:   Methane emissions as a function of soil humidity, determined in laboratory experiments.

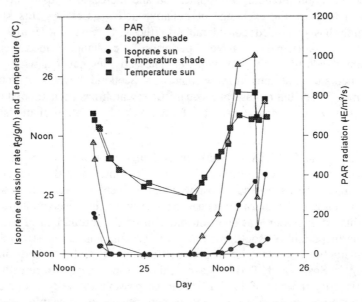

Fig. 9.2.7:   Isoprene emissions from eucalypus trees measured in Portugal.

9.   The assumption at the start of BIATEX that modelling activities in different
      areas, deposition, biogenic emissions and atmospheric reactions, would
      provide a suitable focus to integrate results of the different joint and
      individual experiments has been confirmed in the second half of BIATEX.
      The close combination of experimental work and modelling that has
      emerged in many areas of BIATEX has provided the necessary means for
      integration of the results.

9.2.4 Why did BIATEX not succeed in some areas?

The BIATEX community made a proper diagnosis of a number of quite difficult
problems in the initial phase of the subproject and formulated the research
objectives accordingly.

Good examples are:

> To be able to draw general applicable results from exchange measurements
> at one time and at one location which are valid for other locations and under
> other circumstances, it is necessary to couple research on exchange of trace
> gases with investigations of the underlying mechanisms.

Specific biological mechanisms are mentioned in the description of BIATEX as
of prime importance: this statement has proved to be true. During the course of
BIATEX, attempts have been made to combine the two disciplines, for example
in the systematic model approaches undertaken at the University of Frankfurt. As
already outlined above, the complexity of the problem and the time needed to
become acquainted with the different disciplines, together with a lack of suitable
data, proved more of a barrier than foreseen at the start. On the other hand,
effective collaboration between biologists and researchers who are active in
exchange measurements has been initiated and is leading to new results. Good
examples were provided during the EU/BIATEX workshop on the emissions of
biogenic volatile organic carbon compounds, where biogenic emissions were
linked to biological mechanisms as a first step in providing a more generalised
approach to describe these emissions.

In other areas, for example on the influence of water layers on the deposition
rates of SO_2 and NH_3, which is a simpler problem than some of the biological
issues, acceptable generalised descriptions have been obtained which indeed are a
great asset in generalising the results.

> The description of deposition in non-ideal conditions requires models of a
> total different character compared to the resistance modelling generally
> employed. It had been rightly pointed out at the start of BIATEX that large
> areas of Europe cannot be described in terms of ideal surfaces for classical
> micrometeorological approaches. For this reason emphasis was given to the
> development of 2-D and 3-D models for deposition. As already mentioned,
> the gap between the parameterisation required by modellers to describe
> these processes and available data sets, together with problems connected to
> finding compromises between fundamental meteorological modellers and

the down-to-earth approach favoured by many scientists engaged in exchange measurements, brought these developments to a stand-still during the first half of BIATEX. On the other hand, the confrontation between modellers engaged in fundamental research in these areas, and those scientists who are responsible for providing estimates for deposition loads and biogenic emission fluxes, has made the position clear. Future work in this area now has a good basis and further possibilities can be envisaged.

These types of difficulties will be encountered in any scientific endeavour like BIATEX. The inclusion of ambitious goals in the initial phases should be defended even if it is generally accepted that there is a fair chance that they will not be attained.

Another problem was directly connected to the way EUROTRAC and consequently BIATEX was organised. From the beginning, the generalisation and application of the results from BIATEX in terms of improved descriptions for deposition and biogenic emissions had a high priority. The Steering Group had clear ideas on how to proceed in this matter, but it has been difficult for BIATEX participants, mainly financed by national organisations, to devote sufficient money and manpower to the project. During the last stages of BIATEX these problems were alleviated by important contributions from institutions in the Netherlands and Germany. An important conclusion of the steering group regarding this issue is that the proper means and pathways for generalisation and application must be in place at the start of any research; implementation of generalisation and application halfway through a project is a hazardous operation and can never operate optimally. The late start of this development blocked an extension to, for example, assessments of biogenic NO and reduced sulfur emissions. In all honesty, it must be said that less data were available in these areas than others, but a lack of communal funds was the main impediment.

Summa summarum, central funds for the generalisation and application of results, as well as very rigorous planning, are necessary in order to be able to transfer the results of scientific research into formats which are suitable for supporting the development of environmental policy.

9.2.5 Future developments in atmosphere–biosphere exchange

Research on atmosphere/biosphere exchange will be needed in the future to support the existing and future sulfur, nitrogen and oxidant protocols as well in the area of climate change. Important deficiencies exist in a number of areas:

*   The problems regarding the description of deposition and biogenic emissions over non-ideal terrain have not been solved. It will be necessary to develop and validate new models to tackle this problem. Not only model development, but also validation of deposition over non-homogeneous terrain will require new approaches.

*   Though good progress has been made in the last years in measuring deposition fluxes of aerosols, only very few reliable data are available at

present. It will be necessary to expand the data base on deposition of aerosols considerably in the near future.

* Though participation of southern European groups has been considerably increased in the second half of BIATEX, insufficient data on atmosphere/biosphere exchange are available for Southern Europe. Much effort will be needed to acquire data for southern Europe of the same quality as are available for the northern part.

* Validation of deposition modelling especially for larger areas has not yet been carried out. The uncertainties in estimates of deposition and especially biogenic emissions are still very large (50 to 100 %) and will remain so until methods are developed to verify exchange fluxes over large areas.

* As abatement policy is implemented in Europe sources will become important which have more or less been ignored up to now. Examples are biogenic NO and reduced sulfur emissions. Careful examination is needed to map future needs as a function of changing emission patterns in Europe.

* Due to the abatement policy already mentioned, the contribution of compounds like nitric acid, nitrous acid, organic acids *etc.* in acidification and eutrophication will become much more important in the near future. Study of the exchange of these compounds will be necessary as little reliable data are available at present.

9.2.6 Further information

The BIATEX contribution is shown in context in chapter 4 of this volume.

A complete account can be found in volume 6 in this series (Slanina, 1997) which includes the following reviews from members of the steering group and their co-workers.

Biosphere - Atmosphere Exchange of Ammonia
> *M.A. Sutton, G.P. Wyers, F.X. Meixner, J.K. Schjørring, J. Kesselmeier, G. Kramm and J.H. Duyzer*

Atmospheric Particles and their Interactions with Natural Surfaces
> *M. Gallagher, J. Fontan, P. Wyers, W. Ruijgrok, J. Duyzer, P. Hummelshøj, K. Pilegaard and D. Fowler*

Assessment of Dry Deposition and Total Acidifying Loads in Europe
> *J.W. Erisman and A. van Pul*

Biological Mechanisms involved in the Exchange of Trace Gases
> *J. Kesselmeier, K. Bode, J.K., Schjørring and R. Conrad*

Atmosphere – Surface Exchange of Nitrogen Oxides and Ozone
> *D. Fowler, F.X. Meixner, J.H. Duyzer, G. Kramm and L. Granat*

Exchange of Sulfur Gases between the Biosphere and the Atmosphere
   J. Kesselmeier, P. Schröder and J.W. Erisman

9.2.7 BIATEX: Steering group and principal investigators

Steering group

Sjaak Slanina (Coordinator)	Petten
Jan H. Duyzer	Delft
David Fowler	Edinburgh
Günther Helas	Mainz
Øystein Hov	Bergen
Franz X. Meixner	Mainz
Sten Struwe	Copenhagen

Principal investigators (*) and their contributions to the final report (Slanina, 1997) are listed below.

Measurement techniques

H. Güsten*, G. Heinrich, E. Mönnich, D. Sprung and J. Weppner
   On–line Measurements of Ozone Surface Fluxes: Instrumentation and
   Methodology

H.S.M. de Vries, F.J.M. Harren, G.P. Wyers, R.P. Otjes J. Slanina and J. Reuss*
   Real-time and Non-intrusive Detection of Ambient Ammonia using the
   Photothermal Deflection Technique

G.P. Wyers, R.P. Otjes, A. Wayers, J.J. Möls, A. Khlystov, P.A.C. Jongejan
   and J. Slanina*
   Development of Instrumentation for Concentration Measurements and
   Surface-Exchange Fluxes of Air Pollutants

Exchange of Nitrogen Compounds and Oxygen

S. Cieslik* and A. Labatut
   Ozone deposition on various surface types

J. Duyzer*, H. Weststrate, H. Verhagen, G. Deinum and J. Baak
   Measurements of Dry Deposition Fluxes of Nitrogen Compounds and Ozone

J.W. Erisman, M. Mennen, J. Hogenkamp, E. Kemkers,D. Goedhart, A. van Pul,
   G. Draaijers, J. Duyzer and P. Wyers*
   Dry Deposition Monitoring of SO_2, NH_3 and NO_2 over a Coniferous Forest

L. Granat*, A. Rondon, C. Johansson and R. Janson
   Flux of NOx between Atmosphere and Vegetation – Parameterisation of
   Surface Resistance/Emission

E. Lorinczi*, A. Ritivoiu and T. Camaroschi
   First Emission Inventory of Ammonia in Timis County

F.X. Meixner*, J. Ludwig, H. Müller, F. Böswald and B. Terry
Surface Exchange of Nitrogen Oxides over Different European Ecosystems

U Özer* and S.G. Tuncel
Measurement of Ambient Concentration and Surface Fluxes of O_3, SO_2 NO_x and TSP in Uludag National Park

M.A.H.G. Plantaz, A.T.Vermeulen and G.P. Wyers*
Surface Exchange of Ammonia over Grazed Pasture

S. Walton, M. Gallagher, T.W. Choularton*, J. Duyzer and K. Pilegaard
The Speulderbos Experiment 1993 - Experimental and Modelling Results

Exchange of Sulfur Compounds and Multi-compound Measurements

A. Anghelus, M. Lörinczi, E. Lörinczi* and A. Ritivoiu
Measurements of Wet and Dry Deposition in Timis County, Romania

L. Horváth*, T. Weidinger, Z. Nagy and Ernó Führer
Measurement of Dry Deposition Velocity of Ozone, Sulfur Dioxide and Nitrogen Oxides above Pine Forest and Low Vegetation in Different Seasons by the Gradient Method

W. Jaeschke*, Th. Dietrich, U. Schickedanz and M. Klauer
Biogenic and Anthropogenic Trace Gas Fluxes between the Atmosphere and a Polluted Mixed Forest of Central Europe

J. Kesselmeier*, U. Bartell, S. Blezinger, W. Conze, C. Gries, C. Hilse, R. Hofmann, U. Hofmann, A. Hubert, U. Kuhn, F. Meixner, L. Merk, T.H. Nash III, G. Protoschill-Krebs, F. Velmecke, C. Wilhelm and M.O. Andreae
Exchange of Reduced Sulfur Compounds between the Biosphere and the Atmosphere

Biogenic VOC Emissions

J. Kesselmeier*, C. Ammann, J. Beck, K. Bode, R. Gabriel, U. Hofmann, G. Helas, U. Kuhn, F.X. Meixner, Th. Rausch, L. Schäfer, D. Weller and M.O. Andreae
Exchange of Short Chained Organic Acids between the Biosphere and the Atmosphere

C. Pio*, T. Nunes and A. Valente
Forest Emissions of Hydrocarbons

R. Steinbrecher*, J. Hahn, K. Stahl, G. Eichstädter, K. Lederle, R. Rabong, A.-M. Schreiner and J. Slemr
Investigations on Emissions of Low Molecular Weight Compounds (C_2–C_{10}) from Vegetation

R. Steinbrecher*, H. Ziegler, G. Eichstädter, U. Fehsenfeld, R. Gabriel, Ch. Kolb, R. Rabong, R. Schönwitz and W. Schürmann
Monoterpene and Isoprene Emission in Norway Spruce Forests

J. Reuder, T. Gori, A. Ruggaber*, L. Kins and R. Dlugi
Photolysis Frequencies of Nitrogen Dioxide and Ozone: Measurements and
Model Calculations

Modelling and Validation

C. Cavicchioli, G. Manzi, G. Catenacci, G. Brusasca* and M. Borgarello
Dry Deposition Inferential Measurement in a Rural Region of Northern Italy

M. Ferm*
Improvement and Validation of the Throughfall Technique for Nitrogen
Deposition Measurements to Forest Ecosystems

G.H. Kohlmaier*, P. Ramge, M. Plöchl and F. Badeck
Modelling the Uptake and Metabolisation of Nitrogen Dioxide and Ozone by
Plant Leaves

R. San José*, A. García and H. Casado
Experimental and Modelling Studies of the Deposition Processes in South
European Ecosystems

9.2.8 References

Erisman, J.W., A. van Pul; 1997, Assessment of Dry Deposition and Total Acidifying
Loads in Europe, in: Slanina, J. (ed), *Biosphere-Atmosphere Exchange of Pollutants
and Trace Substances,* Springer Verlag, Heidelberg, pp. 93–116.

Fowler, D., F.X. Meixner, J.H. Duyzer, G. Kramm, L. Granat; 1997, Atmosphere–Surface
Exchange of Nitrogen Oxides and Ozone, in: Slanina, J. (ed), *Biosphere-Atmosphere
Exchange of Pollutants and Trace Substances,* Springer Verlag, Heidelberg, pp. 135–
166.

Gallagher,   M.,   J.   Fontan,   P.   Wyers,   W.   Ruijgrok,   J.   Duyzer,
P. Hummelshøj, K. Pilegaard, D. Fowler; 1997, Atmospheric Particles and their
Interactions with Natural Surfaces, in: Slanina, J. (ed), *Biosphere-Atmosphere
Exchange of Pollutants and Trace Substances,* Springer Verlag, Heidelberg, pp. 45–
92.

Kesselmeier, J., K. Bode, J.K., Schjørring, R. Conrad; 1997a, Biological Mechanisms
involved in the Exchange of Trace Gases, in: Slanina, J. (ed), *Biosphere-Atmosphere
Exchange of Pollutants and Trace Substances,* Springer Verlag, Heidelberg, pp. 117–
134.

Kesselmeier, J., P. Schröder, J.W. Erisman; 1997b, Exchange of Sulfur Gases between the
Biosphere and the Atmosphere, in: Slanina, J. (ed), *Biosphere-Atmosphere Exchange
of Pollutants and Trace Substances,* Springer Verlag, Heidelberg, pp. 167–198.

Slanina, J., (ed); 1994, BIATEX Annual Report, 1993, EUROTRAC ISS, Garmisch-
Partenkirchen.

Slanina, J., (ed); 1997, *Biosphere-Atmosphere Exchange of Pollutants and Trace
Substances,* Springer Verlag, Heidelberg; volume 4 of the EUROTRAC final report.

Sutton, M.A., G.P. Wyers, F.X. Meixner, J.K. Schjørring, J. Kesselmeier, G. Kramm,
J.H. Duyzer; 1997, Biosphere–Atmosphere Exchange of Ammonia, in: Slanina, J. (ed),
Biosphere-Atmosphere Exchange of Pollutants and Trace Substances, Springer
Verlag, Heidelberg, pp. 15–44.

Chapter 10

Laboratory studies

A detailed knowledge of the mechanisms and characteristics of the chemical reactions which take place, or might take place, in the atmosphere is an essential component of atmospheric chemistry. Two subprojects were devoted to laboratory work in EUROTRAC.

10.1	**HALIPP**	Heterogeneous and Liquid Phase Processes
10.2	**LACTOZ**	Laboratory Studies of Chemistry Related to Tropospheric Ozone

HALIPP concentrated on the extraordinarily difficult field of the heterogeneous chemical processes that take place within cloud droplets and on the surface of aerosols, and which are also involved in aerosol and cloud droplet formation. LACTOZ studied the myriad reactions which take place in the free atmosphere.

Also included in this chapter are the results of the **Chemical Mechanism Working Group** (section 10.3), which was set up to encourage exchange between the laboratory experimentalists and the modellers.

10.1 HALIPP: Heterogeneous and Liquid Phase Processes

Peter Warneck (Coordinator)

Max-Planck-Institut für Chemie, Abteilung Biogeochemie,
Saarstr. 23, D-55122 Mainz, Germany

10.1.1 Summary

HALIPP was established as a co-operative research project to meet the need for laboratory studies dedicated to the development of a quantitative description of heterogeneous and liquid phase chemical processes in the troposphere, which include the interaction of trace species with cloud drops, reactions in the aqueous phase of clouds, and heterogeneous reactions occurring on the surfaces of aerosol and ice particles. HALIPP research has resulted in significant progress in the field, especially in the chemistry related to clouds: Aqueous sulfur(IV) oxidation mechanisms have been worked out; the role of metal ions in clouds has been clarified; aqueous reactions of the NO_3 radical have been explored; new techniques for studying gas-liquid mass transfer have been developed and used to determine mass accommodation coefficients; the interactions of nitrogen oxides with components of sea salt and their impact on tropospheric chemistry have been quantified. Progress has been achieved both in *knowledge* and in the *understanding* of heterogeneous and liquid phase chemical processes. A close collaboration and exchange of know-how among research groups with diverse scientific experience were critical in making HALIPP a success.

10.1.2 Aims of HALIPP

Laboratory studies of chemical reaction mechanisms and rate processes are essential in the development of a quantitative and sufficiently detailed description of atmospheric chemistry for use in computer-based simulation models. In contrast to gas-phase reactions, which had been studied for many years with the result that a substantial data base existed even prior to EUROTRAC, heterogeneous and liquid phase reactions had been largely neglected, and the associated mechanisms and reaction rates were poorly known. HALIPP was established in response to a growing recognition that heterogeneous processes cannot be ignored in any reasonably complete description of tropospheric chemistry. In terms of the original work statement the objectives were: (i) to study processes that generate and control oxidants and products in atmospheric water drops; (ii) to study the transfer of free radicals and gases across the gas-liquid interface of atmospheric droplets; (iii) to study the contribution of aerosols to the chemistry of trace gases and radicals, including the photochemical surface production of active species. Major gaps and uncertainties existing at the beginning of HALIPP were related to the mechanisms responsible for the oxidation of sulfur(IV) species in the aqueous phase of clouds; the values of mass accommodation coefficients controlling the gas-liquid transfer of materials; and the role of heterogeneous reactions on surfaces of aerosol particles. In each field of study the aim was to delineate the physico-chemical mechanisms and to quantify the individual rate processes. Although ultimately the data are to be included in computer models simulating atmospheric processes, the implementation of such data in the models was not part of HALIPP activities.

10.1.3 Highlights of HALIPP activities and achievements

a. *Aqueous phase oxidation of sulfur(IV) by non-radical reactions*

*   Stopped-flow techniques have been used to determine rate coefficients, pH dependence and the general mechanism for the non-radical oxidation of sulfur(IV) by hydrogen peroxide, peroxomonosulfate and other hydroperoxides.

*   Rate coefficients for the oxidation of sulfur(IV) by O_3, H_2O_2 and ClO^- have been determined under high ionic strength conditions applicable to deliquescent aerosol particles.

b.   The role of transition metal ions in atmospheric waters

* Sulfur(IV) oxidation in the presence of transition metals (Fe, Mn, Co) has been demonstrated to feature autocatalysis. For iron as a catalyst it has finally been proven (90 years after its discovery) that the oxidation is a radical-driven chain reaction, and a mechanism has been established. An important step in the metal ion redox cycle is the interaction of SO_5^- with the bivalent species to regenerate the trivalent ion, which initiates the chain by formation of a sulfito-complex.

* The catalytic activity of carbon in the oxidation of sulfur(IV) has been demonstrated to arise from traces of iron in soot.

* The predominant fate of the hydroperoxy radical, HO_2, and its anion O_2^-, which do not oxidise sulfur(IV) directly, is reaction with ions of trace metals, copper being most efficient. The reactions produce H_2O_2, which can then oxidise sulfur(IV). The reaction of O_2^- with ozone, which would provide a source of OH radicals, probably cannot compete with trace metals.

* The interaction of nitrogen oxides with sulfur(IV) in the presence of iron or manganese ions produces hydroxylaminedisulfonate, which hydrolyses to form hydroxylamine-monosulfonate under redox cycling of the metal ion. The role of such processes in atmospheric water drops remains to be further explored.

c.   Photochemical sources of radicals in the aqueous phase

* The yields of radicals from the photo-decomposition of NO_3^-, NO_2^-/HNO_2, H_2O_2, $FeSO_4^+$ and $FeOH^{2+}$ have been determined. The Fe(III)-hydroxo complex was found to be the most favourable as a source of OH radicals in terms of quantum yield and abundance in cloud water.

* Photo-decomposition of Fe(III)-oxalato complexes provides an efficient source of HO_2/O_2^- radicals in atmospheric water drops.

d.   Free radical reactions in the aqueous phase

* Pulse radiolysis, steady-state radiolysis and laser flash photolysis have been employed to determine rate coefficients for reactions involved in the chain oxidation of sulfite and hydrogen sulfite initiated by OH radicals and propagated by SO_5^- and SO_4^- radicals.

* The reactivities of SO_5^- and SO_4^- with Fe^{2+} and Mn^{2+}, which are important reactions in the oxidation of sulfur(IV) catalysed by transition metals, have been determined.

* The oxidation of formaldehyde initiated by OH radicals has been studied by steady-state and laser flash photolysis. The products are formic acid and hydrogen peroxide. In the presence of H_2O_2 a chain reaction can occur.

*   The oxidation of hydroxymethanesulfonate (the addition compound of HSO_3^- and formaldehyde), initiated by OH, has been thoroughly investigated and the main reaction steps characterised.

*   The nitrate radical, NO_3, which would enter cloud drops from the gas phase, reacts rapidly with aqueous chloride, HSO_3^- and SO_3^{2-} by charge transfer, and with organic substrates by H atom abstraction. The consequences of these NO_3 reactions to tropospheric chemistry have not yet been fully explored.

e.   Gas-liquid interactions

*   Four experimental techniques, involving a train of water drops, a liquid water jet, a wetted wall flow tube and a water drop floating in an uprising air current, have been developed for the measurement of uptake coefficients, which are required for the calculation of gas-liquid uptake rates. Aqueous phase chemical analysis is used in all cases to quantify the mass transfer. Mass accommodation coefficients are derived by extrapolation to vanishing concentrations and contact times.

*   Mass accommodation coefficients have been determined for important solutes: Values for SO_2, HNO_3, HCl and NH_3 are of the order of 0.1, that is about 10 % of the gas molecules that strike the droplet surface enter into solution; values for formic and acetic acid, HONO and N_2O_5 lie in the range 0.01–0.065. The uptake coefficient for N_2O_5 increases with increasing NaCl salt content from 0.013 to 0.032. The uptake of these gases meets little resistance.

*   The uptake coefficient for ozone is 4.5×10^{-3}, that for CO_2 is by an order of magnitude smaller. The low value for ozone may seriously impede the supply of this important oxidant in cloud drops.

*   Henry's law coefficients for a number of organic nitrates and peroxynitrates that are unstable in water have been determined.

f.   Aerosol surface reactions and photo-catalysis

*   The activity of solids such as silicon dioxide, titanium dioxide, muscovite, hematite, ilmenite, fly ash, in catalysing the oxidation of organic compounds has been explored. TiO_2, which is known to generate OH radicals photo-catalytically from adsorbed water, was found particularly effective in promoting the oxidation of alkanes, simple aromatics, naphthalene and ß-pinene. Fe_2O_3 was shown to be much less photo-catalytically active in this regard.

*   The possibility that photo-catalytic activity arises from the formation of Lewis acid centres was explored. Only TiO_2 and Al_2O_3 showed a noticeable production of reactive sites, the response on SiO_2 and desert sand was minor. In all cases the Lewis acid centres were destroyed in the presence of water. A

higher activity on fly ash was attributed to the presence of carbonaceous material, which reacts with water to produce CO_2 and H_2O_2.

* Compounds such as naphthalene, which are more or less permanently adsorbed on particles, are strongly susceptible to photo-catalytic degradation in sunlight. In the presence of water, Fe_2O_3 forms complexes with organic compounds, which are photolytically decomposed, whereby the dissolution of iron in the aqueous phase is promoted.

g.   Sea salt and nitrogen oxides

* The reaction of nitrogen pentoxide, N_2O_5, with deliquescent sodium chloride to produce $NaNO_3$ and $ClNO_2$ has been studied with submicron droplets suspended in an aerosol chamber and with a flow tube reactor wetted with NaCl solution. The yield of $ClNO_2$ approaches 100 % at high NaCl concentrations. At pH $\leq$ 2, the yield of $ClNO_2$ decreases appreciably and NOCl is formed as a product. The results are explained by a chemical mechanism featuring NO_2^+ and NO_3^- as intermediates.

* The solubility of $ClNO_2$ is much less than that of N_2O_5 so that $ClNO_2$ is released to the gas phase, where it undergoes photo-decomposition to form NO_2 and Cl atoms. The consequences for tropospheric chemistry are an increase in the residence time of NO_x and an enhancement of radical chemistry due to extremely reactive Cl atoms.

* A Knudsen cell reactor with mass spectrometric detection of products was used to study the displacement of chloride and bromide from sodium halide crystals by reaction with HNO_3 and nitrogen oxides. HNO_3 produced 100 % of HCl and HBr from NaCl and NaBr, respectively; NO_2 gave HONO as the main product from NaCl (traces of water were present) , no products were observed with NaBr; N_2O_5 produced 60 % $ClNO_2$, whereas the products with NaBr were Br_2 and HONO; $ClONO_2$ reacting with NaCl produced Cl_2 with high yield, with NaBr the products were Br_2, BrCl and Cl_2. Uptake coefficients ranged from about 10^{-7} for NO_2 reacting with NaCl to 3×10^{-2} for HNO_3 reacting with NaCl and NaBr.

10.1.4   Selected results illustrated

a.   Hydroperoxides as oxidants of sulfur(IV)

The dissolution of sulfur dioxide in cloud water converts it largely to sulfurous acid, hydrogen sulfite (HSO_3^-) and sulfite (SO_3^{2-}). The sum of these species is designated sulfur(IV). Hydroperoxides, especially hydrogen peroxide, H_2O_2, and peroxynitric acid, HNO_4, are considered important oxidants of sulfur(IV) in clouds. The reaction rates exhibit a marked dependence on the pH of the solution. Overall rate coefficients have been measured for a number of such oxidants and a general mechanism has been formulated. Rate coefficients differ for the reactions with hydrogensulfite and sulfite. The reactions with the first species, which is dominant in cloud water, are subject to general acid catalysis. This requires that the effects of buffers used to control the pH must be eliminated by extrapolation to zero concentrations. The remaining pH dependence then can be ascribed to proton catalysis and catalysis by water. Fig. 10.1.1 shows a plot of the logarithm of rate constants as a function of pH, in the HSO_3^- stability regime, for the oxidants hydrogen peroxide, peroxyacetic acid, peroxynitrous acid and peroxynitric acid. The general behaviour is that of a linear decrease of log k with increasing pH until a plateau is reached. The decrease is due to the decline in proton concentration, the plateau is associated with the constant rate coefficient of the water-catalysed reaction. For H_2O_2 the plateau occurs at pH values outside the range shown, in the case of HNO_4 the rate constant associated with the plateau overrules that for proton catalysis. Experimental results obtained from 11 different hydroperoxides including those shown in Fig. 10.1.1 have shown that the third order rate constant for proton catalysis is nearly the same regardless of the nature of the hydroperoxide, whereas the rate constant for the water-catalysed reaction depends strongly on the acid strength of the hydroperoxide.

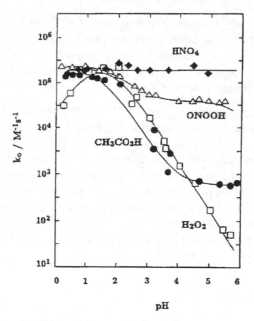

Fig. 10.1.1:   Variation with pH of the buffer-corrected second order rate constant for reactions of sulfur(IV) with H_2O_2, $CH_3C(O)OOH$, ONOOH and O_2NOOH. For details of experimental conditions see Warneck (1996).

Fig. 10.1.2 demonstrates the existence of an essentially linear relation between the logarithm of the rate constant and the pK value for the release of a proton from the hydroperoxide. These results make it possible to transfer the knowledge obtained from the chemical systems studied to other hydroperoxides.

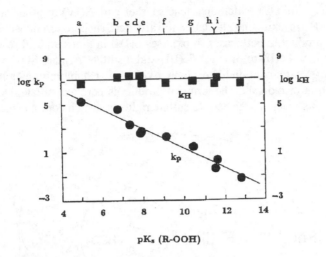

Fig. 10.1.2: Oxidation of HSO_3^- by hydroperoxides: Dependence of rate constants k_H and $k_{HOH}[H_2O]$ on the acid strength of the hydroperoxide; (a) O_2NOOH, (b) $ONOOH$, (c) $HC(O)OOH$, (d) $C_2H_5C(O)OOH$, (e) $CH_3C(O)OOH$, (f) $^-O_3SOOH$, (g) $HOCH_2OOH$, (h) CH_3OOH, (i) H_2O_2, (j) t-C_4H_9OOH, (k) C_2H_5OOH

b.   The role of transition metals in sulfur(IV) oxidation

It has been known for almost a century that transition metal ions can catalyse the oxidation of sulfur(IV) in aqueous solution, but the mechanisms involved have until recently remained obscure. Transition metals occur in cloud water due to the mineral content of aerosol particles that act as cloud condensation nuclei. Collaboration of HALIPP research groups has provided mechanistic and rate data for the metal ion catalysis of sulfur(IV) oxidation. Reasonably complete reaction schemes have been deduced for iron and manganese, which are one-electron oxidants with a favourable abundance in cloud water. It has also been shown that two-electron metal oxidants such as platinum(IV) or thallium(IV) undergo a simple stoichiometric redox reaction with sulfur(IV) without any catalytic effect.

In the presence of Fe(II) and Mn(II) the oxidation of sulfur(IV) displays autocatalytic features, that is the oxidation rate accelerates as the reaction proceeds. This feature is removed when Fe(III) or Mn(III), respectively, are added to the solution. For iron it has been shown that when the reaction is started in the presence of Fe(III) under suitable experimental conditions, Fe(III) is initially almost completely reduced to Fe(II), but in the later stages of the reaction Fe(II) is re-oxidised toward Fe(III). Sulfuroxy radicals have been demonstrated to be formed in the system, so that parts of the mechanism must be identical with the radical initiated chain oxidation of sulfur(IV). This part has been studied separately by kinetic techniques with the result that rate coefficients for chain propagation and termination reactions now are well known. Fig. 10.1.3 shows major parts of the mechanism for the iron-catalysed oxidation of sulfur(IV). The reaction is initiated by the formation of a complex between Fe(III) and sulfur(IV), which partly decomposes under formation of a sulfite radical, SO_3^-. Oxygen converts SO_3^- to SO_5^-, which then reacts either with S(IV) to propagate the chain or with Fe(II) re-oxidising the latter toward Fe(III). Peroxomonosulfate, HSO_5^-, arises as a product in both cases. It oxidises S(IV) in a non-radical reaction, and it also reacts with Fe(II) to produce Fe(III) and a sulfate radical, SO_4^-, which reacts with sulfur(IV) to continue the chain oxidation. Chain termination occurs by recombination of radicals. The terminal product is peroxodisulfate. However, one channel of this reaction regenerates sulfate radicals, which continue the chain.

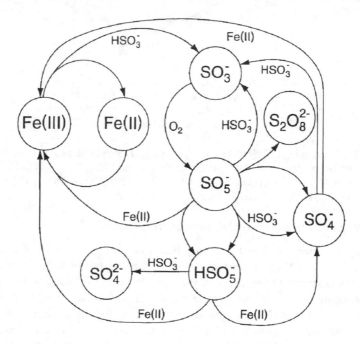

Fig. 10.1.3:   Reaction pathways for the iron-catalysed oxidation of sulfur(IV) in aqueous solution.

Manganese exists in nature mainly in the Mn(II) oxidation state. This contrasts with iron, which occurs largely as Fe(III) in the form of poorly soluble oxides. The principal initiation step generating sulfite radicals in the manganese-catalysed chain oxidation of sulfur(IV) is the formation of a complex between Mn(III) and sulfur(IV). The production of Mn(III) from Mn(II) occurs by reaction with SO_5^- radicals, and in this respect the two mechanisms are similar. In order to generate SO_5^-, however, it is necessary to have a source of SO_3^- radicals, which may either be formed by a mechanism involving a Mn(II)–S(IV) complex or in a synergistic process involving the Fe(III)–S(IV) complex. Traces of iron usually are present in aqueous solution due to unavoidable impurities. Indeed, the addition of Fe(III) to the manganese-sulfur(IV) system accelerates the oxidation rate markedly to values greater than those observed with Mn(II) or Fe(III) alone.

c.   Mechanism of the **OH** induced oxidation of hydroxymethanesulfonate

Hydroxymethanesulfonate (HMS), which is one of the sulfur(IV) compounds present in cloud water, is formed by addition of sulfite/bisulfite to formaldehyde. Because HMS is rather immune to oxidation with ozone and hydroperoxides it was necessary to determine the rate of oxidation by radicals. Rate coefficients for reactions with hydroxyl and sulfate radicals have been determined by means of pulse radiolysis and optical spectroscopy. Additional studies were necessary to determine the complete oxidation mechanism. The results are illustrated in Fig. 10.1.4. Hydroxyl radicals react with HMS by hydrogen abstraction. The intermediate radical thus formed from HMS reacts further with oxygen. The pathway depends on the pH, because the intermediate radical behaves like an acid. At pH 4, where the acid form prevails, a peroxy radical $OOCH(OH)SO_3^-$ is rapidly formed, which decays by a first order process toward a plateau absorption. The dissociation of α-hydroxyperoxy radicals to form HO_2 is an established pathway that would also be applicable here. However, comparison of the observed spectrum with that for HO_2/O_2^- at pH 4 showed that both spectra are different, so that the breakdown of the peroxy radical does not produce solely HO_2. Conductivity experiments confirmed this conclusion. A plausible alternative reaction path is the formation of formic acid and a sulfate radical, which reacts further with HMS to produce a sulfite radical, and ultimately a SO_5^- radical. This pathway oxidises sulfur(IV) and may give rise to a chain reaction. At pH 7, where the anion of the intermediate radical is dominant, the product spectrum corresponds closely to that of O_2^- indicating that it is formed directly without involving an intermediate peroxy radical. The other product, $OHCSO_3^-$, hydrolyses toward formic acid and sulfur(IV). This pathway thus represents an oxidation of formaldehyde rather than sulfur(IV).

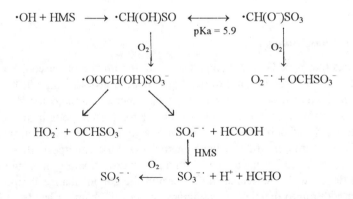

Fig. 10.1.4: Mechanism of the OH radical induced oxidation of hydroxymethane sulfonate (HMS).

d. Radical formation in the aqueous phase by photolysis of solutes

A variety of photochemical processes that are potentially important in the aqueous phase of clouds have been examined to quantify the yields of radicals from these sources. The iron-hydroxo and the iron-oxalato complexes were found most efficient radical sources. Fig. 10.1.5 shows quantum yields for OH production from $FeOH^{2+}$ as a function of wavelength determined with a radical scavenging technique. These data were combined with absorption coefficients and solar actinic fluxes to derive the photo-dissociation frequency $j = 5 \times 10^{-3}$ s^{-1} for mid-latitude summer conditions.

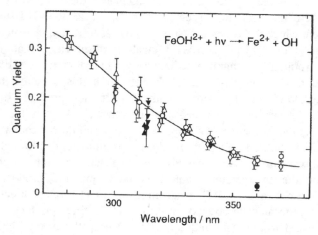

Fig. 10.1.5:   Wavelength dependence of the quantum yield for OH radical formation in the photo-decomposition of $FeOH^{2+}$ in aqueous solution; open points are new measurements, filled data points are previous data.

e. Gas-liquid mass transfer

New techniques that have been developed to study the uptake of gaseous species by liquid water and to measure rates of gas-liquid mass transfer include: (a) a train of newly formed water drops traversing a gas chamber is captured for chemical analysis; (b) a narrow liquid water jet is exposed for a short length to the gas phase, then trapped and subjected to chemical analysis; (c) the loss of a gaseous component flowed through a tube, whose internal walls are covered with a liquid water film, is determined by gas analysis. (c) a water drop floating in an uprising air current is doped with a chemical indicator and the colour change is optically monitored. In addition to these new techniques an aerosol chamber has been used to study the interaction of gases with deliquescent sodium halide particles. Table 10.1.1 compares the main features of the techniques and the working range of uptake coefficients that can be measured. All techniques require a mathematical description of the transport and a comparison of experimental data with model results, which are obtained either in the form of explicit solutions of the differential equations involved or by numerical integration.

Mass accommodation coefficients have been measured by at least two different techniques, mostly the train of droplets and the liquid jet, for important atmospheric trace gases such as sulfur dioxide, ammonia, nitrogen pentoxide, nitrous acid, nitric acid, hydrochloric acid, formic acid, acetic acid, and ozone, with generally excellent agreement. Whereas in most cases the mass accommodation coefficients are large enough to preclude transport limitations, the value for ozone is lower in magnitude so that this important oxidant may be supply-limited.

Table 10.1.1:   Characteristics of the different experimental arrangements used.

	Droplet train technique	Jet technique	Flowtube technique	Wind tunnel experiment	Aerosol Chamber
Working pressure (torr)	10–100	760	760	760	760
Typical exposure time	10^{-3}–10^{-2} s	10^{-4}–10^{-2} s	1 s	> 10 s	> 100 s
Gas phase diffusion limitations	small limitations	limited	limited	limited	very small limitations
Analytical technique	MS-HPLC	DOAS-HPLC	FTIR-HPLC	HPLC	HPLC - gas phase monitoring
Diameter of drops/jet/flow tube(μm)	100	100	6000	40–8000	0.3
Total surface area exposed (cm^2)	10^{-1}	10^{-3}–10^{-2} (a)	10^2	10^{-4}–2	3×10^3 (b)
Nature of the surface	droplet	jet	film	droplet	droplet
Detection limit	$\gamma \geq 10^{-5}$	$10^{-5} < \gamma \leq 10^{-2}$	$10^{-7} \leq \gamma \leq 10^{-4}$		$10^{-7} \leq \gamma \leq 10^{-1}$

(a) not counting liquid jet flow rate,   (b) total particle concentration 5×10^{11} m^{-3}.

f.   The interaction of nitrogen oxides with sea salt particles

Aerosol particles are ubiquitous in the troposphere. When HALIPP was initiated, there were indications that surface reactions on aerosol particles can add significantly to atmospheric chemistry in certain cases. Specifically, the interaction of nitrogen oxides with the liquid water associated with aerosol particles and with dissolved sodium chloride, derived primarily from sea salt, appeared to release chlorine to the gas phase. HALIPP research has demonstrated that the reaction of nitrogen dioxide, NO_2, with NaCl and other halides is rather slow, whereas nitrogen pentoxide, N_2O_5, is very reactive. The precursor to N_2O_5 is NO_3, which is produced by reaction of nitrogen dioxide with ozone. Both NO_3 and its association product N_2O_5 are stable only at night, because they are photo-decomposed in sunlight. Quite unexpectedly it was found that the reaction of nitrogen pentoxide with deliquescent sodium chloride particles as well as with aqueous solutions of sodium chloride generates nitrylchloride, $ClNO_2$, as the main product. This compound is quite volatile so that it is released to the gas phase. Moreover, $ClNO_2$ is photo-decomposed in sunlight whereby a chlorine atom is liberated. A chemical mechanism shown in Fig. 10.1.6 has been advanced to explain this unexpected result. The process begins on the left with the uptake of N_2O_5 by the aqueous phase. The traditional model of hydrolysis leads directly from N_2O_5 to nitric acid formation. However, N_2O_5 may also dissociate to form the ions NO_3^- and NO_2^+, and the latter ion may react either with water or with chloride ion to form HNO_3 and $ClNO_2$, respectively. The uptake of $ClNO_2$ by the aqueous phase is impeded by the presence of chloride, suggesting an equilibrium not only between NO_2^+ and N_2O_5 but also between NO_2^+ and Cl^-. Approximate values for the rate coefficients involved in the reaction scheme of Fig. 10.1.6 have been derived from a comparison of laboratory data with predictions based on computer models describing the chemical system. The formation of $ClNO_2$ from NaCl approximately doubles the source strength of chlorine atoms in the troposphere. There also is evidence for the importance of bromine atoms, which presumably are released in a similar manner from NaBr present in seasalt. A part of the nitrogen dioxide involved in the reaction escapes conversion to nitric acid as it is returned to the atmosphere by the photo-decomposition of nitrylchloride.

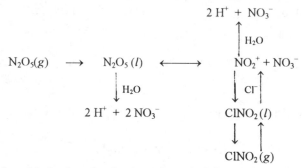

Fig. 10.1.6:   Mechanism for the formation of nitric acid and nitrylchloride from the interaction of nitrogen pentoxide with aqueous sodium chloride solution.

10.1.5 Assessment of HALIPP achievements

Many of the uncertainties that had existed with regard to sulfur(IV) oxidation in the aqueous phase have been removed. In particular, rate coefficients for the reactions involved in the chain oxidation by sulfuroxy radicals have been specified, and the role of iron as a catalyst in the mechanism has been fully delineated. Manganese and cobalt also give rise to autocatalysis, and large parts of the mechanism must be similar. The generation of radicals from photo-labile precursors in clouds has been quantified, and the dominant role of iron complexes as photochemical sources of radicals has been established. Ions of transition metals, especially copper, are scavengers for the HO_2 radical and its conjugate anion, O_2^-, which are converted to H_2O_2. A common mechanism has been established for the reactions of S(IV) with H_2O_2 and other peroxides, specifically peroxomonosulfate, HSO_5^-, which is an important intermediate in the chain oxidation of hydrogen sulphite. The oxidation of formaldehyde by OH radicals in the presence of O_2 forms exclusively formic acid and HO_2 as products. Rate coefficients for the reactions of OH and SO_4^- radicals with hydroxymethanesulphonate and major parts of the reaction mechanisms have been determined. Under atmospheric conditions, SO_4^- radicals interact with chloride to produce chlorine atoms and Cl_2^- radicals. Their reactions have not yet been sufficiently explored to establish the fraction that reacts with water to regenerate OH radicals. The latter reaction determines to some extent the chain length of the radical-driven oxidation of sulfur(IV) in clouds.

Several new techniques for the study of gas-liquid mass transfer have been developed to determine mass accommodation coefficients for a number of trace gases participating in the chemistry of clouds. Some of the techniques will be suitable to study the uptake of radicals by the aqueous phase. This type of investigation has not been performed within HALIPP and remains a task for the future. Some trace gases, $e.g.$ NO_2, may react on the surface of water drops without becoming fully dissolved. Such processes also will need greater attention in the future; the wetted-wall flow reactor might provide a suitable experimental arrangement for this purpose.

With regard to heterogeneous reactions on the surface of materials present in aerosol particles it was recognised from the outset that they required a large exploratory approach. Various studies have shown that minerals can increase the degradation of adsorbed organic compounds under illumination, and that the reaction products often are different from those evolving from gas phase reactions. This result can have significant consequences, especially when product compounds are more toxic than reagent materials. Unfortunately, the concepts advanced so far for surface reactions involving mineral components of the atmospheric aerosol have not been substantiated in the laboratory. For example, it was demonstrated that Lewis acid centres are relatively unimportant as active sites for photo-catalysis. The presence of water on minerals associated with atmospheric particles can have a significant influence on reaction rates, particularly with regard to dissolution of transition metals by the formation of organic complexes. This feature should be given more attention in the future.

The reaction of nitrogen pentoxide, N_2O_5, with deliquescent sodium chloride particles has been shown to produce $ClNO_2$, and a chemical mechanism has been advanced to explain this unexpected result. $ClNO_2$ is transferred to the gas phase where it undergoes photo-dissociation and releases chlorine atoms. The consequences to atmospheric chemistry are twofold: NO_2 is recovered rather than converted to nitrate, and the process represents a source of chlorine atoms additional to that arising from the reaction of OH radicals with HCl.

Most of the original aims of HALIPP have been reached. Remarkable progress has been made in advancing the knowledge and understanding of the aqueous chemistry and gas-liquid mass transfer in clouds. The point has been reached where the new data can be successfully incorporated in simulation models. Good progress has also been made with respect to the interaction of gaseous nitrogen oxides with sodium halides present in sea salt particles, whereas the complexities experienced with surface reactions on minerals associated with atmospheric aerosol particles have defied explanations in terms of mechanistic concepts. Some new knowledge has been obtained nevertheless.

10.1.6   Further information

The HALIPP contribution is shown in context in chapter 3 of this volume. A complete account can be found in volume 2 in this series (Warneck, 1996) which includes the following review.

Review of the Activities and Achievements of subproject HALIPP
P. Warneck, P. Mirabel, G.A. Salmon, R. van Eldik, C. Vinckier,
K.J. Wannowius and C. Zetzsch

10.1.7   HALIPP: Steering group and principal investigators

Steering group

Peter Warneck (Coordinator)	Mainz
Philippe Mirabel	Straßburg
G. Arthur Salmon	Leeds
Christiaan Vinckier	Leuven
Cornelius Zetzsch	Hannover

Principal investigators (*) and their contributions to the final report (Warneck, 1995) are listed below.

Sulfur(IV) Oxidation

T. Amels, H. Elias, U. Götz*, U. Steingens and K.J. Wannowius
Kinetic investigation of the stability of peroxonitric acid and its reaction with sulfur(IV) in aqueous solution

J. Lagrange*, P. Lagrange, C. Pallarès and G. Wenger
Electrolyte effects on aqueous oxidation of sulfur dioxide by strong oxidants

Transition Metal Ions in Atmospheric Waters

J. Berglund and L.I. Elding*
Metal ion catalysed autoxidation of dissolved sulfur dioxide

I. Grgic*, M. Poje, M. Novic, M. Bizjak and V. Hudnik
Catalytic effects of atmospheric aerosol particles in aqueous phase of trace gases

J. Hoigné* and R. Bühler
Kinetics of ozonation processes in cloud and fog water: Effect of transition metal ions on free radical reactions

W. Pasiuk-Bronikowska*, K.J. Rudzinski, T. Bronikowsk and J. Ziajka
Kinetics of S(IV) oxidation in aqueous solution: Impact of transition metal ion
transformations

R. van Eldik*, M. Geißler, C. Brandt and V. Lepentsiotis
The interaction of SO_x and NO_y species in aqueous solution: Photo and metal
ion catalysis

Photochemical Sources and Reactions of Free Radicals in the Aqueous Phase

G.V. Buxton*, G.A. Salmon, S. Barlow, T.N. Malone, S. McGowan, S.A. Murray,
J.E. Williams, N.D. Wood
Kinetics and mechanisms of acid generation in clouds and precipitation

P. Warneck*, H.-J. Benkelberg, U. Deister, M. Fischer, A. Schäfer and J. Ziajka
Chain oxidation mechanisms of S(IV) in aqueous solution: Application of
radical scavenger techniques

R. Zellner*, H. Herrmann, M. Exner, H.-W. Jacobi, G. Raabe and A. Reese
Formation and reaction of oxidants in the aqueous phase

Gas - Liquid Interactions

W. Behnke, M. Elend, A. Frenzel, C. George, H.-U. Krüger, V. Scheer
and C. Zetzsch*
Photo-catalysis of tropospheric chemistry by sea spray

W. Jaeschke*, J. Dierssen, S. Volkwein and J. Wohlgemuth
Phase partitioning and redox processes in fog and cloud systems

P. Mirabel*, C. George, L. Magi and J.L. Ponche
Gas - liquid interactions

U. Schurath*, A. Bongartz, J. Kames, C. Wunderlich, T. Carstens
Laboratory determination of physico-chemical rate parameters pertinent to
mass transfer into cloud and fog droplets

Aerosol Surface Reactions and Photo-catalysis

P. Foster*, I. Denis and V. Jacob
Determination of ß pinene OH radical rate constant – possible impact of solid
particles

J.-C. Petit* and J. Thlibi
Calorimetric study of Lewis acid centres and adsorbed water in the
heterogeneous reactions involving tropospheric particles

P. Pichat*, C. Guillard, C. Hoang-Van, H. Delprat, D. Mas, F. Marme,
T. Bouvier and B. Servajean
Laboratory studies of the photo-transformation of organic compounds in the
presence of water and/or inorganic particulate solids

M.J. Rossi*, F.F. Fenter, K. Tabor, F. Caloz and L. Gutzwiller
   Heterogeneous reactions of nitrogen oxides (NO_2, N_2O_5, HNO_3, $ClONO_2$) with surfaces representative of atmospheric aerosol

C. Vinckier*, N. Van Hoof, S. Ashty and F. Compernolle
   Heterogeneous reactions of air pollutants on various solid surfaces

10.1.8 References

Warneck, P.; 1996, *Heterogeneous and Liquid-Phase Processes,* Springer Verlag, Heidelberg, 254 pp., volume 2 of the EUROTRAC final report.

10.2 LACTOZ: Chemical Processes related to Tropospheric Ozone

Georges Le Bras (Coordinator)

CHRS-LCSR, 1c Av. de la Recherche Scientifique,
F-45071 Orléans Cedex 2, France

10.2.1 Summary

The LACTOZ project has provided an extended kinetic and mechanistic data base that gives much improved quantitative description of the chemical production and loss of ozone from its precursor molecules - nitrogen oxides (NO_x) and volatile organic compounds (VOCs) - in the atmospheric boundary layer and free troposphere. The data base relates to the following classes of reactions:

* reactions of hydroxyl radicals with VOCs; this determines the rates of reaction of VOCs in the sunlit atmosphere;

* reactions of nitrate radicals with VOCs and peroxy radicals and of ozone with alkenes; this controls the non photochemical homogeneous oxidation of VOCs;

* reactions of peroxy radicals (self reactions, reaction with HO_2, other RO_2, NO and NO_2), central to the main ozone generation process;

* reactivity of nitrates and peroxynitrates, which is central to the atmospheric transport of NO_x.

Improved or new photolysis parameters of photo-labile species (*e.g.* carbonyl compounds) have been provided, and significantly new mechanistic information has been established for the oxidation of aromatic and biogenic VOCs (especially isoprene). The available data base has been used to establish structure-reactivity relationships to be used for predicting the reactivity of the very large number of VOCs involved in the generation of tropospheric ozone. As an application of LACTOZ, the data have been incorporated into many atmospheric models describing and predicting ozone and photo-oxidant formation.

10.2.2 Aims and objectives

One of the overall aims of EUROTRAC addressed the problem of transport and transformation of photo-oxidants in north-western Europe, including global, regional and local scale effects involving the atmospheric boundary layer and the free troposphere. Tropospheric ozone is important not only as a photo-oxidant but also for climate because O_3 is a powerful greenhouse gas.

The aims of LACTOZ are to provide, through laboratory investigation of the kinetics and mechanism of relevant gas phase reactions, a quantitative description of the chemical production and loss of ozone from its precursor molecules - nitrogen oxides and volatile organic compounds - in these regions of the troposphere.

The first area of focus was on the reactions involved in the photochemically initiated oxidative degradation of CO and simple hydrocarbons (CH_4, C_2–C_5 alkanes, C_2 and C_3 unsaturated hydrocarbons, *etc.*) and the chemistry of NO_x, relevant for ozone production in the global troposphere. The objective was to provide rate coefficients and reaction mechanisms for a complete model description of the chemistry of NO_x and low molecular weight VOCs in the free troposphere and in air parcels transporting ozone precursors away from source regions.

The second area of focus was the oxidative degradation of more complex volatile organic compounds, (higher alkanes and alkenes; benzene, toluene and higher aromatic compounds; isoprene and other biogenic compounds, oxygenated VOCs), which are important for ozone generation in the atmospheric boundary layer, close to the sources of precursors. In addition to photochemical oxidation initiated by OH, the chemistry of VOC oxidation initiated by the nitrate radical and by reaction with ozone was investigated. The reactions involved in the change in nitrogen speciation resulting from the formation of organic nitrates by the heavier VOCs were also studied.

Within the context of the above foci of the project, the following specific objectives were established for LACTOZ

* to define quantitatively the kinetics and mechanisms for the elementary reactions involved in ozone production in the free troposphere, in particular

  – the oxidative breakdown of simple organic compounds present in the free troposphere (CH_4, HCHO, C_2-C_5 alkanes, *etc.*);

  – the transport and of NO_x into and within the free troposphere and the interconversion of the different NO_x species;

  – the production and loss of odd hydrogen radicals required for accurate calculation of tropospheric free radical concentrations;

* to provide kinetic and mechanistic data necessary for the formulation of models describing the production of ozone in the polluted boundary layer including:

- the determination of the rate constants and mechanisms for OH attack on relatively complex organic compounds including oxygenated species and nitrates;

- the determination of the rate constants and branching ratios for reactions of peroxy and alkoxy radicals, derived from higher molecular weight organic compounds including natural hydrocarbons - isoprene and terpenes;

- the elucidation of NO_3 chemistry with a view to modelling the removal of NO_x and VOCs from the atmosphere during night-time and the impact of this on the overall oxidation of VOCs;

- the investigation of reactions acting as sources and sinks of HO_x (OH, HO_2) and RO_2 radicals in the boundary layer in support of anticipated field measurements of free radical concentrations.

For a given VOC the efficiency of ozone formation is dependent on the details of the degradation pathway. In view of the very large number of VOCs involved in the generation of tropospheric ozone, it was recognised that it would not be feasible to investigate the elementary steps in the oxidation of every VOC. A strategy based on the establishment of structure-reactivity relationships was therefore adopted to provide quantitative rate coefficients and reliable reaction pathways for the diverse reactions of the many individual species.

The nature of the chemical reactions leading to oxidative degradation VOCs and ozone formation in the atmosphere is often determined by the relative rates of competing reaction pathways at several critical points in the sequence. The protocol adopted was therefore to establish these critical points and emphasis was then placed on provision of data allowing the relative importance of the competing processes to be quantitatively defined for a range of VOCs. The critical points were found to involve mainly the reactions of RO_2 and RO radicals.

The considerable achievements of LACTOZ can best be illustrated by mapping the results of the project, as summarised in the next section against the specific objectives that were defined at the start of LACTOZ and that have just been outlined.

10.2.3 Scientific results

a. Reactions of OH radicals

Reaction with the hydroxyl (OH) radical is the most important initiated step in the removal of the majority of VOCs in the troposphere. As a consequence, the rate and extent of formation of secondary pollutants and photochemical oxidants, including ozone, are strongly influenced by the rates of these reactions.

The data base for OH reaction rate coefficients was generally well established, but has nevertheless been improved by providing accurate data for the more complex, higher molecular weight VOCs. As a result of the LACTOZ studies, there is now a much improved data set for aromatic and unsaturated compounds, including isoprene and other biogenic hydrocarbons. Kinetic data have also been obtained for some oxygenated compounds and nitrates for the first time.

Based on the available experimental kinetic data, structure-reactivity relationships have been developed for prediction of the relative yields of OH addition to specific sites in alkenes and poly-alkenes.

Understanding of the mechanism of oxidative degradation following OH attack has improved substantially, in particular for aromatic compounds and for higher (C_5–C_7) alkanes and alkenes.

b.   Reactions of NO_3 radicals

The role of nitrate (NO_3) radicals in the night-time chemistry of the troposphere has been quantified.

Reactions of NO_3 with HO_2 and RO_2 radicals have been shown to be rapid, and to lead to generation of OH and alkoxy radicals respectively. These results provide definitive evidence that the reaction of NO_3 with alkenes can initiate a chain reaction that is a night-time source of OH and other HO_x and RO_x radicals which take part in the chain reaction.

The reactions of NO_3 with volatile organic compounds have been shown conclusively to involve either H-atom abstraction or addition to unsaturated systems. Kinetic data obtained indicate that the reactivity of NO_3 towards saturated compounds is low, and that the reactions may be neglected with respect to the atmospheric removal of these species (Wayne et al., 1991).

Rate-constant measurements for the reaction of NO_3 with alkenes has led to the development of predictive structure-reactivity relationships for these reactions. The results show the importance of NO_3-alkene reactions as a sink for complex alkenes such as biogenic compounds. Carbonyl nitrates, identified as major products of the reactions, are important in long-range transport of odd nitrogen.

c.   Reactions of O_3 with alkenes

Ozone-alkene reactions are an important sink for alkenes and a minor sink for ozone. These reactions provide potentially significant sources of OH radicals, carbonyl compounds, hydrogen peroxide, organic peroxides and acids in the troposphere.

Research within LACTOZ has addressed some of the open questions with respect to reaction rate constants, product formation and general reaction mechanisms. Rate constants have been measured for reactions of ozone with cyclic alkenes. A careful re-measurement of the rate constant for the important reaction of O_3 with ethene has provided a value 30 % lower than that previously recommended.

Studies of the ozonolysis of 2-methyl propene and ethene provide some evidence that the generally accepted mechanism involving biradicals, put forward by Criegee in the 1930s, needs to be complemented by additional reaction pathways.

Quantitative data have been provided for the yield of OH radicals in the reactions of O_3 with a variety of alkenes, including biogenic hydrocarbons. Yields of around 50 % were obtained for typical alkenes, making these reactions a significant source of atmospheric free radicals, a feature which is especially significant at night-time when photolytic sources are absent.

Significant yields of formation of H_2O_2, hydroperoxides and hydroxy-hydroperoxides have been demonstrated for a number of alkenes including biogenic hydrocarbons, and both gas phase and aqueous phase processes are involved.

d.   Reactions of RO_2 radicals

Organic peroxy radicals, RO_2, are the first intermediate species to be formed during the oxidation of organic compounds, and their reactions play a key role in the chain reactions that form ozone in the troposphere. Reaction with nitric oxide (NO), which is the rate determining step for O_3 generation, leads to chain propagation, by forming oxy radicals (RO). This reaction is in competition with terminating processes, in particular $RO_2 + HO_2$ and $RO_2 + NO_2$, so that the relative rates of these reactions are critical parameters in determining the ozone balance.

The work performed within LACTOZ has led to outstanding progress in the state of knowledge of all aspects of RO_2 chemistry, both in terms of reactivity and of reaction mechanisms. The data have been compiled in a review (Lightfoot et al., 1992) and further progress has since been achieved, particularly in defining relationships between structure and reactivity.

One of the principal results concerns the inverse trends in the structure-reactivity relationships that have been observed for $RO_2 + HO_2$ and $RO_2 + NO$ reactions, which are in competition in determining the tropospheric ozone balance. It has been established that increasing the size of alkylperoxy radicals leads to an increase in the rate constant for the former reaction and to a decrease in the rate constant for the latter. In contrast, the presence of electron withdrawing substituents, such as OH or halogens, result in the opposite trend. Since the ratio of rate constants for these reactions can vary by more than a factor of 10 for the most common RO_2 radicals, modelling of the relevant atmospheric chemistry is particularly sensitive to the kinetics of these two processes.

Reactions of RO_2 radicals with NO_2 are efficient terminating reactions when the peroxynitrates formed are stable. This is the case for the reaction forming PAN, one of the most important processes influencing ozone balance and contributing to NO_x transport over long distances. A detailed analysis of the kinetics and thermodynamics of this reaction has been performed, which has led to revision of the currently recommended rate constant values for the forward and reverse reactions.

The thermal decomposition rate constants for 28 peroxynitrates have been studied. A correlation between the activation energies for these reactions with ^{13}C NMR signal shifts of corresponding RX compounds has been found, which can be used to accurately estimate rate constants and activation energies for the thermal decomposition of peroxynitrates.

In remote atmospheres, under low NO_x conditions, peroxy radicals may reach sufficiently   high concentrations for the self reactions $(RO_2 + RO_2)$ and cross-reactions $(RO_2 + R'O_2)$ to occur to a significant extent. These reactions generally possess two reaction channels, a non-terminating pathway forming RO radicals and a terminating pathway forming alcohols and carbonyl compounds. Rate coefficients and branching ratios for the two channels have been obtained for a number of typical radicals. Structure-reactivity relationships are difficult to establish for self reactions, as the rate constants for the different radicals of interest are spread over several orders of magnitude. However, it has been shown that substitution by different types of functional groups in alkylperoxy radicals generally results in a dramatic increase of the rate constant. Consequently, recommendations can be only made for the rate constants for the reactions of several types of primary peroxy radicals that fall in a fairly narrow range and are large enough for such processes to be taken into account in atmospheric reaction schemes. Rate data for various alkyl peroxy cross reactions are reported for the first time. The systems studied were mainly those involving the most abundant peroxy radicals in the atmosphere, CH_3O_2 and $CH_3C(O)O_2$. The most significant result is that all reactions involving acetylperoxy radicals, $CH_3C(O)O_2$, are very fast and must be considered in tropospheric reaction schemes.

Some gaps still remain in our understanding of RO_2 chemistry, but without doubt major advances have been achieved as a result of work performed within the LACTOZ project. The results will also be of considerable value in interpretation of the data from recent instrumental measurements of atmospheric peroxy radicals.

e.   Reactions of RO radicals

Oxy radicals (RO) represent one class of oxidation intermediates for which a variety of alternative degradation channels exist under lower tropospheric conditions; the channels are reaction with O_2, dissociation and/or isomerisation. In the first of these processes, a carbonyl compound with the same number of carbon atoms as the parent VOC is formed together with HO_2 radicals which lead to conversion of NO to NO_2, and hence ozone formation. Dissociation by C–C bond cleavage leads to breakdown of the VOC in the small C chain. The alkyl radicals generated in the dissociative pathways lead to formation of organic peroxy radicals which also give NO to NO_2 conversion and hence ozone formation. In the case of isomerisation, hydroxy substituted alkyl peroxy radicals are formed which ultimately lead to HO_2 formation. The extent of dissociation and isomerisation thus has significant bearing on the total NO to NO$_2$ conversion associated with the oxidative degradation of VOCs.

Within LACTOZ, the reactions of the simpler alkoxy radicals, which are generated in the oxidation of small hydrocarbons, have been quantified using direct techniques. For the study of large oxy radicals, such as those originating from the oxidation of large hydrocarbons in the continental boundary layer, indirect and relative techniques have been applied. It has been shown that reactions of alkoxy radicals with O$_2$ are slow ($k \approx 8 \times 10^{-15}$ cm^3 molecule^{-1} s^{-1}) and essentially independent of the size of the radical. For n-alkoxy radicals, dissociation by C–C bond fission only becomes important for radicals containing at least 4 carbon atoms. For the oxy radicals formed in the oxidation of n-pentane and n-hexane, significant contributions from isomerisation channels involving 1.5 - H atom shifts have been demonstrated. Although the results from three independent studies are largely consistent with theoretical predictions of the relative importance of the various channels open to reaction of RO, a systematic analysis of structure-reactivity relationships for RO radicals has not yet proved possible, and this remains a priority in future work.

f.   Photochemical cross sections and quantum yields

Atmospheric photolysis is the predominant source of free radicals in the atmosphere. Photolysis of NO$_2$ produces O atoms which form ozone; ozone photolysis in the near UV produces O(^{1}D) which reacts with H$_2$O to produce OH radicals. A number of organic species absorb UV light and dissociate to yield peroxy radicals.

Work within LACTOZ has provided refined or new absorption cross section and quantum yield data that can be used to calculate photolysis rates under atmospheric conditions.

Emphasis has been put on: (a) refinement of data for the simple, well characterised molecules (O$_3$, HCHO) involved in HO$_x$ radical production, and (b) determination of data for photolysis of organic compounds formed as products in the degradation of alkanes, alkenes and selected multifunctional VOCs, mainly carbonyl compounds and nitrates. The carbonyl compounds include methylvinylketone, methacrolein and methylglyoxal, formed in the OH and O$_3$ initiated oxidation of isoprene. The atmospheric nitrates studied included ketonitrates formed in the NO$_3$ initiated oxidation of VOCs. The data obtained give information concerning the relative importance of photolysis and attack by OH in the tropospheric degradation of these compounds.

Higher absorption cross section values than previously recommended have been determined for PAN. These new data suggest that the rate of photolysis of PAN in the troposphere is approximately 3–4 times greater than previously calculated.

g.   Oxidation of aromatic compounds

Aromatic hydrocarbons (benzene, toluene, xylene isomers) constitute a major class of organic compounds associated with the urban environment. These compounds are estimated to contribute more than 30 % to photo-oxidant formation in urban areas. Studies within LACTOZ have provided data on many different aspects of the mechanisms of atmospheric oxidation of aromatic compounds, which until recently were poorly characterised.

Studies of the first steps of the oxidation of aromatic hydrocarbons have established that the initial adduct, formed from attack by OH, reacts with O_2 to produce HO_2 on a short time-scale. This observation, together with the rapid reaction of the OH radical with substituted aromatic compounds, accounts for efficient ozone production from these species. However, there is some evidence from product studies that the peroxy radicals formed are very short lived and therefore unable to oxidise NO to NO_2, in contrast to the situation for oxidation of alkane and alkene in the atmosphere. Under atmospheric conditions, muconaldehydes (hexa-2,4-diene-1,6-dial) are proposed as direct products of the reaction of the aromatic-OH adduct with O_2 in the oxidation of toluene and p-xylene.

This conclusion is supported by the observation of similar products (unsaturated 1.4-dicarbonyl compounds, glyoxal, methyl glyoxal, maleic anhydride) in the reactions of OH with muconaldehydes and the OH initiated oxidation of toluene and p-xylene. Smog chamber studies also seem to indicate that cresols (from toluene), and alkylated phenols (from xylene isomers), are very minor products under atmospheric conditions.

The carbon balance found in aromatic compound product studies is presently only approximately 50 % C, however investigations have shown that up to approximately 30 % of the "missing carbon" may be in the form of vicinal polyketones or their hydrates.

h.   Oxidation of biogenic compounds

Terpenes and isoprene are emitted from forests and other vegetation in Europe, the importance of each individual compound depending on tree type and location. Almost all the detailed information obtained so far within LACTOZ refers to isoprene. There are severe experimental difficulties with the study of the terpenes.

In the case of attack on isoprene by OH, and in the presence of oxygen, a peroxy-hydroxy radical is formed; up to six isomers may be produced. Product studies indicate that the initial addition is to one of the terminal carbon atoms. Under conditions where the peroxy-hydroxy radical reacts exclusively with NO, approximately 50 % of the carbon balance is accounted for by three main products: methacrolein, methylvinylketone, and formaldehyde. Attack of NO_3 on isoprene apparently proceeds in much the same manner, but there is considerable controversy about the precise reaction pathway because of the variety of peroxy radicals that can be formed. Reaction of isoprene with ozone follows a rather different course. Ozone adds across one or other of the double bonds, and the ozonide may then fragment in several possible ways, to form methacrolein, methylvinylketone, epoxides and Criegee radicals.

Even for the simplest of the molecules, isoprene, there remain uncertainties about the degradation pathways. One problem concerns the poor carbon balance in almost all of the studies of attack by OH, NO_3 and O_3. Another problem concerns the effect of humidity on product distributions. Yet a further question hangs over the significance of the ozonolysis reactions as a source of radicals.

Almost nothing is known about the mechanisms and specific oxidation pathways for terpenes, and there are substantial experimental obstacles to investigation of these systems. Further work is clearly warranted in view of the potential importance of these processes in atmospheric VOC and ozone chemistry.

10.2.4 Conclusions

The results of the LACTOZ project have provided a much improved quantitative description of the chemical production and loss of ozone from its precursor molecules - nitrogen oxides and volatile organic compounds - in the atmospheric boundary layer and free troposphere.

For the free troposphere, the understanding of the chemistry of oxidation of simple organic compounds and CO under low and high NO_x regimes has been improved. The critical NO_x level that determines whether oxidation leads to depletion or formation of ozone is now better defined.

The kinetic and mechanistic data set established for the oxidation of large VOCs may be used in order to understand ozone formation in the continental boundary layer. The results provide an improved knowledge of the VOCs/NO_x ratios that determine the NO_x-limited and VOC-limited oxidation regimes.

The kinetics and mechanisms for the oxidative breakdown of simple organic compounds present in the free troposphere are now well established and a complete quantitative mechanism is available for the atmospheric oxidation of C_1 and C_2 hydrocarbons. LACTOZ has made a significant contribution to this work, in particular with respect to the reactions of peroxy radicals, which are intermediates in the OH-initiated oxidation of VOCs. The kinetics and mechanisms of the reactions involved in the transport and interconversion of NO_x into and within the free troposphere has also been improved. The project has provided the data necessary for a better evaluation of the role of alkyl nitrates and PAN, which are NO_x reservoirs, in NO_y partitioning in the troposphere. LACTOZ has also shown that the functionalised nitrates produced by the OH, O_3 and NO_3 initiated oxidation of alkenes are potential reservoirs for NO_x. These functionalised nitrates could provide an explanation for the missing species in NO_y field measurements. This possibility requires further investigation.

Significant progress has been made in understanding the reactions of peroxy and oxy radicals, generated in the oxidation of complex VOCs. The extensive kinetic and mechanistic data base obtained for RO_2 reactions within LACTOZ has led to the establishment of structure-reactivity relationships that can be used to predict the behaviour of a large range of peroxy radicals. Despite limitations in the direct experimental detection of oxy radicals, considerable advances in understanding the relative importance of the various possible reaction channels for these species under atmospheric conditions have been made. In particular, the products arising from isomerisation of long-chain RO radicals have been identified for the first time.

The improvement and extension of the photochemical data base for photochemically active species (*e.g.* carbonyl compounds) results in a better assessment of the tropospheric production and loss of odd hydrogen (OH, HO_2) radicals which is required for accurate calculation of tropospheric free-radical concentrations. With a similar aim, the OH source from the ozonolysis of a variety of alkenes, including biogenic hydrocarbons, has been quantified, showing that this source is especially significant at night when photochemical sources of radicals are absent.

A further potentially significant night-time source of OH radicals has been discovered within LACTOZ. Hydroxyl radicals are generated in the NO_3 initiated oxidation of VOC in a chain mechanism involving reaction of NO_3 with peroxy radicals. It has been also established that O_3 reactions with alkenes is a potentially significant source of OH and RO_2 radicals.

Useful advances have been made in the knowledge of the rates and mechanism for OH attack on complex organic compounds, such as aromatic compounds, alkenes, isoprene and other biogenic hydrocarbons, oxygenated compounds and organic nitrates.

The kinetics and mechanisms for the primary reactions of OH radicals with aromatic compounds have been elucidated within LACTOZ. Rapid generation of HO_2 radicals and unsaturated carbonyl compounds in the absence of NO_x has also been demonstrated. These reactive products are expected to further increase ozone formation in the atmosphere.

LACTOZ has carried out work on the atmospheric oxidation of a number of biogenic compounds. Detailed mechanisms for the oxidation of isoprene under conditions of both high and low NO_x have been established.

For ozone-alkene reactions, which are important sinks at least for alkenes (in particular complex biogenic alkenes), the rate constants have been revised downward. Rate data for the reactions of O_3 with alkenes determined within LACTOZ have shown that these reactions are somewhat slower than previously reported; however, it is now apparent that for biogenic species reaction with O_3 provides an important sink. Reaction product investigations from the reactions of O_3 with alkenes have identified the formation of peroxides, which appears to be strongly dependent on the relative humidity.

Various aspects of NO_3 radical chemistry have been elucidated with a view to modelling the removal of NO_x and VOCs from the troposphere under night-time conditions. From the kinetic and mechanistic data obtained for NO_3 reactions with both man-made and naturally produced VOCs, it is now possible to make quantitative estimates of night-time HNO_3 production from NO_x via the homogeneous NO_3 + VOC pathway as opposed to the heterogeneous route via N_2O_5. Night-time production of organic nitrates, which are reservoirs for NO_x, can also be estimated from the data reported.

Applications of the results obtained within LACTOZ to the understanding of ozone formation in both the boundary layer and the free troposphere have received considerable attention throughout the project. LACTOZ has provided a firm basis for an accurate description of the oxidative capacity of the atmosphere and hence for the establishment of realistic NO_x and VOCs emission control strategies. The realisation of the objectives requires the derivation of chemical mechanisms for incorporation in chemical transport models. In the last stages of LACTOZ it has been possible to generate simple mechanistic schemes which begin to satisfy this goal. Further work on these mechanisms should be included in future projects. In addition, LACTOZ has identified a number of scientific uncertainties in our understanding of tropospheric ozone that should be addressed in due course.

10.2.5   Further information

The LACTOZ contribution is shown in context in chapter 2 of this volume. A complete account can be found in volume 3 in this series (Le Bras, 1997) which includes a comprehensive review by the editor and steering group of the activities and achievements of LACTOZ together with the scientific results.

10.2.6   LACTOZ: Steering group and principal investigators

Steering group

Georges Le Bras (Coordinator)	Orléans
Robert Lesclaux	Bordeaux
Karl-Heinz Becker	Wuppertal
R. Anthony Cox	Cambridge
Geert K. Moortgat	Mainz
Howard W. Sidebottom	Dublin
Reinhard Zellner	Essen

Principal investigators (*) and their contributions to the final report (Le Bras, 1997) are listed below.

K. Bächmann*, B. Kusserow, J. Polzer, J. Hauptmann and M. Hartmann
Laboratory Studies of the Atmospheric Decay of Hydrocarbons in the NO_x Free Atmosphere

10.2.7 References

Lightfoot, P.D., R.A. Cox, J.N. Crowley, M. Destriau, G.D. Hayman, M.E. Jenkin, G.K. Moortgat, F. Zabel; 1992, Organic Peroxy Radicals: Kinetics, Spectroscopy and Tropospheric Chemistry, *Atmos. Environ.* **26A**, 1804–1964

Le Bras, G. (ed); 1996; *Chemical Processes in Atmospheric Oxidation*, Springer Verlag, Heidelberg; Vol. 3 of the EUROTRAC final report.

Wayne, W.P., I. Barnes, P. Biggs, J.P. Burrows, C.E. Canosa-Mas, J. Hjorth, G. Le Bras, G.K. Moortgat, D. Perner, G. Poulet, G. Restelli, H. Sidebottom, 1991, *Atmos. Environ.* **25A**, 1.

10.3 The Chemical Mechanism Working Group

D. Poppe[1], M. Kuhn[2] and K.-H. Becker[3]

[1]Institut für Atmosphärische Chemie, Forschungszentrum Jülich,
D-52428 Jülich, Germany
[2]Kempener Straße 125, 50733 Cologne, Germany
[3]Institut für Physikalische Chemie, Bergische Universität-GH Wuppertal,
D-42119 Wuppertal, Germany

10.3.1 Introduction

The Chemical Mechanism Working Group (CMWG) is one of the activities within EUROTRAC to establish interdisciplinary links between the different subprojects, which are basically oriented towards disciplinary work. The scope of the group is to encourage and maintain the scientific communication between the laboratory work in LACTOZ and HALIPP, the modelling in EMEP, EUMAC, TRACT, and GLOMAC, and the field experiments in TOR. The main goal is to communicate information and upcoming new scientific objectives on chemical reaction mechanisms. This includes the methodology of designing chemical mechanisms for chemical transport models, the coupling between gas-phase, heterogeneous and aqueous phase chemistry as well as new reactions of tropospheric relevance. In particular the improved laboratory kinetic data on the oxidation of aromatic hydrocarbons and biogenic alkenes were discussed as well as their implementation into the existing chemical modules. Scientists from in and outside EUROTRAC involved in regional modelling, field studies, and kinetic work participated. Between 1991 and 1995 four workshops were conducted. A list of CMWG members is given in Appendix A, Table 6.

The group executed two major projects. The first concerns the updating of the chemical schemes currently in use in the various chemistry and transport models that are operated in EUROTRAC. Within the community for regional models and air quality studies several types of mechanisms are implemented. While the inorganic chemistry is explicit and basically identical in all implementations, the treatment of the large body of organic reactions is very different. None of the reaction schemes treats all reactions fully explicitly, because such an approach is beyond the capability of today's computers.

10.3.2 EMEP chemical mechanism

The gas-phase chemistry of the EMEP model is an example of a scheme that treats the contributing organic chemistry explicitly. Because kinetic data can be directly used, such type of chemical mechanisms can be easily updated and also scrutinised in a straight forward manner. The EMEP mechanism has been updated by the LACTOZ group (Wirtz *et al.*, 1994). In particular, the most recent results on RO_2 reactions achieved by LACTOZ have been introduced. At present new data on biogenic VOC oxidation including ozonolysis are in preparation for inclusion into the EMEP mechanism.

In carbon bond type schemes the rate constants for the reactions of organic compounds are derived from the chemical reactivity of certain structure elements in that particular compound (Gery *et al.*, 1989). If the reaction rate constant can not be taken directly from the laboratory work, it is a simple linear combination of those. Such chemical mechanisms can be readily updated similar to the explicit schemes.

10.3.3   RADM2 chemical mechanism

The gas-phase chemistry RADM2 of the EURAD model belongs to another type of chemical schemes (Stockwell *et al.*, 1990). For reasons of improved performance with less computational effort the lumping of organic compounds is done by reactivity (for example with respect to the hydroxyl radical OH). Reaction constants for such lumped species are weighted sums of rate constants for individual reactions. The rate coefficients depend in a non trivial fashion on the emission pattern and also on the time scale of the modelling (Middleton *et al.*, 1992). Because of the rather complicated procedure by which kinetic information is incorporated into the scheme, it is not obvious how to scrutinise the mechanism and how to introduce improved reaction rate constants. Even more difficult is the implementation of further reactions. It was realised that the future development of RADM2 should allow for a simpler incorporation of additional and updated kinetic information.

Since RADM2's last update in 1989 rate constants with reduced experimental uncertainties were obtained. Also product distributions and chemical pathways of the degradation of aromatic compounds were investigated. Progress in laboratory work promises a completely different view on the oxidation mechanisms of aromatic hydrocarbons compared to the present state of knowledge. In short, the new results indicate that the ring cleavage, though producing HO_2 radicals, does not involve a reaction step with NO (alkane-type reaction). The products of the ring cleavage, unsaturated dicarbonyls, exhibit a very large reactivity towards OH radicals and photolysis. The improved kinetic data should be incorporated into RADM2 but also into all chemical schemes in particular to account for their potential impact on photo oxidant formation.

10.3.4   Chemical mechanism intercomparison

The second major topic concerns the comparison of chemical schemes for the gas-phase chemistry of the troposphere (Poppe *et al.*, 1995). Several modules that are implemented into chemistry and transport models within and outside EUROTRAC participated in an intercomparison (see Table 10.3.1). Addressing a comparison of the chemical schemes, only time-dependent box model calculations were done. The performances of the mechanisms were tested for five scenarios that simulate chemical processing in the remote atmosphere as well as the moderately polluted planetary boundary layer. To make the intercomparison sensitive to the differences of the chemical schemes the photolysis frequencies for all mechanisms and all cases with one exception were prescribed.

Table 10.3.1: Participants in the intercomparison of chemical mechanisms

Abbreviation	Mechanism	Contributing Author	Reference
EMEP	EMEP MSC-W	D. Simpson	Simpson *et al.*, 1993
UiB	University of Bergen	A. Strand and Ø. Hov	Strand and Hov, 1994
IVL	IVL chemical module	Y. Andersson-Sköld	Andersson-Sköld 1995
ADOM	ADOM II	P. Makar	Lurmann *et al.*, 1986
Ruhnke	Ruhnke mechanism	R. Ruhnke	Ruhnke, 1995
RADM2-IFU	RADM2	F. Kirchner and W. Stockwell	Stockwell *et al.*, 1990
RADM2-FZK	RADM2	B. Vogel, H. Vogel and F. Fiedler	Stockwell *et al.*, 1990
RADM2-KFA	RADM2	M. Kuhn and D. Poppe	Stockwell *et al.*, 1990
Euro-RADM		F. Kirchner and W. Stockwell	Stockwell and Kley, 1994
CBM-IV-LOTOS	CBM4 chemistry	P. Builtjes, M. Roemer and A. Baart	Gery *et al.*, 1989; Builtjes, 1992
CBM-4-TNO	CBM4 chemistry	P. Builtjes, M. Roemer and A. Baart	Gery *et al.*, 1989; Stern, 1994
CB4.1	CB4 Chemistry	M. Das and J. Milford	Gery *et al.*, 1989; Milford *et al.*, 1992

The cases LAND, BIO and FREE deal with rather clean air masses with low burdens of hydrocarbons except the BIO simulation which accounts for the presence of isoprene as an important representative of biogenically emitted hydrocarbons. These three cases consider the ageing of air over five days, no emissions were included. The scenarios are closely related to the equally named cases of the intercomparison of photochemical models conducted by the IPCC group (Prather *et al.*, 1995).

The PLUME1/2 simulations study the chemistry under more polluted conditions with emissions based upon the European emission inventory as compiled by Derwent and Jenkin (1991). Both cases differ in the treatment of the photolytic processes. PLUME1 used the prescribed frequencies, while for PLUME2 the photolysis data of the individual CTM models were to be used.

As an example the diurnal cycles of several trace constituents are discussed. The LAND case (Fig. 10.3.1) models the chemistry of rather clean air with low burdens of pollutant and correspondingly weak photochemical activity. Differences for OH and ozone are small as expected, since the chemistry, that is active in the LAND case, is basically the same in all mechanisms.

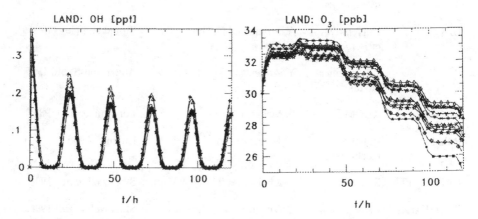

Fig. 10.3.1:   Diurnal cycles of OH and ozone for the LAND case. Results from all participants (Table 10.3.1).

The PLUME1 case simulates the planetary boundary layer under substantial anthropogenic emissions (Fig. 10.3.2). The simulation is sensitive to the different treatment of the hydrocarbon chemistry and their emissions. For ozone, which is formed as a result of many chemical processes, the differences are rather small (Fig. 10.3.2a).

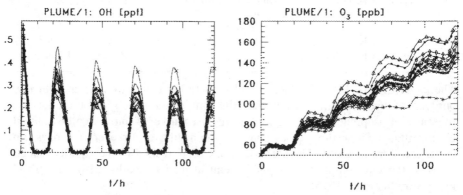

Fig. 10.3.2a: Diurnal cycles of OH and ozone for the PLUME1 case. Results from all participants (Table 10.3.1).

Clustering of models in the central range are evident, however, one model deviates substantially from the others. Enhanced differences are found for the hydrocarbons (Fig. 10.3.2b).

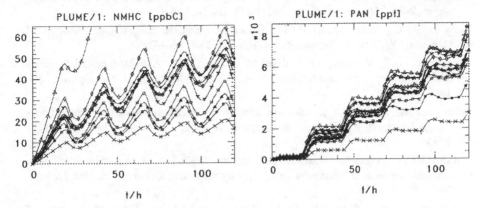

Fig. 10.3.2b: Diurnal cycles of the non-methane hydrocarbons (NMHC) and PAN from the PLUME1 case. Results for all participants (Table 10.3.1).

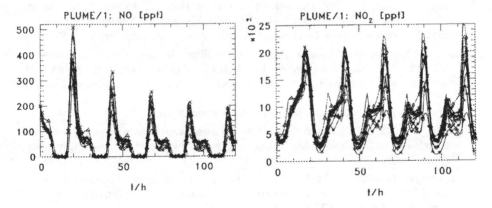

Fig. 10.3.2c: Diurnal cycles of NO and NO_2 for the PLUME1 case. Results from all participants (Table 10.3.1).

Another important result of the intercomparison are the scenarios. They provide with sets of initial conditions for concentrations and with emissions strengths meaningful for the troposphere over Europe. Therefore they have, beyond the scope of the present intercomparison, a much wider field of applications. They can be used, for example, in further comparisons but also for the study of sensitivities toward new reactions or updated rate constants or to check the numerics of the chemical balance equations.

A detailed report of the intercomparison has been published (Poppe *et al.*, 1996).

10.3.5 References

Andersson-Sköld, Y.; 1995, Updating the chemical scheme for the IVL photochemical trajectory model, IVL B-1151, IVL, Goteborg, Sweden.

Builtjes, P.J.H.; 1992, The LOTOS Long Term Ozone Simulation Project, summary report. IMW-TNO Rep. 92/240, TNO, Delft, The Netherlands.

Derwent, R. G., M.E. Jenkin; 1991, Hydrocarbons and the Long-Range Transport of Ozone and PAN across Europe, *Atmos. Environ.* **25A**, 1661–1678.

Gery, M.W., G.Z. Whitten, J.P. Killus, M. Dodge; 1989, A Photochemical Kinetics Mechanism for Urban and Regional Scale Computer Modeling, *J. Geophys. Res.* **94**, 12925–12956.

Lurmann, F.W., A.C. Lloyd, R. Atkinson; 1986, A Chemical Mechanism for Use in Long-Range Transport/Acid Deposition Computer Modeling. *J. Geophys. Res.* **91**, 10905–10936.

Milford, J.B., D. Gao, A.G. Russell, G.J. McRae; 1992, Use of sensitivity analysis to compare chemical mechanisms for air-quality modeling, *Environ. Sci. and Technol.* **26**, 1179–1189.

Middleton, P., W.R. Stockwell, W.P.L. Carter; 1990, Aggregation and Analysis of Volatile Organic Compound Emissions for Regional Modelling. *Atmos. Environ.* **24A**, 1107–1133.

Prather, M., R. Derwent, D.H. Ehhalt, P. Fraser, E. Sanhueza, X. Zhou; 1995, Other Trace Gases and Atmospheric Chemistry, in: J.T. Houghton, L.G. Meira Filho, J. Bruce, Hoesung Lee, B.A. Callander, E. Haites, N. Harris, K. Maskell (eds), *Climate Change 1994 - Radiative Forcing of Climate Change and An Evaluation of the IPCC IS92 Emission Scenarios*, Cambridge University Press, Cambridge, pp. 77-126.

Poppe, D., Y. Andersson-Sköld, A. Baart, P.J.H. Builtjes, M. Das, F. Fiedler, Ø. Hov, F. Kirchner, M. Kuhn, P.A. Makar, J.B. Milford, M.G.M. Roemer, R. Ruhnke, D. Simpson, W.R. Stockwell, A. Strand, B. Vogel, H. Vogel; 1996, *Intercomparison of the Gas-Phase Chemistry of Several Chemistry and Transport Models*, EUROTRAC Special Report, EUROTRAC ISS, Garmisch-Partenkirchen.

Ruhnke, R.; 1995, Ein Verfahren zur Analyse von chemischen Reaktionssystemen und zur Erstellung von Chemiemodulen für atmosphärische Modelle, Ph.D. thesis, Universität Essen, Essen.

Simpson, D., Y. Andersson-Sköld, M.E. Jenkin; 1993, Updating the chemical scheme for the EMEP MSC-W oxidant model: current status, EMEP MSC-W Note 2/93, Norwegian Meteorological Institute, Oslo.

Stern, R.M.; 1994, Entwicklung und Anwendung eines dreidimensionalen photochemischen Ausbreitungsmodells mit verschiedenen chemischen Mechanismen, in: Meteorologische Abhandlungen **8**, Institut für Meteorologie der Freien Universität Berlin, Verlag von Dietrich Reimer, Berlin.

Stockwell, W.R., P. Middleton, J.S. Chang, X. Tang; 1990, The second generation regional acid deposition model chemical mechanism for regional air quality modelling, *J. Geophys. Res.* **95**, 16343–16367.

Stockwell, W.R., D. Kley; 1994, The Euro-RADM Mechanism: A Gas-Phase-Chemical Mechanism for European Air Quality Studies, Forschungszentrum Jülich GmbH (KFA), Jülich.

Strand, A., Ø. Hov; 1994, A two-dimensional global study of tropospheric ozone production, *J. Geophys. Res.* **99**, 22877–22895.

K. Wirtz, C. Roehl, G.D. Hayman, M.E. Jenkin; 1994, *LACTOZ re-evaluation of the EMEP-MSC-W photo-oxidant model*, EUROTRAC Special Report, EUROTRAC ISS, Garmisch-Partenkirchen.

Chapter 11

Modelling and Emissions

Modelling is at the heart of applied atmospheric chemistry: the computer codes try to encapsulate both the understanding of the atmospheric processes and the detailed knowledge of the many contributing meteorological, chemical and biochemical processes. Some modelling was carried out in all the subprojects but three subprojects were specifically concerned with model development or the generation of emission data for modelling.

    11.1   **EUMAC**   European Modelling of Atmospheric Constituents

    11.2   **GENEMIS**   Generation of European Emission Data

    11.3   **GLOMAC**   Global Modelling of Atmospheric Chemistry

EUMAC was devoted to the development of regional scale chemical transport models. In the formation of GLOMAC was the recognition that Europe exists in a global context and that our emissions contribute to the global burden and that this also determines our background. Indeed the problem of incoming ozone from the North Atlantic had to be taken into account in the formulating the forthcoming European Commission ozone directive to establish acceptable ozone levels for Europe.

GENEMIS was a Cinderella project of EUROTRAC. In the early considerations, although pressed from some quarters, the committees decided against having a subproject devoted to emissions. However it was soon realised that for the model results to have any credibility at all reliable emission data was required and GENEMIS was formed. The prince has now certainly arrived to marry Cinderella and emissions have been accorded an overall priority in EUROTRAC-2, both from a scientific point of view and in the view of the environmental agencies. (ISS; 1999; EUROTRAC-2: Project Description and Handbook; EUROTRAC-2 ISS, München, 1998. See also: *www.gsf.de/eurotrac*; p20 and p32).

11.1 EUMAC: European Modelling of Atmospheric Constituents

Adolf Ebel (Coordinator)

Universität zu Köln, Institut für Meteorologie und Geophysik, EURAD,
Aachener Straße 201-209, D-50931 Köln, Germany

11.1.1 Summary

The EUROTRAC subproject EUMAC (European Modelling of Atmospheric Constituents) continued its work on the development of an advanced model hierarchy for air pollution dispersion simulations. The applicability of the long-range transport model system EURAD (European Air Pollution Dispersion Model System) to episodic simulations has been demonstrated for a variety of cases. Evaluation studies focusing on the meteorological as well as chemical parts of the system were conducted. Methods for nesting smaller-scale three-dimensional transport models into larger-scale ones were further improved. Applications of the former model type, in particular the EUMAC Zooming Model (EZM), are published in volume 7 of this series (Ebel *et al.*, 1997a).

The chemical transport simulations were accompanied by sensitivity and process studies also using box models. First results from large-eddy simulations reveal the importance of concentration perturbations for their chemical transformation in the atmosphere. The role of biogenic emissions, aerosols and fog for chemical transformations were topics of further process studies. A major issue of this branch of research in EUMAC was cloud effects on the budgets of atmospheric pollutants.

Co-operative work with other EUROTRAC subprojects and other projects and programmes was continued, for instance, by pursuing comprehensive comparisons of long-range transport models. An important field of collaboration continues to be the generation of reliable emission models and scenarios for three-dimensional chemical transport simulations.

11.1.2 Aims of the subproject EUMAC

The main focus of EUMAC was the development of a three-dimensional Eulerian model system for the simulation of air pollution in the troposphere over Europe and its application to scientific problems, environmental planning and the support of experimental studies. The work on the European scale was complemented by the development and application of smaller-scale chemical transport models and by sensitivity as well as process studies employing a hierarchy of models. The EURAD system was developed as the core model of

EUMAC. One of its tasks was to support also other EUROTRAC subprojects (*e.g.* ALPTRAC, TOR, TRACT) through episodic simulations.

The model hierarchy was assembled in a way that zooming of chemical transport calculations to smaller areas embedded in the larger European domain became possible. The treatment of effects due to the presence of clouds and precipitation should be improved and the role of fog and aerosols for chemical transformations explored.

An essential goal was the evaluation of the meso-scale transport models used in EUMAC by applying observations made in the framework of EUROTRAC and data from other sources, in particular meteorological data from the European Centre for Medium-Range Weather Forecast (ECMWF) and air quality data from EMEP and national monitoring networks.

11.1.3 Structure of EUMAC

In the course of its development EUMAC selected six major scientific tasks on which it would focus its activities. EUMAC Specialist Groups (ESGs) were formed to carry out the tasks. The respective themes treated by them were:

1)   long-range transport modelling,

2)   model evaluation,

3)   smaller-scale models,

4)   chemical mechanisms,

5)   clouds, and

6)   emissions.

This chapter is structured according to the listed topics. Further information on scientific results obtained through EUMAC can be found in the special issue on the subproject in *Meteorology and Atmospheric Physics* **57**, No. 1–4, 1995.

11.1.4 Long-range transport modelling

a.   Approaches to regional modelling

The original plans of EUMAC contained the idea to explore and compare different existing and new approaches to chemical transport modelling on the larger regional, *i.e.* European, scale. In this way, the reliability and accuracy of the core model EURAD should be checked and improved. Different meteorological drivers, *e.g.* the Eta model of the University of Belgrade (Mesinger *et al.*, 1990), and chemical mechanisms were under consideration. The ideas could partly be realised through contributions from the Finnish Meteorological Institute (Hongisto and Joffre, this issue) focusing on sea-land interactions and from the Max-Planck-Institute for Meteorology connecting the regional climate model HIRHAM (the Hamburg version of HIRLAM (High

Resolution Limited Area Model)) with the EURAD-CTM (EURAD Chemical Transport Model; Langmann and Graf, see Volume 7). Furthermore, a comparison exercise involving five different long-range transport models could be initiated as described in the section on model evaluation and will soon be completed.

b.   Simulation of episodes

Various episodes have been chosen for simulation of chemical transformation and transport of pollutants. Their selection was mainly determined by specific applicational reasons. For instance, one of the first simulations was carried out to study the spread of the radioactive cloud after the Chernobyl nuclear reactor accident. Other cases were selected according to the dates of field campaigns conducted in the framework of EUROTRAC. As far as the EURAD model system is concerned several episodes were simulated in support of other ongoing environmental projects (e.g. on the changing tropospheric pollution in East Germany and the impact of aircraft emissions). They cover a broad spectrum of meteorological situations.

Joint EUMAC cases

Two of the episodes have been treated as cases for joint studies within the subproject EUMAC, namely the episodes from 25th April to 7th May 1986 (so-called joint wet case, JWC) and from 17th to 20th July 1986 (so-called joint dry case, JDC) . The model data base was made accessible to all who wanted to participate in the joint investigations of a great variety of issues related to long-range transport calculations. In this way it was possible to avoid dispersion of EUMAC modelling activities to a great deal.

Collaboration with other subprojects

Several cases were chosen for joint analysis with other subprojects. A EUMAC/TOR case (episode in July/August 1990) was defined using the availability of sufficient TOR observations during a photo-smog episode as the criterion for selection. The TRACT case (Sept. 1992) covers the period of the TRACT campaigns, whereas the ALPTRAC case (March 1990) was chosen from several attractive episodes with campaign data from the ALPTRAC network (Ackermann et al., 1994).

c.   Model improvement and development

The connection of the EURAD-CTM with HIRHAM established a new link between meso-scale air pollution and climate modelling. In cooperation with the cloud group of EUMAC new cloud parameterisations were tested in EURAD and an improved cloud chemistry mechanism was prepared. Also modifications of the gas-phase chemical mechanism and the photolysis module were studied. The EURAD Emission Model (EEM) was adapted to episodic emission input data from the subproject GENEMIS. It was demonstrated that inclusion of an aerosol module parameterisation ammonium sulfate and nitrate formation is essential for reliable estimates of the budgets of nitrogen compounds.

Through a nesting option which became available with the improved meteorological model (MM5) of EURAD the range of applicability of the system increased considerably.

d. Application to research and environmental assessment

Long-range transport modelling was applied to a variety of research and environmental problems. Whereas the HIRHAM/CTM system aims at the treatment of regional climate issues and the Finnish contribution deals with coastal processes which are typical and important for European land type properties controlling pollutant transport, the EURAD system was used to treat a relatively large spectrum of problems. In a few cases this was achieved by applying EURAD in other environmental projects which originated after the start of EUROTRAC. Occasionally, this was done in collaboration with other groups contributing to EUMAC, e.g. with those running smaller scale models or developing dry and wet chemical mechanisms.

To demonstrate the range of topics to which the EURAD model system was applied some of them are listed here without going into details: impact of air traffic emissions on the tropopause region (Petry et al., 1994; Lippert et al., 1994), dynamics of the tropopause including cross-tropopause ozone transport (Ebel et al., 1991, 1996); Kuwait oilfield burning; nitrogen deposition to the North Sea (Bock, 1994; Memmesheimer and Bock, 1994); monitoring network optimisation and transport model initialisation (Petry, 1993); emission reduction scenarios (Memmesheimer et al., 1991); assessment of air quality changes in East Germany since 1989 (Jakobs et al., 1995); and the impact of biogenic emissions.

It is an essential experience originating from EUMAC work that it is crucial to carry out joint and strongly interactive development of all components of an air quality model system. Otherwise it may easily happen that inconsistencies of independently designed parts of the system hamper the progress of model improvement and application.

11.1.5 Model evaluation

Model evaluation in atmospheric science faces the problem that the model output is generally far more comprehensive than the observational data available for validation. The model output is superior than the observational data both in terms of the variables represented and in terms of time and space resolution. On the other hand, any scientific theory (and any model based on such theory) must be scrutinised with independent data taken from reality in order for the model to be considered relevant and acceptable for the user.

Model evaluation can principally be made in a variety of approaches. A straightforward way is to compare a quantity which had been forecast by the model with the equivalent observed quantity. It is generally assumed that the model is the more credible the smaller the difference between the forecast and observed quantity. An indirect way is to compare quantities forecast by one model with those derived by another mode. Here the underlying hypothesis is that both models gain more credibility the closer their results are to each other. The models run by the EUMAC community are characterised by coupling meteorological and chemical modules. Evaluation can put emphasis upon either the meteorological or the chemical part. In order for a tracer concentration to be correctly forecast both the meteorological and the chemical module must be independently as correct as possible. Based upon these considerations the model evaluation group has put emphasis on two different but closely related aspects. In a first approach the meteorological component of EUMAC was analysed with specific emphasis upon evaluation of the water budget. In a second series of studies the trace gas predictions by EURAD were diagnostically evaluated by employing independent observations.

a.   Evaluation of hydrological components within EURAD

The transport of atmospheric trace constituents is intimately related to the hydrological cycle. Several of the trace species which are routinely simulated in the EURAD model are chemically coupled to water. Furthermore, clouds can modify the actinic flux. Thus, a high quality simulation of the hydrological cycle is a necessary prerequisite for successful modelling of trace species in the free atmosphere.

The diagnostic model DIAMOD of YANAI type designed by Hantel (1987) and used as a tool for evaluation has been implemented at the University of Vienna. Input is the gridscale routine data (obtained from ECMWF) representative for atmospheric columns with 100 km horizontal scale. DIAMOD yields, among other quantities, the vertical convective flux h and the rain flux r in the free troposphere; h is the correlation between moist enthalpy (proportional to equivalent temperature) and vertical motion. Both fluxes h and r are considered functions of pressure in an atmospheric column; they are calculated as sub-gridscale residuals from the averaged budget equations for sensible and latent heat.

While most evaluation methods employ field quantities (e.g. concentrations, like specific humidity) DIAMOD uses fluxes (i.e. the sub-gridscale fluxes h and r). The reason is that these are considerably more sensitive to insufficient parameterisation. The complete evaluation philosophy has been discussed by Hantel et al. (1995).

An example is shown in Fig. 11.1.1 for a case of severe precipitation in southern Germany and northwestern Austria. The case is characterised by a meteorological summer situation with weak pressure gradients, almost of tropical type (Dorninger et al., 1992). Fig. 11.1.1 demonstrates that the EURAD model depicts the location of the maximum of the fluxes correctly but underestimates the amplitudes. This result is also typical for other cases, for example, the Chernobyl event which was investigated by Wang (1994).

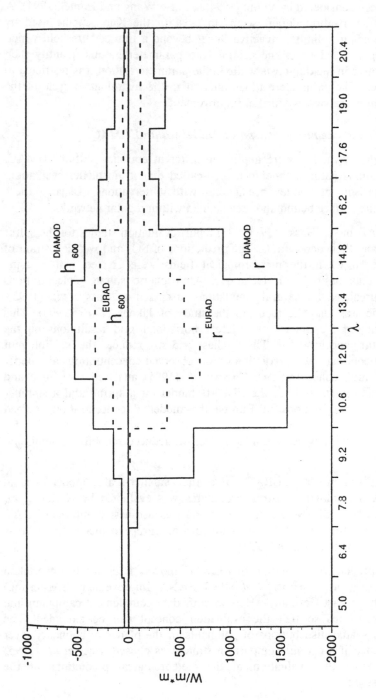

Fig. 11.1.1: East-West section (in about 48 degrees northern latitude) of vertical convective flux h (across level 600 hPa) and of rain flux r (across earth = B4s surface). Horizontal axis: Geographical longitude, centred at individual atmospheric column, increment 1.4 degrees. Both fluxes in energy units W/m^2, positive in downward direction; a rain flux of 30 W/m^2 corresponds to 1 mm precipitation per day. Full curves: h and r diagnosed by DIAMOD; dashed: h and r forecast by EUMAC. Fluxes are 12-hour averages centred on 1st August, 1991.

Possible reasons for the underestimation of the free atmosphere sub-gridscale fluxes have been considered by Wang (1994; see also Wang and Hantel, 1994). A relevant parameterisation deficit seems to exist in the Kuo scheme used by EURAD. It tends to inhibit convective fluxes beyond a limited threshold which results in negative values of the central Kuo parameter b; this quantity then becomes negative in the diagnosis while in the parameterisation it is restricted to positive values. The parameterisations implicit in the hydrological cycle of the meteorological model deserve further improvement.

b.   *Evaluation of chemistry transport modules within EUMAC*

Several air quality models were applied to different episodes within EUMAC. These applications naturally included a so-called diagnostic model evaluation, *i.e.* a comparison of model predictions with observations. Usually these observations are surface bound and are obtained within regular networks.

For the EURAD model Hass *et al.* (1993) report evaluation studies for the sulfur components and their depositions. The predictions of SO_2 and wet deposition of sulfate improved when the meteorological fields were created with a four-dimensional data assimilation technique. Atmospheric sulfate concentrations were under predicted for several simulations discussed by Hass *et al.* (1993). Sulfur dioxide was also the focus in the study of Jakobs *et al.* (1995). This evaluation showed the impact of different horizontal grid resolutions on the predicted sulfur concentrations. The smallest grid size reduced the bias inherent in Eulerian modelling: over prediction of low observed concentrations and under prediction of high concentrations. Hass *et al.* (1995) analysed the ozone and nitrogen dioxide predictions of the EURAD model for a spring and a summer episode. For several regions in Europe the modelled concentrations agreed reasonably well with the observations. Larger discrepancies were found *e.g.* for NO_2 in eastern Europe which seems to reflect the reduced reliability of emission data for that region.

A joint application of EURAD and the smaller-scale model system KAMM/DRAIS employing nesting procedures was evaluated by Nester *et al.* (1995). The EZM model (Moussiopoulos, 1994) was also nested into the EURAD model. The evaluation of EZM results for several European metropolitan areas can be found in Moussiopoulos (1994).

A more comprehensive evaluation of EUMAC models is currently undertaken within the SANA project (Ebel *et al.*, 1995; SANA: Improvement (SANierung) of Air Quality in East Germany). Here aircraft data complement campaign and network observations so that the evaluation concept can be extended. The temporal and spatial distributions of pollutants in the planetary boundary layer can be evaluated if a careful comparison strategy is chosen. Ebel *et al.* (1995) also report about the evaluation of the meteorological predictions of the modelling systems.

Of special interest is a study comparing EURAD, the LOTOS model (applied in TOR), EMEP models and a model from the Free University of Berlin (REM3).

The study involving all four models is focusing on photo-oxidants. A comparison dealing with the sulfur components and their wet deposition is carried out between the EURAD model and EMEP's acid rain model. The performance and behaviour of the models for a specific period in July/August 1990 is studied. Fig. 11.1.2 shows results from the photo-oxidant intercomparison study.

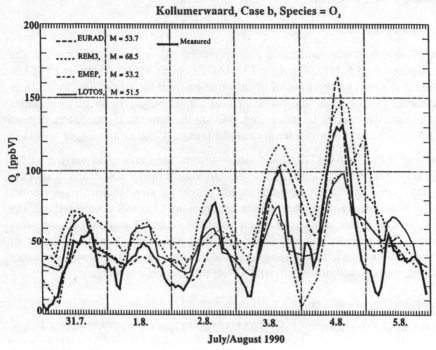

Fig. 11.1.2:   Ozone time series for the TOR station Kollumerwaard in the Netherlands. Episode: July 31st–August 5th, 1990. The measured ozone concentrations (ppb) are shown and the results from 4 long-range transport models used in Europe.

11.1.6   Smaller-scale models

a.   Goals of implementation and application

A group of transport models for the smaller meso-scale was improved and applied in the subproject EUMAC. The major goal was to provide tools for zooming chemistry transport calculations from the continental down to the local scale. The reasons to apply different types of smaller-scale models in EUMAC are multi-fold ranging from availability and resources for the individual models over flexibility, differences in their physical and chemical content, and design for application to domains with specific characteristics, *e.g.* alpine terrain, coastal areas or regions with peculiar emission conditions. Such tasks not only need specialisation of models but also experienced groups being capable to deal with specific local and regional problems of the polluted atmosphere.

In the course of the project an alternative to smaller-scale modelling with existing β and γ meso-scale models was developed, namely direct nesting of domains of decreasing size with the EURAD model system. This method has mainly been applied in projects which co-operate with EUROTRAC, *e.g.* SANA, whereas EUMAC also continued to follow its original plan to develop zooming or nesting capabilities for a broader class of models which can and have also been used as stand-alone model systems.

b.   Nesting methods

Special investigations addressed the problem of optimum nesting of the meteorological part (MEMO) of the EUMAC Zooming Model system to EURAD. Different schemes for the transfer of information from the coarse grid to the finer grid model (matching at the boundaries or for the whole mesh of the smaller scale model) were tested in connection with studies how these procedures depend on the refinement of topography in complex terrain (Gantner and Egger, 1994).

Coupling of the KAMM/DRAIS model system to the coarse-grid version of the EURAD system was carried out. The meteorological model KAMM is nested via the basic state of the coarse-grid model providing large-scale forcing to the smaller scale model. The chemical transport model DRAIS is coupled via the initial and boundary conditions taken from the EURAD simulations (one-way nesting). This method allows better handling of lateral boundary fluxes of chemical species and, therefore, enables more accurate simulations for smaller domains under conditions where such fluxes are particularly relevant.

c.   Specific applications

The interactive application of the KAMM/DRAIS and the EURAD model system was forming an essential link between the subprojects EUMAC and TRACT where both models were applied with the aim to support the interpretation of TRACT campaign measurements. Another joint application of both model systems was carried out in the framework of the SANA project focusing on extreme air pollution conditions in eastern Germany.

Studies of special situations of air pollution included environmentally "hot spot" problems related to areas with dense population and intensive industrial activity. Applications to the Greater Lisbon Area (Borrego *et al.*, 1995), Athens and Barcelona are among them. Furthermore, studies were carried out to explore the role of aqueous phase chemistry, and to test alternative radiation schemes and the consequences of changing update intervals for meteorological fields used for transport calculations in smaller-scale models (Giovannoni *et al.*, see Ebel *et al.*, 1997a).

A particularly appealing development within EUMAC towards smaller scale model application is described in the following section.

d.   The EUMAC zooming model

An important product of the EUMAC Specialist Group on submodels and parameterisation is the EUMAC Zooming Model (EZM; Moussiopoulos, 1994b; Moussiopoulos, 1995), which is a contemporary model system capable of addressing problems related to air pollutant transport and transformation at the local-to-regional scale. In the last four years the EZM has been successfully utilised in a number of case studies, primarily for large European urban agglomerations. Although initially designed as a tool to be used in conjunction with a continental scale model, the EZM evolved to be now a stand-alone model system which may be driven directly with measured data. Core models of the EZM are the non-hydrostatic prognostic meso-scale model MEMO (Kunz and Moussiopoulos, 1995a) and the photochemical dispersion model MARS (Moussiopoulos et al., 1995a).

Several recent EZM applications have a strong environmental policy background. Three examples are given below.

1)   The study launched by the Greek Ministry of the Environment aiming to assess the environmental impact of constructing the New Athens Airport. For this purpose, nested grid simulations were performed to describe the dispersion of inert air pollutants in the Athens basin and the adjacent Mesogia plain for different meteorological conditions (Moussiopoulos et al., 1995b).

2)   The "Heilbronn Ozone Experiment" which was performed in the period 23rd–26th June, 1994 with the intention to prove to what extent peak ozone levels may be reduced by the aid of short term local scale interventions. Nested grid simulations were performed with the EZM for the situation prevailing during the experiment in order to analyse the detailed meteorological and air pollutant measurements carried out during this period. Dispersion calculations were also performed for the hypothetical case of no emission reductions during the period of the experiment. By comparing the results of the latter calculations with those of the standard ones conclusions were drawn with regard to the effectiveness of the emission reduction interventions (Moussiopoulos et al., 1995c).

3)   The Auto/Oil study commissioned by DG-XI of the European Union. This joint study of several EU Directorates, ACEA and Europia is supposed to serve as the scientific basis for setting future European motor fuel qualities and motor vehicle exhaust limits. The role of the EZM was to analyse air quality in the local-to-regional scale (Moussiopoulos and Kallos, 1995). In the above cases and in numerous previous applications (Moussiopoulos, 1994a), the results of the EZM were found to agree fairly well with available observations.

e.   The APSIS model intercomparison study

The Athenian Photochemical Smog Intercomparison of Simulations (APSIS) was performed in the frame of EUMAC, with the objective to check the capabilities of contemporary models to describe wind flow and ozone formation in Athens (Moussiopoulos, 1993). Models employed in EUMAC like EZM and KAMM/DRAIS participated among several others not directly linked to this EUROTRAC subproject. Although APSIS was rather a demonstration exercise than an in-depth evaluation project, its findings lead to the following conclusions.

a)   Contemporary prognostic meso-scale models are capable of reproducing most of the wind field features which significantly affect air pollutant concentrations in complex airsheds like Athens.

b)   Statistical analyses of deviations between model results and presently available measured data may lead to useful assessments of the wind model performance (Kunz and Moussiopoulos, 1995b).

c)   Available experimental datasets are hardly sufficient for a conclusive evaluation of photochemical dispersion models. The consequence from the latter conclusion is that detailed field campaigns are needed in order to build a comprehensive experimental database for the conclusive evaluation of present and future air pollution models at the local-to-regional scale. Such campaigns should primarily aim to check the model accuracy with regard to budgets for all major compounds. In addition, the experimental evidence should allow representing the interaction between anthropogenic and natural emissions.

f.   Large eddy simulations

Research started on the problem of the influence of eddy mixing on chemical transformations in the atmosphere during the end phase of EUROTRAC. A large-eddy model was complemented with a simplified NO_x chemistry scheme to demonstrate the principal effects of eddies in the convective boundary layer. As a first result it could be demonstrated that updraft regions are areas of enhanced chemical activity. It is expected that the results of more detailed studies of this phenomenon will lead to significant consequences for the parameterisation of chemical transformation processes in Eulerian transport models (Nieuwstadt, see Ebel *et al.*, 1997a).

11.1.7   Chemical mechanisms

After the EUROTRAC Working Group on Chemical Mechanisms was formed in 1991 the EUMAC Specialist Group on Chemistry was mainly operating in the framework of the new group to avoid duplication of work. Within EUMAC sensitivity studies regarding the chemical mechanisms as integral part of the whole chemical transport system were carried out in connection with long-range transport simulations. The role of ammonium nitrate and sulfate formation, transport and deposition for the nitrogen budget of the planetary boundary was studied leading to far-source concentration and deposition patterns which were markedly different from calculations without the aerosol phase and in better agreement with available observations (Ziegenbein *et al.*, 1994). The chemical mechanism of EURAD was analysed with respect to the photochemical ozone creation potential of various VOCs under changing meteorological and emission conditions (Heupel, 1995). The dependence of chemical calculations on initial conditions was tested. And several studies focusing on changes of VOC/NO$_x$ ratios were conducted in connection with emission sensitivity tests. The RADM2 gas-phase chemistry code used in EURAD was expanded to enable the treatment of aircraft emissions in the upper range of the model domain (Lippert *et al.*, 1994). A new module for calculating photolysis frequencies was tested (Hass and Ruggaber, 1995).

a.   Validation of gas phase chemistry using field measurements

A particularly important contribution to EUMAC has been the validation study of the RADM2 gas-phase mechanism employing data from field measurements (Poppe *et al.*, 1994). Experimental data on the hydroxyl radical, which is the key substance in the photochemistry of the troposphere, was compared with model calculations using RADM2. Due to the short chemical lifetime of OH a zero-dimensional treatment suffices to model OH. The calculations overestimate the atmospheric OH concentration on the average by 20 % in the continental planetary boundary layer under rural and moderately polluted conditions. Based on this results a comparison for the longer-lived intermediates ozone and other photo-oxidants was conducted. Measurements were carried out during a period in August 1990 with a high photochemical activity and substantial burdens of primary and secondary pollutants. Simulations of the diurnal cycles are in good agreement with the field data for ozone, formaldehyde and PAN after improving the chemical mechanism with respect to the isoprene degradation mechanism (Zimmermann and Poppe, 1994).

b.   Future needs

As regards the RADM2 scheme an assessment for remote areas with low non-methane hydrocarbon mixing ratios and also for the free troposphere should be carried out. A similar conclusion applies to other condensed mechanisms for gas phase reactions in the polluted atmosphere.

11.1.8 Clouds

Clouds play a crucial role with regard to the development of the weather and climate and the associated changes in our environment. They are not only responsible for a redistribution of water substances and heat in the atmosphere, thereby affecting the water content and temperature of the soil, but also act as chemical processors. The cloud processes occur on the relatively small-scale of an individual cloud or a cloud system. However, they have a large impact on the regional and global scale.

a.   Aims

As regards the EURAD model system cloud effects including wet phase chemical processes enter at various points in the meteorological and chemical part. The aim of the ESG on clouds was to improve the treatment of cloud processes in the EURAD model and to adapt the associated parameters to European conditions. Consequently, the physical content and performance of the model as a whole have improved.

The group which assembled consists of experts from different domains connected to cloud/fog dynamics, microphysics and chemistry, thus reflecting the broad spectrum of problems to be tackled: Peter Builtjes and Jan Matthijesen from TNO (NL); Nadine Chaumerliac and Sylvie Cautenet from LaMP (F); Renate Forkel and Ralph Dlugi from the University of Munich (D); Hermann Jakobs and Manfred Laube from the University of Cologne (D); Jean-Michel Giovannoni and Frank Müller from ETH, Lausanne (CH); Günther Mauersberger from the University of Cottbus (D); Andrea Flossmann, Andreas Bott and H.R. Pruppacher from the University of Mainz (D); Robert Rosset from the University of Toulouse (France), and Chris Walcek from SUNY (USA). Herein, R. Rosset and C. Walcek acted as advisors while the other members of the group actually contributed to improve the EURAD model. H. Jakobs was the link to the EURAD model and helped with all questions concerning the EURAD model and the implementation of codes.

b.   Results

The groups at TNO, LaMP and University of Cottbus have worked on schemes concerning the aqueous phase chemistry mechanisms. Here, they have developed different degrees of complexity. The most sophisticated scheme came from the University of Cottbus. It is based on the comprehensive liquid-phase system compiled by Müller and Mauersberger (1995). Even though this project dealt with the development of special fast numerical solver, the cloud chemical module is nevertheless time consuming. The group has started to develop a reduced mechanism, however, has not yet finished with identifying the dominant pathways as they found out that the local sensitivity analysis of Pandis and Seinfeld (1989) is not sufficient for this purpose. TNO and LaMP developed simplified aqueous phase chemical modules, focusing mainly on ozone effects. The module of LaMP considers the main oxidation chains of ozone in the presence of NO_x and takes into account the role of drop size for mass transfer. With this module little effect of a cloud on ozone evolution was found, but a strong modification of the distribution of NO_x due to an efficient scavenging of radicals that modifies the conversion of NO to NO_2.

The group at the ETH in Lausanne investigated the structure of the EURAD model which consists of two components, a meteorological part and a chemical part. The chemical transport model CTM is driven by the stored meteorological fields. In the work of the ETH group the CIT (CALTECH / Carnegie Mellon Institute of Technology) photochemical transport model was extended with respect to the representation of clouds. An aqueous phase reaction mechanism and the STAR radiation transfer model (Ruggaber *et al.*, 1994) was also implemented. The group of the University of Munich investigated chemical reactions in fog events, focusing especially on sulfate production and photochemical reactions. They studied the EUMAC joint cases (April/May and July 1986) and found that the liquid phase sulfate production in the fog sometimes exceeds the gas phase production during 24 hours within a layer of the same height. Consequently, fog events which are sub-gridscale phenomena for meso-scale models can nevertheless have a noticeable effect on the larger scale.

The group of the University of Mainz worked on the convection parameterisation including scavenging of aerosol particles and wet deposition. A new mass conservative scheme based on the work of Fritsch and Chappel (1980) was developed and incorporated into the two parts of EURAD. It has been applied to one of the EUMAC Joint Cases and the change in the precipitation and wet deposition was studied. First tests have indicated that less cloud water than in the Kuo scheme is formed due to convection and that less convective precipitation is consequently calculated. As a result less wet deposition is obtained. These effects go in the right direction as the Kuo scheme is suspected to over-predict convective precipitation. Further tests are necessary to thoroughly judge the performance of this new scheme. The group at the University of Cologne has worked on the improvement of the large scale cloud treatment and the associated scavenging and wet deposition (Mölders *et al.*, 1994). They have developed an ice parameterisation scheme to be coupled to the warm cloud scheme. Results of sensitivity experiments predicted that not only the cloud structure and dynamics but also the distribution of trace gas species in the atmosphere and wet deposition of sulfate at the earth surface are strongly affected by turning on or off the ice parameterisation or the riming process.

c.   Problems to be addressed in future research

Work on problem areas attacked by the ESG on Cloud Processes has well advanced not only by indicating weak points in the current understanding and parameterisation of clouds and the associated multiphase processes but also by pointing out the directions for an improvement and taking first steps into this direction. The problem areas, however, are too complex to expect a comprehensive solution within the period EUROTRAC has been existing. Further work is needed to harvest the fruits of the research done so far. From the work of the aqueous phase chemistry group a significant improvement in the in-cloud reactions can be expected. Also, the effects of the separated parameterisations of aqueous-phase processes in the meteorological and chemical part of air pollution model systems have to be further looked into. Fog episodes need to be parameterised in large scale models. Furthermore, the coupling of the cumulus convection parameterisation and the ice parameterisation scheme for meso-scale clouds promises a substantial improvement of the understanding of the wet deposition properties of clouds. In future, the work should be more intimately linked with aerosol research.

11.1.9 Emissions

Regarding air pollution emissions form the starting point for the chemical reactions and the transport phenomena in the atmosphere. Emission inventories are generated on an annual basis. Emission values are the result of calculations using emission factors relating certain activities, *e.g.* combustion, to the release of certain trace substances, *e.g.* sulfur dioxide. The emission factors stem from selected measurements and are chosen to represent typical conditions of the activity, *e.g.* road traffic in a country. The emission factors need to be constantly updated and adjusted to the prevailing situation. Equally, the changing nature of sources and activities needs to be tracked over time.

The study of atmospheric transport and deposition of trace substances on regional scales requires a high temporal resolution of pollutant concentrations. The generation of an emission data base with hourly data was started in the framework of EUMAC by research groups at Essen University, subsequently Rostock University, and Stuttgart University. After GENEMIS was established the major responsibility for emission work was transferred to this EUROTRAC subproject. Yet, the design of the model linking the data base generated in GENEMIS and also EMEP, namely the EURAD Emission Model (EEM) remained a task of EUMAC. In this context it is gratefully acknowledged that an emission scenario for 1982 was made available by the Dutch-German PHOXA Project (Stern and Scherer, 1989) for first applications of the EURAD model before EUMAC and GENEMIS could design their own emission data bases.

a.   Aims

The emission work for the EURAD model aimed at producing an improved emission data base with a high temporal resolution and European coverage. The high resolution is achieved by disaggregating the annual emission to hourly values. The activities like the operation of a power plant have to be modelled for each month of the year for typical weekdays and hours of the day. Specifically, emission files for specified time episodes have to be generated as input to the EURAD CTM. Besides anthropogenic emissions, biogenic emissions also have to be taken into account. Specifically, the goal of present activities has been to change from preliminary emission scenario calculations using EMEP inventories (Memmesheimer *et al.*, 1991) to more detailed and, as VOCs are concerned, also more sophisticated emission modelling based on CORINAIR and LOTOS (Friedrich *et al.*, 1994).

b.   Results

The CORINAIR inventory contains annual emissions from 1985 for the countries of the European Union. The data have been updated to 1986. For non-EU countries the annual emissions from LOTOS have been used. The 1990 CORINAIR inventory was not released in time for use within the project. For 1990 and subsequent years the LOTOS 1990 inventory was adapted for model work.

The two modelling groups at Essen and Stuttgart have developed computer codes for the various source types and activities in close cooperation. The former group has been responsible for large point sources and has modelled the temporal behaviour of power plants, large combustion units, refineries, iron and steel mills, etc. The Stuttgart group has been responsible for area sources entailing the diverse activities in the industrial sector, the residential sector and the transport sector. It has also coordinated the build-up of the data bank and linking of the computer models for emission estimates. Biogenic emissions have been treated at the University of Cologne in the framework of the EURAD project.

The emission scenarios now available allow us to calculate emission episodes for any given time interval ranging from days to months in selected years. The results are formatted as input for atmospheric transport models like EURAD. Emission episodes generated are:

Joint Wet Case	(JWC)	25th April – 7th May	1986
Joint Dry Case	(JDC)	17th July – 20th July	1986
EUMAC-TOR Case	(ETC)	1st July – 10th July	1990
		29th July – 5th Aug.	1990
EUMAC-ALPTRAC		7th March – 11th March	1990
		21st March – 31st March	1991
SANA Episode 1990		10th Oct. – 20th Oct.	1990
SANA Episode 1991		28th Aug. – 6th Sept.	1991
TRACT Episode		14th Sept. – 22nd Sept.	1992
GLOMAC		1st Jan. – 31st Dec.	1990

c.   Problems for future work

The emission modelling work proved the feasibility of disaggregation of annual emission data to hourly values. The quality of the episode data especially depends on the quality of the annual emission inventories. It is a constant task to improve the inventories. Not only the emissions themselves but also the technical data appending to the type and nature of the emission sources need permanent revision. The new CORINAIR 1990 represents an important step in this direction and will form the basis for future episode calculations.

Extensive information of the technical, social, economic and meteorological factors affecting the temporal behaviour is required. Future work will have to ascertain the quality of the information available outside the non-EU area, especially in eastern Europe. Ammonia emissions need to be investigated more thoroughly all over Europe. Biogenic hydrocarbon, nitrogen and sulfur emissions are not yet fully understood regarding their dependence on meteorological and land type parameters.

The computer software and data format need to be streamlined in order to enable better communication with application models. It also needs to be improved in order to more easily adapt to changing data bases like the new CORINAIR-EMEP 1994. In general, the logistics of inventory design, emission scenario modelling and transfer to air quality models need to be developed further in the light of the experience gained through the co-operation of EUMAC and GENEMIS.

11.1.10   Conclusions

When comparing the original plans of EUMAC with its achievements at its end one finds considerable successes which were not expected at the beginning but also areas where the subproject did not reach its original objectives. In most cases this can be attributed to a lack of necessary resources often enhanced by inconsistent funding policy on a national basis. Thus it became clear in the very early phase of EUMAC that the establishment of a EUROTRAC community model in the form of a pure service project would not be possible under the given conditions and that other ways for supporting the EUROTRAC community had to be sought. Therefore, direct links to two subprojects of major importance for model application, namely TOR and TRACT, were formed through specific contributions of the EURAD modelling group to their work programme. Later on, collaboration was also sought with other subprojects, e.g. ALPTRAC and TESLAS. EUMAC considerably benefited from the work of the EUROTRAC Working Group on Chemical Mechanisms, since this aspect of research was not well enough covered by subproject contributions, and the WG on Clouds linking modelling activities with field and laboratory work in EUROTRAC.

A useful and not expected development occurred in the area of smaller-scale modelling regarding community support and application. A versatile model system, the EUMAC Zooming Model (EZM), recently became available. It has been used by several groups for treatment of local to smaller regional air pollution problems. At the same time a nesting version of the regional model EURAD was designed which extended its range of application to smaller spatial scales through increasing the horizontal resolution by two, in future three, orders of magnitude. This is also an achievement beyond what was expected in the planning phase of EUMAC.

Major problems originated from the reduced weight which was put on emission work in the early phase of EUROTRAC. Though EUMAC supported emission research from the beginning of the subproject on it took some time before an independent subproject on emissions (GENEMIS) could be started. Air quality simulations for Europe are still suffering from this late start since the design of emission data bases suitable for meso-scale air pollution simulations and implementation in chemical transport models is more tedious than usually believed.

Considerable improvements of the deposition and biogenic emission modules are expected to result from relevant work of the subproject BIATEX. Of course, transfer of results achieved there again needs its time and remains a task for the post-EUROTRAC era. Also, more comprehensive use of TOR results for model evaluation would be desirable regarding future work. On the other hand, model results from EUMAC were able to provide essential support for ozone budget studies carried out in TOR. Special topics where model results became essential are fluxes between the free troposphere and polluted boundary layer and stratosphere-troposphere exchange. The latter issue has also been linking EUMAC and GLOMAC activities and became an important aspect for aircraft emission research where the EURAD model system found a stimulating field for application.

As outlined in the previous section evaluation studies have been carried out using existing observational data bases, in particular those from monitoring networks (EMEP, national) for air pollution. Yet, as has been shown by EUMAC, it is absolutely necessary to also evaluate the meteorological part of the model system driving the chemical transport model (Tilmes et al., 1995). Extension of the evaluation of the chemistry part of the model systems which presently mainly focuses on ground-based observations to upper levels using air-borne measurements is essential. First approaches are based on TRACT and SANA field campaigns. More work is needed here so that the accuracy and reliability of meso-scale air pollution dispersion modelling with the EURAD system and other components of the EUMAC model hierarchy will further increase in future.

11.1.11   Further information

The EUMAC contribution is shown in context in chapter 5 of this volume, a set of papers were also published in *Atmospheric Environment* (Ebel *et al.*, 1997b).

A complete account of the subproject can be found in volume 7 of this series which includes an overview of air quality models and modelling.

11.1.12   EUMAC: Steering group and principal investigators

Steering group

Adolf Ebel (Coordinator)	Köln
Karl-Heinz Becker	Wuppertal
Carlos A. Borrego	Aveiro
Remi Bouscaren	Paris
Peter Builtjes	Delft
Andrea Flossmann	Clermont-Ferrand
Ulf Hansen	Rostock
Michael Hantel	Vienna
Heinz Hass	Aachen
Nicolas Moussiopoulos	Thessaloniki
Dirk Poppe	Jülich
Robert Rosset	Toulouse

Contributions from principal investigators (*) to the final report (Ebel *et al.*, 1997a), together with some of their co-workers, are listed below.

Larger Scale Modelling of Air Pollutant Transport, Transformation and Deposition in Europe

A. Ebel*, H. Elbern, H. Hass, H.J. Jakobs, M. Memmesheimer, M. Laube, A. Oberreuter and G. Piekorz
Simulation of Chemical Transformation and Transport of Air Pollutants with the Model System EURAD

B. Langmann and H.-F. Graf*
Another Meteorological Driver (HIRHAM) for the EURAD Chemistry-Transport Model: Validation and Sensitivity Studies with the Coupled System

M. Hongisto and S.M. Joffre*
Transport Modelling over Sea Areas

D. Poppe*, J. Zimmermann and M. Kuhn
Validation of the Gas-Phase Chemistry of the EURAD Model by Comparison with Field Measurements

M. Hantel* and Y. Wang
Diagnostics of Vertical Subscale Fluxes (SUBFLUX)

Smaller Scale Modelling of Air Pollutant Transport, Transformation and Deposition in Europe

K. Nester*, F. Fiedler and H.J. Panitz
Coupling of the DRAIS Model with the EURAD Model and Analysis of Subscale Phenomena

N. Moussiopoulos*, G. Ernst, Th. Flassak, Ch. Kessler, P. Sahm, R. Kunz, C. Schneider, T. Voegele, K. Karatzas, V. Megariti and S. Papalexiou
The EUMAC Zooming Model, a Tool Supporting Environmental Policy Decisions in the Local-to-Regional Scale

L. Gantner* and J. Egger
Nesting of Mesoscale Models in Complex Terrain

C. Borrego* and N. Barros
Development and Application of a Mesoscale Integrated System for Photochemical Pollution Modelling

J.M. Giovannoni*, F. Müller, A. Clappier and A.G. Russell
An Air Quality Model that Incorporates Aqueous-Phase Chemistry for Calculating Concentrations of Ozone, and Nitrogen and Sulfur Containing Pollutants in Europe

F.T.M. Nieuwstadt*
Subgrid Modelling of Atmospheric Chemistry with the Aid of Large-Eddy Modelling

Cloud and Fog Effects and their Parameterisation in Regional Air Quality Models

G. Mauersberger*
Cloud Chemistry Modelling

A.I. Flossmann* and H.R. Pruppacher
Parameterisation of Wet Deposition by Convective Clouds

W.R. Stockwell* and T. Schönemeyer
Modelling Cloud Chemical Processes

R. Forkel*, A. Ruggaber, W. Seidl and R. Dlugi
Modelling and Parameterisation of Chemical Reactions in Connection with Fog Events

N. Chaumerliac* and S. Cautenet
New Developments in Regional Cloud Chemistry Modelling

P.J.H. Builtjes* and J. Matthijsen
Regional-Scale Modelling of Cloud Effects on Atmospheric Constituents

11.1.13 References

Ackermann, I., H. Hass, H.J. Jakobs, M. Schwikowski, U. Baltensperger, H.W. Gäggeler, P. Seibert; 1994: Observation and simulation of air pollution at Jungfraujoch: A joint ALPTRAC/EUMAC case study, Paul Scherrer Institute, Villingen, Switzerland, Annual Report, p. 18.

Bock, H.J.; 1994: Parametrisierung der trockenen Deposition mit dem EURAD Modell: Stickstoffeintrag in die Nordsee durch die Atmosphäre. Diploma thesis, University of Cologne, Inst. Geophys. and Meteor., 50923 Cologne, Germany.

Borrego, C., M. Coutinho, N. Barros; 1995: Intercomparison of two mesometeorological models applied to the Lisbon Region. *Meteor. Atmos. Phys.* **57**, 21–30.

Dorninger, M., M. Ehrendorfer, M. Hantel, F. Rubel, Y. Wang; 1992: A thermodynamic diagnostic model for the atmosphere. Part I: Analysis of the August 1991 rain episode in Austria. *Meteorolog. Zeitschr.* N.F. **1**, 87–121.

Ebel, A., H. Hass, H.J. Jakobs, M. Memmesheimer, M. Laube, A. Oberreuter; 1991: Simulation of ozone intrusion caused by a tropopause fold and cut-off low. *Atmos. Environ.* **25A**, 2131–2144.

Ebel, A., H. Elbern, J. Hendricks, R. Meyer; 1996: Stratosphere-troposphere exchange and its impact on the structure of the lower stratosphere. *J. Geomagn. Geoelectr.* **48**, 135–144.

Ebel, A., H. Feldmann, F. Fiedler, H.Hass, H.J. Jakobs, O. Klemm, K. Nester, E. Schaller, A. Schwartz, J. Werhahn; 1995: Contributions to the evaluation of chemical transport models within the SANA project, in: A. Ebel, N. Moussiopoulos (eds), *Proc. Conf. Air Pollution '95 III*, **4**, Computational Mechanics Publication, pp. 103–110.

Ebel, A., R. Friedrich, H. Rodhe (eds); 1997a, *Tropospheric Modelling and Emission Estimation*, Springer Verlag, Heidelberg, 440 pp.

Ebel, A., *et al.*; 1997b, EUMAC: European Modelling of Atmospheric Constituents, *Atmos. Environ.* **31**, 3117–3264.

Friedrich, R., M. Heymann,Y. Kasas; 1994: The GENEMIS inventory - The estimation of European emission data with high temporal resolution. *EUROTRAC Annual Report 1993, Part 5, Section II*, EUROTRAC ISS, Garmisch-Partenkirchen, pp. 1–19.

Fritsch, J.M., C.F. Chappel; 1980: Numerical prediction of convectively driven meso-scale pressure systems. Part I., *J. Atmos. Sci.* **37**, 1722–1733.

Gantner, L., J. Egger; 1994: Nesting of mesoscale models in complex terrain, in: P.M. Borrell, P. Borrell, T. Cvitaš, W. Seiler (eds.) *Proc. EUROTRAC Symp. '94*, SPB Academic Publishing bv, The Hague, pp. 886–888.

Hantel, M.; 1987: Subsynoptic vertical heat fluxes from high-resolution synoptic budgets. *Meteor. Atmos. Phys.* **36**, 24–44.

Hantel, M., H.-J. Jakobs, Y. Wang; 1995: Validation of parameterized convective fluxes with DIAMOD. *Meteor. Atmos. Phys.* **57**, 201–226.

Hass, H., A. Ruggaber; 1995: Comparison of two algorithms for calculating photolysis frequencies including the effects of clouds. *Meteorol. Atmos. Phys.* **57**, 87–100.

Hass, H., A. Ebel, H. Feldmann, H.J. Jakobs, M. Memmesheimer; 1993: Evaluation studies with a regional chemical transport model (EURAD) using air quality data from the EMEP monitoring network. *Atmos. Environ.* **27A**, 867–887.

Hass, H., H.J. Jakobs, M. Memmesheimer; 1995: Analysis of a regional model (EURAD) near surface gas concentration predictions using observations from networks. *Meteor. Atmos. Phys.* **57**, 173–200.

Heupel, M.K.; 1995: Ozone creating potentials of different hydrocarbons as calculated with a one-dimensional version of EURAD. Air Pollution Studies with the EURAD Model System (2), *Mitteilungen Institut für Geophysik und Meteorologie*, University of Cologne, 50923 Cologne, Germany, No. 105, pp. 29–56.

Jakobs, H.J., H. Feldmann, H. Hass, M. Memmesheimer; 1995: The use of nested models for air pollution studies: an application of the EURAD model to a SANA episode. *J. Applied Met.* **34**, 1301–1319.

Kunz, R., N. Moussiopoulos; 1995a: Simulation of the wind field in Athens using refined boundary conditions. *Atmos. Environ.* **29B**, 3575–3591.

Kunz, R., N. Moussiopoulos; 1995b: Statistical analysis of prognostic mesoscale model results in the frame of APSIS, in: H. Power, N. Moussiopouos, C.A.. Brebbia (eds), *Air Pollution III, Vol. 1*, Computational Mechanics Publications, pp. 175–182.

Lippert, E., J. Hendricks, B.C. Krüger, H. Petry; 1994: The chemical history of air parcels from an airplane flight path. in: U. Schumann, D. Wurzel (eds), *Impact of emissions from aircraft and spacecraft upon the atmosphere, Proc. Inform. Sci. Coll.*, Cologne, pp. 348–355.

Memmesheimer, M., H.J. Bock; 1995: Deposition of nitrogen into the North Sea for air pollution episodes calculated with the EURAD model. In Air Pollution Studies with the EURAD Model System (2), *Mitteilungen Institut für Geophysik und Meteorologie*, University of Cologne, 50923 Cologne, Germany, No. 105, p. 57–70.

Memmesheimer, M., J. Tippke, A. Ebel, H. Hass, H.J. Jakobs, M. Laube; 1991: On the use of EMEP emission inventories for European scale air pollution modeling with the EURAD model. *Proc. EMEP workshop on Photo-oxidant Modelling for Long-Range Transport in Relation to Abatement Strategies*, Berlin, pp. 307–324.

Mesinger, F., T.L. Black, D.W. Plummer, J.H. Ward; 1990: Eta model precipitation forecasts for a period including tropical storm Allison. *Weather Forecast* **5**, 483–493.

Mölders, N., H. Hass, H.J. Jakobs, M. Laube, A. Ebel; 1994: Some effects of different cloud parameterizations in a mesoscale model and a chemistry transport model. *J. Appl. Met.* **33**, 527–545.

Moussiopoulos, N.; 1993: Athenian photochemical smog: intercomparison of simulations (APSIS), background and objectives. *Environ. Software* **8**, 3–8.

Moussiopoulos, N.; 1994a: Urban air pollution modelling in Europe. in: J.M. Baldasano, C.A. Brebbia, H. Power, P. Zannetti (eds), *Computer Simulation*, Computational Mechanics Publications, pp. 341–348.

Moussiopoulos, N. (ed); 1994b: The EUMAC Zooming Model (EZM), model structure and applications. *EUROTRAC Report*, EUROTRAC ISS, Garmisch-Partenkirchen.

Moussiopoulos, N.; 1995: The EUMAC Zooming Model, a tool for local- to-regional air quality studies. *Meteor. Appl. Phys.* **57**, 115–134.

Moussiopoulos, N., G. Kallos; 1995: Study of a model for the quality of air concerning non-reactive pollutants in selected European cities. *Final Report to CEC DGXI*, Contract No. B4-3040/94/000142/MAR/B3.

Moussiopoulos, N., P. Sahm, Ch. Kessler; 1995a: Numerical simulation of the photochemical smog formation in Athens, Greece - A case study. *Atmos. Environ.* **29B**, 3619–3632.

Moussiopoulos, N., P. Sahm, A. Gikas, A. Karagiannidis, K. Karatzas and S. Papalexiou; 1995b: Analysis of air pollutant transport in the Athens basin and in the Spata area with a three-dimensional dispersion model in: N. Moussiopoulos, H. Power, C.A. Brebbia (eds), *Air Pollution III, Vol. 3*, Computational Mechanics Publications, pp. 141–152.

Moussiopoulos, N., P. Sahm, Ch. Kessler, T. Voegele, Ch. Schneider; 1995c: Interpretation of the "Heilbronn Ozone Experiment" by the aid of the EUMAC Zooming Model. in: H. Power, N. Moussiopoulos, C.A. Brebbia (eds), *Air Pollution III, Vol 1*, Computational Mechanics Publications, pp. 395–401.

Müller, D., G. Mauersberger; 1995: An aqueous phase chemical reaction mechanism. EUROTRAC special report: *Cloud-Models and Mechanisms*, EUROTRAC ISS, Garmisch-Partenkirchen.

Nester, K., H.J. Panitz, F. Fiedler; 1995: Comparison of the DRAIS and EURAD model simulations of air pollution in a mesoscale area. *Meteor. Atmos. Phys.* **57**, 135–158.

Pandis, S.N., J.H. Seinfeld; 1989: Sensitivity analysis of a chemical mechanism for aqueous-phase atmospheric chemistry. *J. Geophys. Res.* **94**, 1105–1126.

Petry, H.; 1993: Zur Wahl der Anfangskonzentrationen für die numerische Modellierung regionaler troposphärischer Schadstoffelder. Mitteilungen Institut für Geophysik und Meteorologie, No. 89, Universität zu Köln, 50923 Köln, Germany.

Petry, H., H. Elbern, E. Lippert and R. Meyer; 1994: Three dimensional mesoscale simulations of airplane exhaust impact in a flight corridor. in: H. Schumann, D. Wurzel (eds), *Impact of emissions from aircraft and spacecraft upon the atmosphere*, *Proc. Inform. Sci. Coll.*, Cologne, pp. 329–335.

Poppe, D., J. Zimmerman, R. Bauer, T. Brauers, D. Brüning, J. Callies, H.-P. Dorn, A. Hofzumahaus, F.-J. Johnen, A. Khedim, H. Koch, R. Koppmann, H. London, K.-P. Müller, R. Neuroth, C. Plaß-Dülmer, U. Platt, F. rohrer, E.-P. Röth, J. Rudolph, U. Schmidt, M. Wallasch, D.H. Ehhalt; 1994: Comparison of measured OH-concentrations with model calculations. *J. Geophys. Res.* **99**, 16633.

Ruggaber, A., R. Dlugi, T. Nakajima; 1994: Modelling of radiation quantities and photolysis frequencies in the troposphere. *J. Atmos. Chem.* **18**, 171–210.

Stern, R., B. Scherer; 1989: Application of a complex acid deposition/photochemical oxidant model to an acid deposition episode over north-western Europe within the PHOXA program. *Technical PHOXA Report,* Free University Berlin, Institute for Geophysics.

Tilmes, S., H. Hass, H.J. Jakobs; 1995: Transport schemes in Eulerian Mesoscale Models. Air Pollution Studies with the EURAD Model System (2), *Mitteilungen Institut für Geophysik und Meteorologie,* University of Cologne, 50923 Cologne, No. 105, pp. 1–28.

Wang, Y.; 1994: Semiprognostischer Test der Cumulusparametrisierung mit dem Kuo-Schema über Europa. Phd thesis, University of Vienna, 109 pp.

Wang, Y., M. Hantel; 1994: Convective fluxes and cumulus parameterization in the EURAD model. in: P.M. Borrell, P. Borrell, T. Cvitaš, W. Seiler (eds) *Proc. EUROTRAC Symp. '94,* SPB Academic Publishing bv, The Hague, pp. 850–855.

Ziegenbein, C., I.J. Ackermann, A. Ebel; 1994: The treatment of aerosols in the EURAD model: Results from recent developments. in: J.M. Baldasano, C.A. Brebbia, H. Power, P. Zannetti (eds), *Air Pollution II, Vol. I,* Computational Mechanics Publications, pp. 229–237.

Zimmermann, J., D. Poppe; 1994: A supplement of the RADM2 chemical mechanism: The photo-oxidation of isoprene, Jül-report 1938, Research Center Jülich, 52425 Jülich, Germany.

11.2 GENEMIS: Generation of European Emission Data for Episodes

Rainer Friedrich (Coordinator)

Institut für Energiewirtschaft und Rationelle Energieanwendung (IER)
Universität Stuttgart, Heßbrühlstraße 49a, D-70569 Stuttgart, Germany

11.2.1 Summary

For the analysis of atmospheric processes and the application and evaluation of atmospheric models, reliable emission data with high temporal and spatial resolution are needed. Within the GENEMIS subproject, such data have been generated. To achieve a high temporal resolution, the processes that cause the emissions have been analysed, and parameters which can be used to describe the temporal variation of the underlying activities have been identified and used to describe the temporal course of the emissions. It was shown that strong variations of emissions with time occur. For example, emissions are approximately 30 % lower on weekend days than during working days. Also, the ratio between anthropogenic VOC and NO_x emissions is higher during the day than in the night, furthermore it is much higher in urban than in rural areas. Progress was also achieved in estimating NH_3 emissions and their temporal distribution. Another achievement is the development of methods for updating annual inventories. Know-how from national experts of several countries contributed to improve the database, especially for central and eastern Europe. In an experiment at the Gubrist tunnel in Switzerland, measurements of the concentrations of pollutants at the entrance and the exit of the tunnel were made to derive the emissions in the tunnel. The resulting emissions were compared to the calculated emissions. A good agreement between measurements and calculated emissions could be shown for gasoline cars.

11.2.2 Aims of the research

The quality of annual emission inventories has experienced large progress throughout the last years. The same holds for atmospheric models, which have been developed for the simulation of transport and deposition of pollutants. In contrast to that, not much information has been available about the temporal distribution of emissions. Particularly photo-oxidant formation is correlated to processes, which depend on time, like daily or hourly emissions. As a survey of Heymann (1993) shows, emission data are the major source of uncertainty for model results. Hence, a scientific understanding of photo-oxidant pollution and the recommendation of efficient measure to reduce the air pollution problems in Europe require sufficient knowledge about the temporal variation of emissions.

Therefore the EUROTRAC subproject GENEMIS (Generation of European Emission Data for Episodes) has been established to serve the following purposes:

*   harmonisation and combination of different annual emission inventories,

*   transformation of given annual inventories into arbitrary grids and cells,

*   collection and evaluation of meteorological, socio-economic and technical data, that can be used as indicators for the temporal and spatial disaggregation of emission data,

*   development of models for the temporal disaggregation of annual emissions,

*   development of models for the update of annual inventories,

*   improvement of the data base by the implementation of national expert know-how,

*   split of VOC emissions into species and classes of species,

*   preparation of actual land use data for an enhanced calculation of natural emissions,

*   generation of European emission data for episodes in different years,

*   evaluation of emission data.

11.2.3 Principal scientific results

The generation of emission data with high temporal and spatial resolution is based on a combination and harmonisation of the annual CORINAIR and LOTOS inventories. A suitable check for the quality of emission factors applied for the preparation of these annual inventories can be obtained from comparisons with experiments.

Staehelin *et al.* (1997) performed measurements for NO_x, CO and VOC species in the Swiss Gubrist tunnel. A good agreement between the measured and the calculated emission factors can be shown for gasoline cars and for the total transport in the tunnel (see Table 11.2.1). For diesel vehicles the statistical basis was too small to come to significant results. The simulation model developed by John for the calculation of emission factors (Obermeier *et al.*, 1995) considers the gradient of the road and the different flow of the air within a tunnel.

In addition to the total VOC emissions, a large number of VOC components was measured and compared to the currently used emission factors. A remarkable good agreement was found for most of the analysed components.

Also other emission data and emission factors for other sectors have been investigated and improved, *e.g.* power plants (Karl *et al.*, 1997; Fliszár-Baranyai *et al.*, 1997).

The temporal resolution of the annual emission data is derived from the evaluation of actual statistical data, so-called *indicator data*, and from appropriate simulation models developed for the estimation of the temporal course of emissions (*e.g.* Friedrich *et al.* (1993); Lenhart *et al.* (1995a, 1995b)).

Table 11.2.1: Comparison of emission factors from experiment and model simulation

type	pollutant	tunnel experiment		simulation
		factor	stand. deviation	factor
gasoline	NO_x	623	52	851
	VOC	444	25	292
	CO	4146	230	4185
diesel	NO_x	13135	313	9443
	VOC	259	119	847
	CO	1065	1040	2307
all	NO_x	3564	113	2870
	VOC	401	47	422
	. CO	3422	420	3744

The GENEMIS exercise has proved that for all important sectors sufficient indicators are available, and that a large number of realistic time factors can be provided. Table 11.2.2 contains a list of the different emission source categories in the left column and the according indicators in the right column.

Table 11.2.2: Indicators for the temporal disaggregation of annual emission inventories

Sector	Indicator data for the temporal disaggregation
public power	fuel use, load curve
refineries	fuel use, working times, holidays
small consumers	fuel use, degree days, production, user behaviour
industrial combustion	fuel use, temperature, degree days, production, working time, holidays
production processes	production, working times, holidays
extraction and distribution of fuels	production, working times
solvent use	production, working times, holidays
road transport	traffic counts, road statistics
gasoline evaporation	temperature, traffic counts
air traffic	LTO cycles, passengers, freight
mobile sources and machinery	working times, user behaviour
waste treatment and disposal	time factors
agriculture	use of fertiliser, animal breeding
nature	temperature, land use

Actual economic developments, holidays, working times, climatic conditions, and changes of behaviour are considered by these indicators, so a generation of reliable time factors is possible. On the other hand, a huge amount of data has to be collected, harmonised and evaluated before the generation of time factors can be conducted.

Useful indicator data have been delivered by Fudala *et al.* and Pazdan *et al.*, for Poland, Fliszár-Baranyai *et al.* for Hungary, Rode for Slovenia, Jelavic for Croatia, Milyaev *et al.* for Russia, Bogdanov for Bulgaria, and Lorinczi *et al.* for Romania. For Poland, Slovenia, Hungary, and Bulgaria other general statistical data have also been collected. For more details see Ebel *et al.* (1997).

Adolph *et al.* developed models for the temporal resolution of large point source emissions. Fig. 11.2.1 indicates that base load, medium load and peak load power plants have to be distinguished. Moreover, load curves differ from country to country due to different working times and user behaviour.

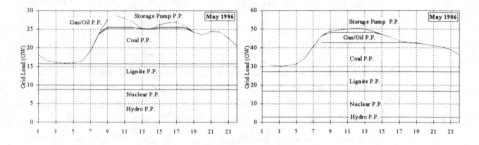

Fig. 11.2.1: Hourly loads for a Wednesday in May 1990 (Germany left, Italy right).

Information about the temporal variation of NH_3 emissions were provided by Asman and Wickert (1995). Fig. 11.2.2 displays the daily variation of ammonia emissions in Kastrup, Denmark and Be Bilt, the Netherlands.

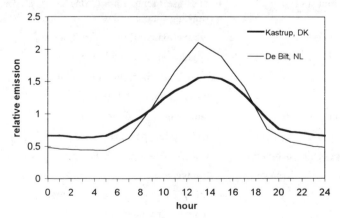

Fig. 11.2.2: Hourly distribution of ammonia.

The collection of indicator data often turns out to be difficult. Especially for central and eastern European countries a lack of suitable information was found. Therefore, the GENEMIS subproject CERES (Central and Eastern European High Resolution Emission Data Calculation) was established in 1994 to incorporate the expert know-how of scientists from these countries into the GENEMIS data base. In the results of the CERES project the economic decline in the years 1990–1992 in eastern Europe (see Table 11.2.3) and the resulting decrease of emissions have been described.

Table 11.2.3: Changes of Gross Domestic Product and industrial production in eastern Europe.

country	changes from 1990–92 in percent	
	GDP	industrial production
Bulgaria	−28	−44
Croatia	−39	−39
Hungary	−14	−19
Poland	−8	−8
Romania	−25	−37
Russia	−26	−25
Slovenia	−25	−24

The land use mapping carried out by Köble and Smiatek was an important input for the calculation of biogenic emissions. Fig. 11.2.3 illustrates the forest map of Europe. A combination of land use with adequate emission factors yields biogenic emission data with sufficient spatial and temporal resolution (e.g. Winiwarter; Moussiopoulos et al.).

Based on all these models and information a number of emission data sets with high temporal and spatial resolution have been generated and passed on to several modeller groups in Europe. Table 11.2.4 gives a list of the emission data currently available.

Table 11.2.4: Available emission data with high temporal and spatial resolution.

Area investigated	Europe
Annual data source	CORINAIR, LOTOS, CERES, IER
Spatial resolution	territorial units (NUTS code), EURAD grid (80×80 km, 60×60 km)
Pollutants	SO_x, NO_x, VOC, CH_4, CO, CO_2, N_2O, NH_3, VOC split into 32 species
Periods investigated	episodes (1–2 weeks in the years 1985, 1986, 1990, 1991 and 1992), the whole year 1990
Temporal resolution	hourly for episodes, daily for the year 1990

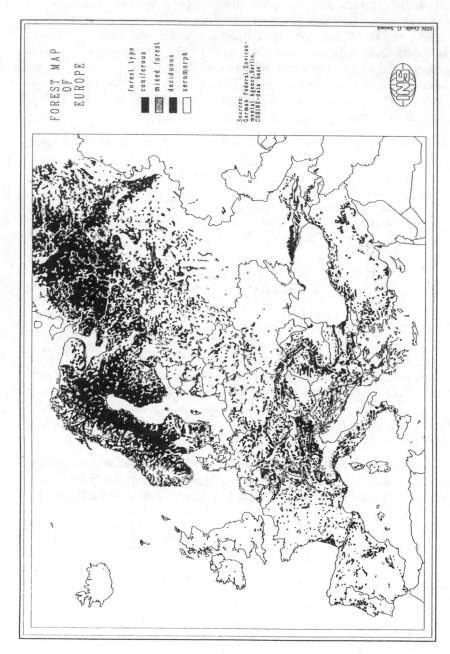

Fig. 11.2.3:   Forest map of Europe.

As an example, Fig. 11.2.4 represents the temporal variation of VOC emissions during a summer episode in 1990 for the Stuttgart area. Large differences between day and night and maxima at the rush hour (4–5 p.m.) caused by road traffic can be detected. Moreover, Fig. 11.2.4 shows that emissions from different sectors have a different temporal distribution.

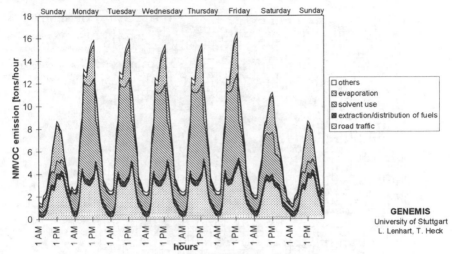

Fig. 11.2.4:   VOC emissions during a summer episode 1990 in the Stuttgart area.

The spatial distribution of emissions can be seen in Fig. 11.2.5 where the hourly NO_x emissions in Europe are plotted for Monday, 30th July 1990 at 2–3 a.m. In this early morning hour the emissions are relatively low, but even then significant regional differences occur.

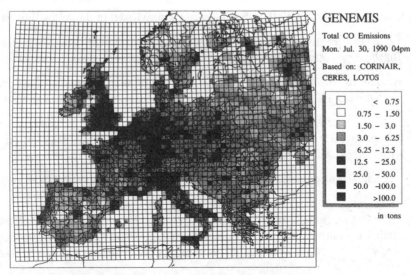

Fig. 11.2.5:   Total NO_x emissions in Europe on Monday, 30th July, 1990 at 4 p.m.

Major emissions stem from point sources, especially power plants. During the day the NO$_x$ emissions from road transport dominate. Fig. 11.2.6 shows the emission maximum at about 4 p.m. Densely populated areas with high transport activities can be clearly identified.

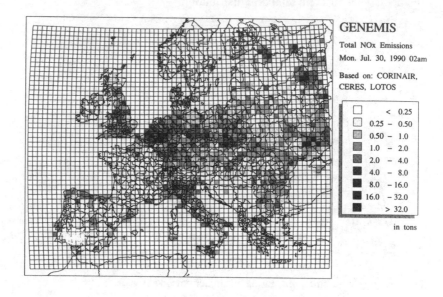

GENEMIS

Total NOx Emissions
Mon. Jul. 30, 1990 02am

Based on: CORINAIR,
CERES, LOTOS

☐	< 0.25
☐	0.25 – 0.50
▨	0.50 – 1.0
▨	1.0 – 2.0
▨	2.0 – 4.0
■	4.0 – 8.0
■	8.0 – 16.0
■	16.0 – 32.0
■	> 32.0

in tons

Fig. 11.2.6:   Total NO$_x$ emissions in Europe on Monday, 30th July, 1990 at 4 p.m.

11.2.4 Summary of the results

The main results of the GENEMIS project are summarised in the following list.

* Relevant meteorological, socio-economical and technical data are available for the description of the temporal course of emissions.

* Models for the temporal disaggregation of annual emission data base have been developed. Their use leads to reasonable estimations of the variation of emissions.

* Emissions from most sectors, like road traffic, gasoline evaporation or biogenic emissions, show a strong time dependency.

* The emissions deviate considerably from country to country and from region to region.

* Different pollutants follow rather distinct time curves and have a different spatial resolution. This fact also has an impact on the ratio of anthropogenic NMVOC to NO$_x$, which is of importance for the formation of photo-oxidants.

*   The ratio of anthropogenic VOC to NO_x emissions is higher in urban areas than in rural regions,

*   The ratio of anthropogenic VOC to NO_x emissions is higher in summer than in winter,

*   The ratio of anthropogenic VOC to NO_x emissions is higher during the day than at night.

*   The emissions from single sources and the total emissions of some pollutants vary up to several hundred percent during the week.

*   Emissions are higher, by several factors, during daytime compared to night-time

*   Anthropogenic emissions are approximately 30 % higher on working days than on weekends or holidays

*   The temporal behaviour of emissions calculated with the GENEMIS methodology shows considerable differences compared to the simple patterns applied in the past for the temporal disaggregation.

*   The tunnel experiment in Switzerland could confirm the assumed emission factors of NO_x, CO and VOC species for gasoline cars.

GENEMIS provides plausible estimations of European emission data with high temporal and spatial resolution. This results are of importance for the simulations with atmospheric models. Especially the regional dependency of emissions and the changing ratio of VOC to NO_x have a large influence on concentration, transport and deposition of pollutants. The reduction in emissions during the weekend can be seen as a sort of natural experiment for analysing the effects of short term emission reduction.

11.2.5   Further information

The GENEMIS contribution is shown in context in chapter 5 of this volume. A complete account of the subproject can be found in volume 7 of this series (Ebel et al., 1997) which includes a chapter on assessment, improvement, and temporal and spatial disaggregation of European emission data

11.2.6   GENEMIS: Steering group and principal investigators

Steering group

Rainer Friedrich (Coordinator)	Stuttgart
Jan Berdowski	Delft
Adolf Ebel	Cologne
Janina Fudala	Katowice
Ulf Hansen	Essen
Johannes Staehelin	Zürich
Szabina Török	Budapest

Principal investigators (*) together with their contributions are listed below.

Temporal Disaggregation of Emission Data

L. Lenhart, T. Heck and R. Friedrich*
The GENEMIS Inventory: European Emission Data with High Temporal and
Spatial Resolution

D. Adolph, U. Hansen* J. Kim and B. Zinsius
Temporal Variation of Emissions from Large Point Sources in Europe

W. Winiwarter *
Development of Methods for the Temporal Disaggregation of NMVOC and
NO$_x$ Emission Data for Austria

W. Asman, H. Nielsen, H. Langberg, S. Sommer, S. Pedersen, H. Kjaer
and L. Knudsen*
Diurnal and Seasonal Variation in the NH$_3$ Emission Rate

Evaluation of Emissions and Emission Factors

M. Memmesheimer, H.J. Jakobs, A. Oberreuter, H.J. Bock, G. Piekorz, A. Ebel
and H. Hass*
Application of the EURAD Model to a Summer Smog Episode using
GENEMIS Emission Data

J. Staehelin* and K. Schläpfer
Emission Factors from Road Traffic from a Tunnel Study (Gubrist Tunnel,
Switzerland) ·

Land Use Data

R. Köble and G. Smiatek*
Mapping Land Cover for Europe

Emissions in Central and Eastern Europe

M. Heymann*
Effects of the Economic Crisis on Atmospheric Emissions in Central and
Eastern Europe 1990–1992: Results of the CERES Project

J. Fudala*, M. Cenowski and M. Zeglin
Development of the Indexes for Temporal and Spatial Disaggregation of Air
Pollutant Emissions in Poland

W. Pazdan*, R. Pazdan and M. Krzeminski
Results of the CERES Project in Poland

V. Milyaev* and A. Yasenski
Data Collection and Evaluation in Russia for GENEMIS

S. Bogdanov*
Data Collection and Evaluation in Bulgaria for GENEMIS

Generation of Emission Inventories

N. Moussiopoulos*, P. Tsilingiridis and P. Pistikopoulos
Evaluation of Biogenic Non-methane Volatile Organic Compound Emissions
from Greek Forests

A. Müezzinoglu*, M. Tiris, A. Bayram and T. Elbir
Emission Inventory of Major Pollutant Categories in Turkey

A. Anghelus, T. Camaroschi, E. Lörinczi*, M. Lörinczi and A. Ritivoiu
Emission Data Bank for Timis County, Romania

Power Plant Emission Factors

U. Karl*, D. Oertel, C. Veaux and O. Rentz
Determination of NO_x and Heavy Metal Emissions from Coal-Fired Power
Plants

R. Fliszár-Baranyai, J. Fekete, J. Osán, B. Török and S. Török*
Determination of Specific Emission Factors and Diurnal Cycles for Fossil-
Fuel Burning Power Utilities

11.2.7 References

Ebel, A., R. Friedrich, H. Rodhe (eds); 1997, *Troposphere Modelling and Emission Estimation*, Springer Verlag, Heidelberg.

Fliszár-Baranyai, R., J. Fekete, J. Osán, B. Török and S. Török; 1997, Determination of Specific Emission Factors and Diurnal Cycles for Fossil-Fuel Burning Power Utilities, in: Ebel, A., R. Friedrich, H. Rodhe (eds); *Troposphere Modelling and Emission Estimation*, Springer Verlag, Heidelberg, pp. 346–352.

Friedrich R.; Generation of Time-dependant Emission Data, in: P.M. Borrell, P. Borrell, T. Cvitaš, W. Seiler (eds.) *Proc. EUROTRAC Symp. '92*, SPB Academic Publishing bv, The Hague 1993, pp. 255–268.

Heymann M., A. Trukenmüller, R. Friedrich; Development Prospects for Emission Inventories and Atmospherical Transport and Chemistry Models (DEMO), *Report to the Institute for Prospective Technological Studies*, Joint Research Centre Ispra 1993.

Karl, U., D. Oertel, C. Veaux and O. Rentz; 1997, Determination of NOx and Heavy Metal Emissions from Coal-Fired Power Plants, in: Ebel, A., R. Friedrich, H. Rodhe (eds); *Troposphere Modelling and Emission Estimation*, Springer Verlag, Heidelberg, pp. 337–345.

Lenhart L., Friedrich R.; European emission data with high temporal and spatial resolution, *Water, Air & Soil Pollution* **85** (1995) 1897–1902.

Lenhart L., Friedrich R., European emission data with high temporal and spatial resolution, poster presentation held at the 5th International Conference on Acidic Deposition, Göteborg, Sweden, 26–30 June 1995

Lenhart L., Friedrich R., European emission data with high temporal and spatial resolution, *Air Pollution III, Vol. 2*, Computational Mechanics Publications 1995, pp. 285–292.

Obermeier A., Friedrich R., John CH., Seier J., Vogel H., Fiedler F.,Vogel B., Photosmog - Möglichkeiten und Strategien zur Verminderung des bodennahen Ozons, *ecomed*, Landsberg 1995

Staehelin, J., K. Schläpfer; 1997, Emission Factors from Road Traffic from a Tunnel Study (Gubrist Tunnel, Switzerland), in: Ebel, A., R. Friedrich, H. Rodhe (eds); *Troposphere Modelling and Emission Estimation*, Springer Verlag, Heidelberg, pp. 249–260.

Wickert B., *et al.*, Ermittlung von Luftschadstoffemissionen in den neuen Bundesländern - Endbericht, in press 1995.

11.3 GLOMAC: An Overview of Global Atmospheric Chemistry Modelling

Henning Rodhe

Stockholm University, Department of Meteorology,
S-106 91 Stockholm, Sweden

11.3.1 Summary

The basic aim of the GLOMAC Subproject has been to develop and apply 3-D models of the global troposphere and lower stratosphere for simulation of transport, transformation and removal of chemical species relevant to the occurrence of ozone in the troposphere and to the dispersion and deposition of acidifying pollutants.

At the outset of the subproject it was realised that the current computers would not allow the application of complex chemistry schemes in the most meteorologically advanced General Circulation Models (CMS). Two parallel approaches were then followed to fulfil the aim:

1. To use a meteorologically relatively simple model (MOGUNTIA), based on monthly mean climatological data, in combination with a more complex chemistry scheme and

2. To use an advanced GCM with a detailed description of meteorological processes (ECHAM) in combination with simple chemistry schemes.

Both these lines, have been successful and we are now approaching the time when relatively complex chemical schemes can be handled in models like ECHAM. In the future, we also plan to run the chemistry and the meteorology interactively in ECHAM so that changes in the climate brought about by greenhouse gases and aerosols feed back on the emission transport and removal of these species.

The MOGUNTIA model has been extensively tested and refined within the subproject and applied to a variety of situations. Despite its obvious limitations (*e.g.* poor spatial and temporal resolution), it has proven to be very useful for first order estimates of global distributions. The model has been made available to several groups outside GLOMAC.

In addition to the two models mentioned above (MOGUNTIA and ECHAM), many simulations have been made using an off-line version of a GCM, *i.e.* the TM model developed by Martin Heimann in Hamburg. This model can be viewed as a compromise between MOGUNTIA and ECHAM, in it that allows a higher temporal resolution than MOGUNTIA but retains much of its simplicity.

Specific results related to the various chemistry applications are summarised below (section 11.3.3).

11.3.2 The original aims of the subproject

The scientific objectives outlined in the founding document, approved by EUROTRAC's Steering Committee 23rd July, 1987, were as follows:

> The objectives of this project are to develop a 3-D model of the global troposphere and lower stratosphere for simulation of transport, transformation and removal of chemical species relevant to the occurrence of ozone in the troposphere and its possible long-term changes. Once the model is operational it will be used to clarify questions such as:

> *   What are the sources and sinks of tropospheric ozone and its precursors NO_x, CH_4, CO and NMHC?

> *   How do anthropogenic processes (industrial emissions, agriculture *etc.*) in Europe and elsewhere influence the composition of the background troposphere and the global climatic system?

> *   How far away from the industrial regions are sulfur compounds and other acidifying species transported?"

11.3.3 Principal scientific results

Substantial achievements have been made in the development of global models for simulation of the distribution of various chemical species in the global troposphere and lower stratosphere. The climatological tracer transport model, MOGUNTIA, developed at the MPI for Chemistry in Mainz, has been further tested, improved and applied to a variety of situations. During the latest years of the GLOMAC subproject, more and more emphasis has been placed on the use of the ECHAM model (a GCM, developed at the MPI for Meteorology in Hamburg, based on the model used by the European Centre for Medium Range Weather Forecast (ECMWF)) for the tracer transport simulations. The ECHAM model includes a much more detailed description of meteorological processes (than models like MOGUNTIA), which makes it possible to treat transport, mixing and removal processes in a more realistic fashion. In addition to GLOMAC and ECHAM an off-line transport model (TM) has been used to test chemistry and transport parameterisations, *e.g.* to improve the emissions of CH_4, to calculate stratosphere-troposphere exchange rates and to develop a scheme for desert dust.

Specific scientific results are summarised below. More details are given in the reports from the individual contributions to GLOMAC and in the published papers listed at the end of this part of the volume.

a.   Improved estimates of sulfur and nitrogen species and methane emissions

Model simulations of the global distribution of chemical species in the atmosphere require realistic estimates of the emissions of the species in question and their precursors. Considerable efforts have been made in the subproject to improve such emission estimates.

The most important reactive sulfur compounds in gaseous form in the atmosphere are sulfur dioxide (SO_2) and dimethylsulfide (DMS). The largest source (> 60 % of the total emission) of sulfur in the atmosphere is SO_2 released by fossil fuel combustion and industrial processes. Previous weaknesses in the description of anthropogenic SO_2 emissions included the lack of seasonal variations. The second largest source is the oceanic emission of DMS, which is due to production by phytoplankton. The abundance of DMS depends on marine ecosystem dynamics, but the mechanisms are not yet well understood. The study so far has indicated the need for an improved description of the geographical and seasonal variations of the emission of DMS from the ocean surface. A discrepancy between observed and simulated aerosol sulfate concentration in the upper troposphere, particularly over the central Pacific, has highlighted a weakness in the description of the volcanic emission of SO_2. The following improvements have been introduced in the sulfur scheme used in the GLOMAC models:

*   a seasonal cycle in the anthropogenic sulfur emissions,

*   a height dependence in the anthropogenic emission,

*   a DMS emission parameterisation based on observed concentration in ocean surface water and model surface wind speeds,

*   a separation of the volcanic emissions into one part due to continuous, non-eruptive emission and another part due to eruptive volcanoes,

*   anthropogenic sulfur emission inventories have been established for the years 1860, 1870 ..... 1980 to be used in time dependent climate simulations in the ECHAM model. The inventory is based on a 5 by 5 degree grid.

The sources of methane are poorly understood. The most promising technique presently used is based on inverse modelling whereby the sources are inferred from a combination of model calculations of the full methane cycle and comparisons with actual observations. There are still considerable inaccuracies in these estimates due to the fact that many different source-sink configurations may produce rather similar mixing ratio distributions thus creating an ill-conditioned problem. Nevertheless, a reasonably accurate estimate can be done.

There is also progress in the determination of the nitrogen emission, though there are difficulties partly due to limitations in our knowledge of the chemistry involved. It appears, for example, that the present assumption of the oxidation of NO_x to HNO_3 by OH is too fast, and the nitrate concentration in remote boundary layers is under predicted. With respect to nitrogen generated by lightning – one of the most uncertain sources – it is found that the simulated concentrations are sensitive to the vertical distribution of the lightning source but that there are not enough observations to infer the most realistic such distribution. Using the annual nitrate wet deposition and the observations as a constraint it is estimated that the total lightning source strength is unlikely to be higher than 20 Tg N/yr.

b.   Simulation of the sulfur cycle and the climate forcing by anthropogenic sulfate aerosols

A sulfur cycle has been implemented in the MOGUNTIA and ECHAM models. The model treats the three sulfur species as prognostic variables: dimethylsulfide (DMS), sulfur dioxide (SO_2), both as gases, and finally sulfate SO_4^{2-} as aerosols. The emission occurs as DMS and SO_2. Four types of natural emission are considered: emissions from oceans, soils, vegetation and inactive volcanoes. All biogenic emissions are assumed to occur as DMS. The model incorporates both dry and wet deposition.

Simulated distributions of sulfur compounds in air and precipitation are broadly consistent with observations in and around industrialised regions in the northern hemisphere. In more remote areas the number of observations is too limited to enable a detailed evaluation of the model performance. Systematic comparisons of sulfur simulations using the ECHAM model with observations have also revealed some shortcomings. One such problem is a systematic under prediction of the concentration of aerosol sulfate at high latitudes in winter, especially in the polluted parts of the world (Europe and North America). It seems as if the oxidation of SO_2 by H_2O_2 and O_3, as described in the model, is too slow under such conditions.

The concentration of aerosol sulfate in surface air over the northern hemisphere is estimated to have increased by more than a factor of two due to anthropogenic emissions. The increase over central and northern Europe may be as high as a factor 100 during the winter months. Model calculated sulfate distributions have been used to estimate the effect of back scattering of solar radiation by sulfate aerosols. The estimated radiative forcing by anthropogenic sulfate aerosols was estimated to be almost as large, but of opposite sign, as the forcing caused by anthropogenic emissions of greenhouse gases to date. Within the most polluted regions the negative aerosol forcing may well have offset the whole positive forcing due to greenhouse gases. The estimated distributions of anthropogenic aerosol sulfate have been used by several modelling groups to study the corresponding radiative forcing and the climate response to the combined forcing by greenhouse gases and aerosols. The negative anthropogenic climate forcing caused by CCN induced changes in cloud optical properties has been estimated to be on the order of 1 W/m^2 as a global average. The largest forcing is likely to occur over oceans adjacent to the polluted continents.

Global temperature trend data for the past 75 years have been analysed in order to look for a possible aerosol forcing signal. A systematic cooling during the past 50 years in Europe, the eastern parts of North America and China is roughly consistent with the climate response expected from the increased emissions of anthropogenic sulfate during this period.

The MOGUNTIA has been used to translate future sulfur emission scenarios in the Asian region into expected deposition rates. These simulations indicate that parts of southern and eastern Asia are likely to be affected by acidification problems during the next few decades.

c.   Implementation of aerosol processes in global transport models

Efforts have been made to improve the description of aerosols in the global models. This is required for realistic simulations of the climate forcing due to anthropogenic aerosols. Three activities were undertaken:

*   The dynamics of aerosol formation from the gas phase, through nucleation, condensation, coagulation and cloud-processing was studied with a box model (AERO2) applied to the clean marine boundary layer. Correlations between sulfate aerosol mass and number of cloud condensation nuclei were calculated that agreed with observations, and that could be used as parameterisations in large-scale models.

*   A simplified two-mode aerosol model (IMAD-MOG) was developed for inclusion in global transport models. This model was compared with the full aerosol model (AERO2) using the whole range of input parameters that can be encountered in the global atmosphere. This is still pointing at deficiencies in the simple model which must be further studied.

*   A global emission inventory was developed for black carbon aerosols and implemented in the global model, MOGUNTIA. The black carbon annual emissions of industrial activities and biomass burning are estimated to be 5.5 and 7.9 Tg yr^{-1}, respectively. The calculated fields were compared with measurements of BC in air and in precipitation around the world, showing agreement within a factor of two.

d.   Simulations of the nitrogen cycle

Simulations have been carried out of the global distributions of reactive nitrogen compounds, including NO, NO_2, NO_3, HNO_3, N_2O_5 and PAN, both to study the nitrogen cycle and to investigate its influence on the distribution of O_3 and other important oxidants. Comparisons with observations of nitrogen compounds have revealed certain systematic discrepancies especially over the remote Pacific indicating that the lightning source may not be well represented in the model.

By using the MOGUNTIA model we calculated that the heterogeneous reactions of NO_3 and N_2O_5 on aerosol particles have a substantial influence on the concentrations of NO_x, O_3 and OH. In the northern hemisphere, where the anthropogenic perturbation of the atmosphere is largest, aerosols are an important sink for NO_x. In the southern hemisphere, heterogeneous removal of NO_x on clouds is most significant. The inclusion of the aerosol reactions into the model reduced the calculation annual average global abundance of NO_x by about 50 %, particularly owing to strong NO_x reductions during winter, when night-time N_2O_5 formation is relatively efficient. Calculated O_3 concentrations are reduced by about 15 % during winter time in the subtropics of the northern hemisphere.

The chemistry routine from MOGUNTIA has been implemented in the TM model to simulate the effects of nitrogen oxide emissions from subsonic aircraft. The maximum increase due to aircraft in the zonal mean ozone concentration was found to be on the order of 3 % in winter and 7 % in summer for 1990.

e.   Simulations of the methane cycle

The sources of methane have been determined by the application of an inverse technique to the global methane budget using the 3-D atmospheric transport model TM2 together with atmospheric observations of the CH_4 mixing ratios and its $^{13}C/^{12}C$ isotope ratios. The study has focused on the time period 1983–1989 in which it has been assumed that the global methane budget has been in a quasi-stationary state. The sources and sinks of CH_4 are represented by a linear combination of source components whose spatial and seasonal distribution is based on the statistics of land use, agriculture etc. Global magnitudes of the source components are determined by minimising the composite of the simulated concentration and the observations at the different monitoring stations. A similar calculation is performed for the methane isotopes.

The result shows that an improved scenario of sources and sinks can be obtained through 3-D atmospheric transport modelling combined with observations of atmospheric mixing ratios. Using the inversion technique, the a priori uncertainties of the magnitudes of CH_4 from rice paddies, bogs, swamps, biomass burning and emissions from Siberian gas sources have been more than halved. The improvement from using isotopic ratio information is rather small because a relatively small number of isotopic measurements are currently available.

f.   Simulation of ozone

Catalysis by NO_x in the CO and hydrocarbon chains plays a decisive role in O_3 production. Because of the strongly different atmospheric lifetimes of NO_x, CO, CH_4 and non-methane hydrocarbons (NMHC), the production of O_3 is not a linear function of the concentrations of these compounds. Using the global 3-D model of the troposphere MOGUNTIA, we evaluated the uncertainties in the O_3 simulations related to the NO and NMHC input rates and its geographical distribution at up to 50 %. This non-linear dependence of O_3 production on its precursor concentrations is crucial when past and future atmospheres are simulated. It partly explains why earlier two-dimensional studies overestimated the importance of photochemical production of O_3 in the troposphere and changes in O_3 and OH concentrations since the pre-industrial period.

The impact of higher hydrocarbon chemistry on O_3 concentrations which is experimentally and theoretically proved is also a factor of uncertainties in O_3 global simulations due to the complexity of NMHC emissions, the NMHC with both natural and anthropogenic sources are shown to contribute to about half the net chemical production of O_3 in the troposphere.

It was also shown that close to source areas up to 30 % of the NO_x available in the boundary layer is captured during NMHC oxidation to form organic nitrates. Organic nitrates are then transported and decomposed over oceanic areas thus increasing NO_x levels over the north pacific and Atlantic oceans by up to 100 %. Ozone production rates in the troposphere driven by NO_x levels are therefore affected.

g.   *Ozone chemistry changes in the troposphere and consequent radiative forcing of climate*

The MOGUNTIA model has been applied to simulate changes in the chemical composition of the troposphere since pre-industrial times. Our computations representing the contemporary troposphere in the northern hemisphere confirm earlier calculations that the annually rather constant pristine O_3 concentrations are overwhelmed by anthropogenic photochemical O_3 production, causing a characteristic summertime O_3 maximum. Most photochemical O_3 production occurs in the lower troposphere between 30° N and 60° N, while net O_3 destruction dominates over the remote tropical oceans, particularly in the northern hemisphere. The latter is associated with relatively high ozone volume mixing ratios advected southward from the mid-latitude production areas. Tropospheric ozone has an average lifetime of about 1 month, while the NO_x needed for O_3 formation has a lifetime of about 1–2 days, so that by the time the mid-latitude emitted pollutants reach the tropical oceans, the ozone previously formed reaches an NO_x-poor environment where efficient O_3 destruction predominates.

Wintertime boundary layer O_3 concentrations at middle north latitudes may have increased by about a factor of 2 to 3 since 1850, while summertime O_3 concentrations may be more than three times higher than in the pre-industrial troposphere. The latter may even be an underestimate considering that our model does not simulate episodic summer smog O_3 formation. The O_3 increase is not only limited to the boundary layer; calculated altitude profiles indicate that the anthropogenic influence on O_3 extends throughout the troposphere, especially in northern mid-latitudes. The total tropospheric O_3 burden in the northern hemisphere may have increased by almost a factor of two, while in the southern hemisphere this increase may be about 50 %. The latter is mainly caused by biomass burning CO and NO_x emissions. Thus, the catalytic role of NO_x in atmospheric oxidation processes and the strong industrial and biomass burning CO and NO_x sources are the major causes of the anthropogenic O_3 increase.

h.   *Simulations of the fate of HCFCs and HFC-134a in the troposphere*

The MOGUNTIA model has been applied to simulate the fate of several HCFCs and HFC-134a – halogenated hydrocarbons replacements of chlorofluorocarbons (CFCs) – in the troposphere. Based on the most recent kinetic data, phase photochemical reactions, describing the oxidation chemistry of HCFC-22 ($CHClF_2$), HCFC-123 (CF_3CHCl_2), HCFC-124 (CF_3CHClF), HFC-134a (CF_3CH_2F), HCFC-141b (CH_3CCl_2F) and HCFC-142b (CH_3CClF_2) in the troposphere, and heterogeneous removal processes determining the concentrations of the oxidation products of these halogenated hydrocarbons in the troposphere are taken into account for this study. Published past and future emissions scenarios were adopted for these halogenated hydrocarbons.

We calculate that HCFC-22 levels will reach a maximum of 190 ppt around year 2005, the levels of all other HCFCs remaining below 100 ppt. Lifetimes of HCFC-22, HCFC-123, HCFC-124, HCFC-141b, HCFC-142b and HFC-134a are evaluated to be shorter than CFCs: 13.7, 1.3, 6.0, 9.5, 20.3 and 13.9 years respectively, with the main loss due to reaction by OH. The chlorine loading potentials (CLP) of these substitutes range from 0.014 for HCFC-123 to 0.166 for HCFC-142b, whereas HFC-134a has by definition a zero CLP.

i.   Development of meteorological part of the models

Stratosphere - troposphere exchange

The vertical transport in the TM model has been tested in simulations of the transport of radioactive tracers like ^{90}Sr, ^{14}C, ^{7}Be and ^{10}Be, and of tropospheric and stratospheric "air". The comparison with measured deposition of radioactive tracers indicated a problem with the input ECMWF data in the Antarctic region.

The vertical exchange rates derived from simulations of the transport of air agree to within 30 % with independent estimates of the residual mean meridional circulation. It was also shown that a horizontal resolution of at least about 4–5° is needed to include the bulk of the vertical transport between the stratosphere and the troposphere. At a resolution of 8–10° the model was not able to capture most vertical transport by extra tropical cyclones. The simulations using ECMWF data as input showed evidence of effects of changes in the ECMWF model formulation on the vertical fluxes. This is a strong argument for the use of reanalysed meteorological data in the future.

Greenhouse effects of ozone calculated with the radiation code of ECHAM

A radiative-convective model containing the radiation code of ECHAM has been used in a number of pilot studies:

The effect of ozone hole deepening over Antarctica has been studied by calculating the changes in the temperature profile over Antarctica from the observed depleted ozone profile. The result, an 18 K cooling maximising at 70 hPa, is in agreement with observations.

The 3-dimensional ozone distributions calculated with the TM2 model and MOGUNTIA have been used as input for the radiation code in order to calculate a 2-dimensional geographical distribution of radiative forcing. This has been done for the ozone perturbation due to aircraft emissions and for the lower tropospheric ozone perturbation due to anthropogenic activities since pre-industrial times.

Nudging

A procedure was developed for 4-D assimilation of analyses from ECMWF in the ECHAM model. With this so-called nudging technique a number of parameters in the ECHAM model are relaxed towards analysed observations. The best choice for the nudging coefficient and for the set of parameters to be nudged has been determined. Among others it was found that the combination of ECMWF analyses and ECHAM physics by nudging gives an improvement of the precipitation in the inter-tropical convergence zone relative to the ECMWF forecasts. The nudging technique offers the opportunity to use the ECHAM model in the future for the study of episodes (*e.g.* field campaigns) with a realistic description of the day-to-day weather. This will also offer the possibility to validate the climate model.

Precipitation scavenging

Efforts have been made to improve the description of cloud transformation and precipitation scavenging in the models. These processes have been shown to have a major influence on the distribution of not only the aerosol components and soluble gases like HNO_3, H_2O_2 and SO_2, but also O_3, CH_2O and OH. In order to simulate cloud processes in MOGUNTIA a global cloud data net was developed based on synoptic observations.

Several different schemes for precipitation scavenging has been tested both in MOGUNTIA and ECHAM. ^7Be and ^{210}Pb have been used as tracers to validate these scavenging schemes.

j.   Model validation

The transport and chemistry performances of the models have been tested by comparisons with available trace gas measurements. Particularly important in this respect are comparisons with measurements of ozone and methylchloroform (MCF) in the background troposphere. Simulations of MCF yield indications about the model OH field since MCF (about 6 years) is relatively long (close to that of methane) so that its proper simulation also validates the global model transport properties. Also, aircraft measurements of PAN and higher hydrocarbons have been used to test MOGUNTIA, which is the only model that currently includes the calculation of the chemistry of these gases.

*   MOGUNTIA model: In this model emissions of methane, carbon monoxide and nitrogen oxides are calculated explicitly, while the other two models simulate NO_x emissions and fix the concentrations of other trace gases at the surface. The simulated global north-south gradient of methane concentrations corresponds closely to measurements. However, the seasonal cycle of methane in the model seems to be somewhat weak, particularly in the southern hemisphere. Simulated MCF concentrations agree within about 10 % with measurements, suggesting that OH concentrations and thus methane removal are calculated adequately. However, it is likely that the seasonal cycle in methane emissions is underestimated. Model predicted ozone concentrations in the lower troposphere are in approximate

agreement with measurements in regions not directly polluted. Calculated ozone concentrations in the ITCZ over the central Pacific and Indian Oceans are up to two times higher than indicated by observations. This may be due to the coarse grid scale of the model, thus artificially transporting too much NO_x from the source regions. The model matches well with the relatively strong surface ozone gradient over the Atlantic Ocean between 20 degrees north and south latitude. Ozone concentrations in polluted regions during summertime are generally underestimated because the model does not properly represent "smog" chemistry.

* The TM2 model has been used in combination with a chemical module similar to that used in MOGUNTIA. The OH concentrations thus calculated have been validated by modelling the increase of atmospheric CH_3CCl_3 mixing ratios during the 1980s. It should be noted here that reaction with OH in the troposphere is not the only sink of atmospheric CH_3CCl_3, but photolysis in the stratosphere and hydrolysis in the oceans are additional sinks of minor, but still significant importance, which were both included in the TM2 model. OH concentrations were multiplied with a global scaling factor in order to optimise the agreement of modelled and observed CH_3CCl_3 mixing ratios during the 1980s. The root mean squared deviation of these modelled and observed CH_3CCl_3 mixing ratios at the ALE/GAGE sites obtained using the optimal scaling factor for the OH concentrations (0.81) was only 3 ppt. The calculated effective tropospheric CH_3CCl_3 lifetime due to OH reaction alone was 5.1 years, and the corresponding methane lifetime due to OH reaction of 8.3 years is well in agreement with calculations performed by Prinn et al. (1995). We are further using the TM2 model in an (ongoing) "inverse modelling" study to deduce information on methane emissions from temporal and spatial variations in observed atmospheric methane mixing ratios. Thereby a scenario of methane emissions could be constructed, that well reproduces the main features (north-south gradients, and seasonal cycles in particular) of methane seen in the NOAA/CMDL measurements.

* ECHAM model: In the validation of the ECHAM model particular emphasis has been given to the simulation of ozone distributions in the troposphere. Regarding surface O_3 concentrations, the model appears to perform qualitatively well, showing a realistic seasonal variability for 13 measurement sites in background locations. The model captures the ozone spring maximum and summer minimum in remote marine locations. The model quantitatively reproduces ozone concentration patterns in background tropical locations, but underestimates ozone at the surface and the free troposphere in mid-latitudes. There appear to be two main reasons for this. First, practically all stratospheric ozone down-mixing in the ECHAM model occurs at 30 degrees latitude, where it is efficiently destroyed by photochemical reactions. The model grid resolution is too coarse to resolve tropopause folding at mid-latitudes, where a significant amount of ozone is expected to enter the troposphere, associated with active synoptic

disturbance. Second, the model underestimates net chemical ozone production. This may partly be due to insufficient mixing of NO_x from the polluted boundary layer into the free troposphere by entrainment. Up to now higher hydrocarbon chemistry has been neglected in ECHAM, which may significantly influence the ozone production over polluted regions, while transport of PAN is expected to lead to increased ozone formation in remote locations. Furthermore, we have compared calculated and observed annual average nitrate wet and total deposition values. For most sites (50 in total) in Europe and South and North America the modelled deposition agrees within about 50 %, which is satisfactory considering that the nitrate deposition is strongly dependent on variable NO_x sources, chemistry and precipitation patterns.

* Model intercomparison: The ozone mixing ratios at thirteen background stations have been used as "standards" for an intercomparison of the three models used in the project. In general it can be concluded that the differences between the models are not very large, while deviations from observations are sometimes significant, particularly at high latitudes. A general conclusion may also be that qualitative agreement with measurements is quite good, especially in low latitudes; seasonal cycles are represented well, although the calculated surface O_3 mixing ratios are generally too low. It is encouraging that O_3 levels are quite well reproduced in the tropics and springtime O_3 maxima in remote regions agree qualitatively with observations. However, all models underestimate surface ozone by about a factor of two in high latitudes, especially over Central Antarctica. The major reason for this is the lack of effective downward transport of O_3 from the stratosphere at mid- and higher latitudes due to insufficient resolution of the 3-dimensional atmospheric circulation. Preliminary results from tests with a higher resolution version of ECHAM (type 2) indicate strong improvements of these deficiencies.

11.3.4 Achievements with respect to original aims

The subproject has, by a broad margin, achieved the aims set up at the outset. Very substantial progress has been made in the development and application of tracer transport models of the global troposphere and lower stratosphere. Several important unexpected results also have come out of the subproject. These include

* the important role anthropogenic sulfate aerosols play in climate forcing and

* the influence of chemical processes in clouds for the balance of ozone and related species in the troposphere.

11.3.5 Further information

The GLOMAC contribution is shown in context in chapter 5 of this volume. A complete account of the subproject can be found in volume 7 of this series (Ebel *et al.*, 1997).

11.3.6 GLOMAC: Steering group and principal investigators

Steering group

Henning Rodhe (Coordinator)	Stockholm
Paul Crutzen	Mainz
Maria Kanakidou	Paris
Hennie Kelder	de Bilt
Jos Lelieveld	Mainz
Frank Raes	Ispra
Erich Roeckner	Hamburg

Principal investigators and their contributions are listed below.

H. Rodhe*, L. Gallardo, U. Hansson, E. Kjellström and J. Langer
  Modelling Global Distribution of Sulfur and Nitrogen Compounds

P. Crutzen*, B. Brost, F. Dentener, H. Feichter, R. Hein, M. Kanakidou, J. Lelieveld and P. Zimmermann
  Development of a Time-Dependent Global Tropospheric Air Chemistry Model "GLOMAC" based on the Weather Forecast Model of the "ECMWF"

M. Kanakidou*, P. David and N. Poisson
  Impact of VOCs of Natural and Anthropogenic Origin on the Oxidising Capacity of the Atmosphere

F. Raes*, J. Wilson and W. Cooke
  Implementation of Aerosol Processes in Global Transport Models

H. Kelder*, M. Allaart, J.P. Beck, R. van Dorland, P. Fortuin, L. Heijboer, A. Jeuken, M. Krol, T.H. The, P. van Velthoven and G. Verver
  Contributions to Global Modelling of Transport, Atmospheric Composition and Radiation

11.3.6 References

Ebel, A., R. Friedrich, H. Rodhe (eds); 1997, *Troposphere Modelling and Emission Estimation,* Springer Verlag, Heidelberg.

Chapter 12

Instrument Development

Progress in science has always gone hand in hand with the development and improvement of instrumentation. EUROTRAC itself relied on instruments development in previous work, and EUROTRAC has left a legacy of improved techniques and instrumentation for the next generation. As is indicated in chapter 6 and elsewhere in this volume (see *instruments* in the index) instruments were developed in many subprojects, but three subprojects were formed specifically to concentrate on particularly promising techniques.

12.1 **JETDLAG** Joint European Development of Tunable Diode Laser Absorption Spectroscopy for the Measurement of Atmospheric Trace Gases

12.2 **TESLAS** Tropospheric Optical Absorption Spectroscopy

12.3 **TOPAS** Tropospheric Ozone Research

The techniques chosen offered, in two cases, rapid specific measurements of trace-substance concentrations at high sensitivity; the other could be used to make high-altitude measurements by vertical sounding from the ground.

Two of the instrument development subprojects had a difficult history and, after review by the Scientific Steering Committee, had to be re-organised; TOPAS in 1991 and JETDLAG in 1993. A critique of instrument development within EUROTRAC, which indicates the problems faced by such projects with the EUREKA context, is given in chapter 6.4 of this volume.

12.1 JETDLAG: Joint European Development of Tunable Diode Laser Absorption Spectroscopy for the Measurement of Atmospheric Trace Gases

David J. Brassington (Coordinator)

Atmospheric Chemistry Research Unit,
Imperial College of Science Technology and Medicine, Silwood Park, Ascot, UK

12.1.1 Introduction

The JETDLAG subproject was concerned with the development of tunable diode laser absorption spectroscopy for measurements of trace atmospheric species. In this introduction we give a short overview of the TDLAS technique. The following sections then outline the aims and achievements of JETDLAG between 1988 and 1995.

The origins of TDLAS lie in the development of tunable lead salt diode lasers in the mid 1960s. These lasers provided the first convenient high-resolution tunable source for laser spectroscopy throughout the mid-infrared. The first use of tunable diode lasers for atmospheric measurements employed a long open-path, with a retro-reflector to return the beam to the instrument (Ku and Hinkley, 1975; Hinkley, 1976) but the most important application of tunable diode lasers to atmospheric measurements has turned out to be their use in conjunction with a long-path cell to provide high sensitivity local measurements (Brassington, 1995; Schiff et al., 1994). This type of measurement was pioneered in the late 1970s (Reid et al., 1978a; 1978b; 1979; 1980; El-Sherbiny et al., 1979) and TDLAS has since developed into a very sensitive (down to 20 ppt for some species) and general technique for monitoring most atmospheric trace species. The only requirement is that the molecule should have an infrared line-spectrum which is resolvable at the Doppler limit. In practice this includes most molecules with five or fewer atoms together with some larger molecules (Schiff et al., 1983).

The principle of TDLAS is absorption spectroscopy using a single isolated absorption line of the species. However typical line-strengths are such that, even for strongly absorbing species, a typical atmospheric concentration of 1 ppb produces an absorption of only 1 part in 10^7 over a 10 cm path-length. Conventional absorption spectroscopy techniques cannot measure such small absorptions. TDLAS overcomes this problem firstly by using a multi-pass cell to give path-lengths of 100 m or more. Such cells achieve the long path by using mirrors to fold the optical path, giving typically 100 passes of a 1 m base-length cell. Secondly various types of modulation spectroscopy are employed in which the diode-laser wavelength is modulated over the absorption linewidth at frequencies of anywhere between 100 Hz and 2 GHz. These modulation techniques allow absorptions as low as 1 part in 10^5 to be measured with a 1 Hz bandwidth. In combination these two techniques give detection limits of around 20 ppt for the most strongly absorbing species (*e.g.* NO_2, HCl, NH_3) and better than 1 ppb for almost all species of interest in atmospheric chemistry (provided they have resolvable lines).

TDLAS is typically used in a continuous sampling mode in which air is continually drawn through the multi-pass cell at a pressure of about 30 mbar (achieved with a pressure-reducing valve on the inlet). Operating at reduced pressure narrows the absorption lines so avoiding possible interferences from other species and also reduces the range over which the laser needs to be modulated, which in turn reduces the effects of laser noise.

The most critical component of a TDLAS system is the diode laser. The JETDLAG subproject was concerned exclusively with instruments using the lead-salt family of semiconductor lasers, which produce wavelengths between 3 and 30 μm. An individual laser of this type could typically be tuned over a 100 cm^{-1} range and would need to be operated at liquid nitrogen temperatures. TDLAS instruments can also use InGaAsP diode lasers (which do not need cooling) operating at wavelengths below 2 μm, but in this wavelength region absorption lines are due to overtone or combination bands with strengths typically two orders of magnitude lower than the fundamental bands. Thus these lasers cannot normally be used for measurements of trace species.

The main features of TDLAS which make it such a valuable technique for atmospheric measurements are as follows:

* As a high resolution spectroscopic technique it is virtual immune to interferences by other species - a problem that plagues most competing methods. This ability to provide unambiguous measurements leads to the use of TDLAS as a reference technique against which other methods are often compared.

* It is a general technique. The same instrument can easily be converted from one species to another by changing the laser and calibration cells. Similarly it is easy to construct an instrument which will measure several species simultaneously by multiplexing the outputs of several lasers (commonly up to four) through the multi-pass cell.

* It offers automated measurements at time-constants of a minute or so. (This compares with the bubblers, diffusion, or denuder tubes which for many species are the main alternative and which require sampling periods of at least several hours.) The time-constant of TDLAS can be traded off against sensitivity and this allows flux measurements of relatively abundant species such as CH_4 and CO by the eddy-flux correlation technique. For this application time-constants of 70 ms have been achieved.

Against these advantages must be set the facts that TDLAS instruments are complex and expensive, and they currently require expert operators. The diode lasers themselves can be unreliable and each laser is unique, leading to the need to re-characterise the instrument whenever a new laser is installed.

12.1.2   JETDLAG aims and history

The JETDLAG subproject was approved by the IEC in November 1987. The subproject had four specific objectives.

* To develop and test a standard ground based TDLAS instrument capable of unattended field operation. The intention was to develop an instrument which would then be commercialised and thus be available for use by other EUROTRAC participants.

* To construct special purpose instruments for other EUROTRAC programmes such as TOR, GCE, TRACT, ALPTRAC, and BIATEX.

* To develop new TDLAS components and techniques. In particular high temperature diode lasers, fibre optic components, high speed detectors, FM spectroscopy and signal processing.

* To obtain high resolution IR spectral data for atmospheric species where these were missing or unreliable.

We shall see in the following sections that, with the exception of the first objective, all these aims have been achieved. As approved in 1987 JETDLAG had 26 proposed projects and represented 26 separate institutions. Inevitably many of the proposed projects had funding problems. By February 1990 only eight projects had received funding with a further 12 projects still expecting to receive funds. However by May 1992 work was actually in progress on 17 projects. Thus after a slow start JETDLAG eventually developed into a substantial subproject.

The fact that the first objective, the development of a standard instrument suitable for commercial production, could not be achieved was due largely to problems in interfacing between commercial organisations and the independent research groups within JETDLAG. Inconsistency of funding of the various projects which were between them all necessary to the development of the instrument was a further factor. In retrospect in seems inevitable that there should have been problems with this objective. It is worth noting that over the eight year course of JETDLAG commercial instruments have become available, developed entirely by the companies concerned. However these are still bespoke systems which require some operator expertise. A standard off-the-shelf instrument is still not available.

When it became clear in 1992 that it would not be possible to achieve the aim of developing a standard instrument, the SSC decided to review JETDLAG. As a result of this review JETDLAG was reorganised into a much smaller subproject with only six individual contributions, all except one focused on using TDLAS for field measurements. The exception was a contribution aimed at further development of the diode lasers. Because of this reorganisation the individual final reports largely cover only those contributions which survived the reorganisation and so do not give a true picture of the totality of the work of JETDLAG.

12.1.3 Principal results

a. Laser development

Diode laser quality is critical to many aspects of TDLAS performance. Laser noise, power, mode structure, and beam quality can all directly affect instrument sensitivity. Also the difficulty of producing instruments which can be easily used by non-expert operators is mainly due to non-reproducibility in the laser behaviour. If lasers could be produced that would reliably give a single mode output at a known wavelength when operated at a given set of operating conditions the usability of TDLAS instruments would be transformed. The ability to operate at higher temperatures would also reduce the complexity of instruments. Because of these factors laser development was an important component of JETDLAG.

The group at the Fraunhofer Institute (IPM) in Freiburg undertook this work. They developed test methods for studying laser beam patterns, mode behaviour and linewidth and improved their techniques, including MBE, for fabricating buried heterostructure and double heterostructure lasers using PbEuSe, PbSe, PbSrSe, and PbSnSe materials (Schlereth *et al.*, 1990a; 1990b; Lambrecht *et al.*, 1991; 1993) The test instrumentation allowed studies of the properties of these lasers (Fischer *et al.*, 1991; Agne *et al.*, 1992). Some work on DBR and DFB lasers was undertaken (Fach *et al.*, 1994) but the main project to develop these types of laser cavity remained unfunded. Overall the work at IPM has resulted not only in improved lasers with higher temperature operation, better mode patterns and beam quality, but also in an increased understanding of the laser physics and of the fabrication processes. This will provide a basis for future laser development.

b. High resolution spectroscopy

High resolution spectroscopic studies, including measurements of line strengths, pressure broadening coefficients and pressure shifts, were undertaken by groups at Ispra and at Université Libre de Bruxelles. Tunable diode laser spectroscopy was used at Ispra (Baldacchini *et al.*, 1989; 1990; 1991; 1992; 1993) to study NH_3, N_2O, CO_2, CH_3Cl, and HCOOH. FTIR measurements at Bruxelles (Herman *et al.*, 1991; Herman and Vander Auwera, 1992; Mélen *et al.*, 1992; Mélen and Herman, 1992a; 1992b) included HONO, C_2H_2, NO, NO_2, N_2O_4 and C_2H_6, C_3O_2 and CH_3CHO.

Tunable diode laser spectroscopy was also used as a tool in laboratory studies of atmospheric chemistry (Hjorth *et al.*, 1989) mainly in support of the LACTOZ subproject.

c.   Component development

TDLAS systems are difficult to align and once aligned are susceptible to thermal and mechanical drift. Use of fibre optic components could have benefits both in overcoming these alignment problems and also in achieving more compact and robust systems by replacing lenses mirrors and beam-splitters by fibre components. This was investigated by a project at the Fraunhofer Institute (IPM) in Freiburg. Zirconium fluoride fibres, with good transmission at 4 μm, were developed in conjunction with a commercial company and evaluated for use with Tunable diode lasers (Grisar *et al.*, 1992). Results were encouraging but more development would be needed before these components could be used in a TDLAS system.

Signal processing forms an important aspect of any TDLAS system but some groups paid special attention to novel or improved techniques. These included studies of improved signal averaging and fitting methods (Werle; 1994a; Werle *et al.*, 1994) optimum strategies for increasing sensitivity by long term averaging (Werle *et al.*, 1993), and comparisons of different techniques for multiplexing several lasers (Mücke *et al.*, 1991). Attention was also paid to the important topic of quality assurance of TDLAS measurements (Mücke *et al.*, 1994a).

At the start of JETDLAG high frequency modulation techniques were just beginning to be applied to TDLAS. They appeared to offer significantly improved sensitivity and faster response. A project at the Fraunhofer Institute in Garmisch-Partenkirchen (IFU) was aimed specifically at developing such an FM instrument. An instrument was developed using single-tone FM spectroscopy with a modulation frequency of 100 Mhz (Werle *et al.*, 1992; 1994a). This achieved a sensitivity of around 1.7×10^{-6} with a path length of 36 m at a 6 Hz bandwidth. This should be equivalent to 7×10^{-7} over a 60 s average. Compared to the best WMS systems using 100 m path-lengths this represents an improvement of nearly three in sensitivity. However such performance proved difficult to obtain reliably under field conditions. As part of this program significant contributions were made to understanding the influence of high frequency laser noise on TDLAS performance (Werle *et al.*, 1989a) and quantum limited performance was demonstrated (Werle *et al.*, 1989b) (although not in a field usable system).

High frequency instruments were also developed by the group at the Max Planck Institute in Mainz and these are mentioned in the next section.

d.   Instrument development

The development and application of instruments for field measurements of atmospheric species was the central aim of JETDLAG. Five groups contributed to this: Fraunhofer Institut für Atmosphärische Umweltforschung, Garmisch-Partenkirchen (IFU); Max Planck Institut, Mainz (MPI); Forschungszentrum Jülich (KFA); Service d'Aéronomie, Paris; and Imperial College, London (ACRU).

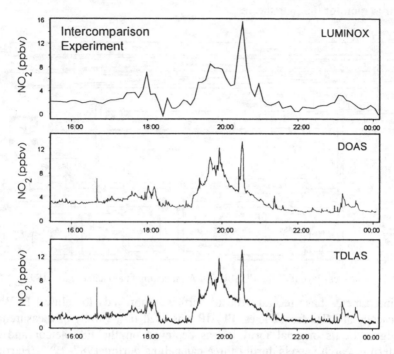

Fig. 12.1.1:   Intercomparison of NO_2 measurements using TDLAS, Luminox and DOAS instruments (Mücke *et al.*, 1994a).

The work of the Garmisch group on developing FM techniques has already been mentioned. The instrument was initially used for ambient NO_2 measurements (Werle *et al.*, 1993) (30 ppt sensitivity) and was further developed to allow simultaneous measurements of two species using frequency modulation multiplexing of the two lasers. In this form the instrument made measurements of background levels of HCHO and H_2O_2 on the MV Polarstern during a 1994 cruise in the Atlantic Ocean. The group has also carried out intercomparison studies of TDLAS NO_2 and HCHO measurements (Mücke *et al.*, 1994a). For NO_2 the TDLAS was compared with DOAS and Luminox methods (see Fig. 12.1.1), for HCHO the comparison was with the enzymatic/fluorimetric method. In both cases convincing agreement was obtained. The possibility of measuring HO_2 radical by TDLAS was investigated (Mücke *et al.*, 1994b). This is considerable challenge since sensitivities of better than 10^{-7} would be needed. Although not yet achieved, studies indicated that a sensitivity of 4×10^{-8} might be achievable, corresponding to 10 ppt of HO_2.

The main instrument developed by the MPI group was a four laser system (FLAIR) which was used both for ground ship and aircraft measurements. The instrument as originally developed used a 1.5 m base-length White cell with 28 L volume. A later version (Fig. 12.1.2) was made more compact and rugged for aircraft studies. This used a 1 m White cell with 7 L volume and 126 m total absorption path. Both systems used time division multiplexing between the lasers in order to monitor the four species.

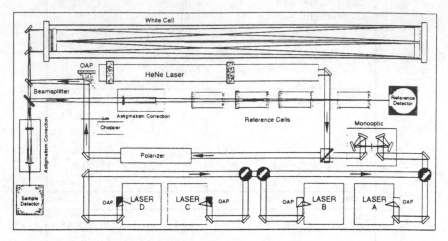

Fig. 12.1.2:   Optical layout of the FLAIR TDLAS instrument (Fischer *et al.*, 1997).

The instrument has recently been further improved to allow HFWMS spectroscopy at MHz frequencies. FLAIR was used in a variety of measurement campaigns. In its original form it was deployed on the FS Meteor and RV Polarstern research vessels during three campaigns during 1987–1988 (Harris *et al.*, 1989; 1992a; 1992b). During TROPOZ II in 1991 it made airborne measurements involving 28 flights measuring vertical profiles of HCHO, H_2O_2, NO_2 and CO. In 1992 FLAIR was used on a DC3 for airborne measurements in the SAFARI 92 campaign studying biomass burning. Species measured were NO_2, HCHO, N_2O and CH_4. Since then a series of ground based campaigns have been undertaken including two FIELDVOC campaigns and an OCTA campaign at the Izaña TOR station.

In addition to FLAIR the MPI group have constructed a fast response instrument using two-tone FMS which has been used for eddy-flux measurements of N_2O and CO (Wienhold *et al.*, 1994; 1996). A further instrument has been constructed to measure isotope ratios ($^{12}C/^{13}C$ and H/D) in atmospheric CH_4 (Bergamaschi *et al.*, 1994).

At KFA an instrument was developed specifically for HCHO measurements as part of the TOR project. It is housed in a van and uses a 1 m White cell. Sensitivity was 300 ppt for a 1 minute measurement. The instrument has undertaken a range of field studies with particular emphasis on the influence of HCHO from car exhausts on photo-chemical ozone production.

The work at Service d'Aéronomie was directed towards developing an instrument for eddy-flux correlation measurements of O_3, CO, and CH_4 on board a Fokker 27 research aircraft (Mac Leod *et al.*, 1994). The instrument was developed in collaboration with the Société Bertin with the Fraunhofer Institute Freiburg providing the diode lasers and electronics. It uses wavelength modulation spectroscopy and the beams from the three lasers are combined using dichroic beam-splitters (see Fig. 12.1.3).

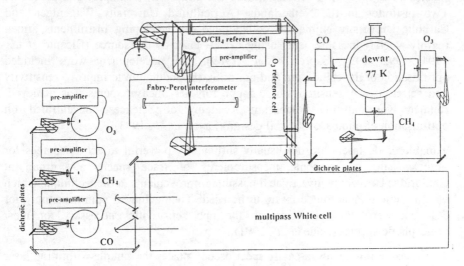

Fig. 12.1.3:   Airborne TDLAS system for measurements of O_3, CH_4, and CO vertical fluxes (Mac Leod *et al.*, 1994).

Different modulation frequencies allow the signals from the three lasers to be separated in the electronics. The White cell exchange time is < 1 s so allowing measurements at a 1 Hz rate. The instrument has successfully measured CO and CH_4 profiles and development continues.

The group at Imperial College developed an instrument which was originally employed on studies of the transformation of fossil-fuelled power station emissions. It was housed in a van and capable of mobile operation. HCl was measured in a power station plume and NH_3 measurements were made during two cloud chemistry experiments at Great Dun Fell (Brassington, 1989). Originally a single species instrument, it was later modified to measure two species using time division multiplexing and in this form was employed for simultaneous measurements of HNO_3 and NH_3. The instrument is of conventional design using a 1 m base-length White cell and 50 kHz modulation frequency. A high pumping speed (20 slpm at the inlet) is used to reduce sampling problems which are known to be severe for all the three species studied. Further, airborne, instruments are currently under development for tropospheric and stratospheric measurements.

e.   *Photoacoustic and intracavity spectroscopy*

JETDLAG provided a home for projects aimed at developing two non-TDLAS laser spectroscopic techniques, which offer promise for measurements of trace species.

Photoacoustic spectroscopy uses a sensitive microphone to detect the sound pulses produced by the absorption of a pulsed or chopped laser beam, tuned to the species absorption, in a cell containing the atmospheric sample (Sigrist, 1994). Two institutes in the Netherlands (Agricultural University Wagenigen and Catholic University Nijmegen) collaborated on developing instruments aimed mainly at NH_3 measurements using a CO_2 laser as the source (Bicanic et al., 1991; 1992; 1993; Sauren et al., 1989). The development work included optimisation of the cell design and use of Stark modulation to improve sensitivity and selectivity. A sensitivity of 0.5 ppb for NH_3 measurements was achieved. Related work at the Hungarian Academy of Sciences concentrated on optimisation of photoacoustic cell design (Favier et al., 1993).

Tunable diode lasers are not usually sufficiently powerful to act as sources for photoacoustic spectroscopic measurements of trace species. A group at Cambridge University investigated possible improvements to the technique which would allow tunable diode lasers to be used. They achieved ppm sensitivity for SO_2 but were not able to reach the ppb sensitivities necessary for most atmospheric species (Bone et al., 1991).

As the name implies intra-cavity spectroscopy places the sample within the laser cavity. Because the light within the cavity makes many passes very high effective path lengths and hence absorption sensitivities can be achieved. With a dye laser an absorption sensitivity of 3×10^{-10} cm^{-1} has been seen. Work at Hamburg University attempted to extend intra-cavity spectroscopy to Tunable diode lasers (Baev et al., 1992a; 1992b). This involves anti-reflection coating the lasers so that they may be operated in an external cavity. Using GaAlAs lasers absorption sensitivities of better than 10^{-6} cm^{-1} were obtained using linear and ring cavities. Use of a fibre-laser allowed sensitivities of 10^{-7} cm^{-1}. Thus the technique shows promise but has not yet produced a usable instrument.

12.1.4 Achievements

At the start of the JETDLAG subproject TDLAS was just beginning to make an impact on measurements of trace atmospheric species. As can be seen from section 12.1.3, the TDLAS technique has now developed to a point where it is extensively used to measure the concentrations, and increasingly also the vertical fluxes, of a wide range of species. The JETDLAG subproject, representing as it does most of the European research in this field, has made major contributions in advancing TDLAS to this point. The only disappointment is that TDLAS has not developed into a routine instrument which can easily be operated by non-specialists. The reason for this lies mainly in the lasers which are still non-reproducible in their characteristics so that whenever a laser is changed or cycled to room temperature the instrument must be re-characterised. This problem is unlikely to be overcome in the near future since the market for these lasers is not large enough to support the development needed.

Specific areas in which JETDLAG has made important contributions to TDLAS development include:

* Improvements to lasers, including higher temperature operation, better mode quality and improved beam patterns. The JETDLAG work in this area represents a major part of the world-wide effort.

* Developments of FM spectroscopic methods for higher sensitivity. The development of three field usable FM systems within JETDLAG is at the forefront of research in this area. The use of an FM system for flux measurements is also a significant achievement.

* Deployment of multi-species instruments using time and frequency multiplexing methods. This has resulted in instruments able to monitor up to four species simultaneously.

* Construction of robust and compact instruments capable of reliable operation in aircraft and ships.

* Spectroscopic studies of a range of species. Spectroscopic data is essential to the optimum design of TDLAS systems.

* Development of improved signal processing strategies leading to increased sensitivity and better quality assurance.

* Demonstration, through a large array of contributions to field measurement campaigns both ground, ship and aircraft based, of the usefulness of TDLAS in atmospheric chemistry studies.

Thus JETDLAG can be considered a success and the techniques developed within it will certainly find continuing application with the projects of EUROTRAC-2.

12.1.5 Further information

The JETDLAG contribution is shown in context in chapter 6 of this volume. A complete account of the subproject can be found in volume 8 of this series (Bösenberg et al., 1997) which includes a general scientific report by the subproject coordinator.

12.1.6 JETDLAG: Steering group and principal investigators

Steering group

David Brassington (Coordinator)	Ascot
Horst Fischer	Mainz
Dieter Klemp	Jülich
Hélène Mac Leod	Paris
Maurus Tacke	Freiburg im Breisgau
Peter Werle	Garmisch-Partenkirchen

Principal investigators (*) and their contributions to the final report (Bösenberg *et al.*, 1997) are listed below.

A. Lambrecht, H. Böttner and M. Tacke*
Development of infrared tunable diode lasers specifically suited for spectroscopy applications

H. Fischer, J. Bonifer, J.P. Burrows*, D. Klemp, U. Parchatka, J. Roths, C. Schiller, T. Zenker, R. Zitselsberger and G.W. Harris*
Development and application of multi-laser TDLAS Instruments for ground-based, shipboard and airborne measurements of trace species in the troposphere

P. Werle*, R. Mücke and F. Slemr*
High frequency modulation spectroscopy with tunable diode lasers

G. Baldacchini* and F. D'Amato
High resolution molecular spectroscopy: pressure broadening and shift of ammonia

R. Grisar*, J. Anders, M. Knothe and W.J. Riedel
IR-fibre optical components for trace gas analysis equipment

D. Brassington*
TDLAS measurements of HCl, NH_3 and HNO_3 in the troposphere

H. Mac Leod*, H. Poncet, G. Ancellet, I. Carrasco, O. Lubin, G. Mégie, F. Huard and W. Riedel
Development of an airborne tunable diode laser spectrometer for fluxes measurements by eddy correlation technique

12.1.7 References

Agne M., U. Schiessl, A. Lambrecht, and M. Tacke; 1992, in: R.Grisar, H. Bottner, M. Tacke, G. Restelli (eds), *Proc. Int. Symp. Monitoring of Gaseous Pollutants by Tunable Diode Lasers,* Kluwer Academic Publishers, pp. 93–101.

Baev V.M., J. Eschner, M. Paeth, R. Schüler, P.E. Toschek; 1992a, Intracavity spectroscopy with diode lasers, *Appl. Phys.* **B55**, 463–477.

Baev V.M., J.H. Eschner, J. Sierks, A. Weiler, P.E. Toschek; 1992b, Dynamics of multi-mode dye lasers,in: *Proc. XVIII Int. Quantum Electronics Conference (IQEC '92),* Vienna.

Baldacchini G., A. Bizzarri, L. Nencini, G. Buffa, O. Tarrini; 1989, New measurements of self broadening and shift of ammonia lines, *J. Quant. Spectros. Radiat. Transfer* **42**, 423–428.

Baldacchini G., A. Bizzari, L. Nencini, V. Sorge, G. Buffa, O. Tarrini; 1990, Pressure broadening and shift of ammonia transition lines in the v_2 vibrational bands, *J. Quant. Spectrosc. Radiat. Transfer* **43**, 371–380.

Baldacchini G., F. D'Amato, E. Righi, F. Cappellani, G. Restelli; 1991, Applicazioni di spettroscopia molecolare in fase gassosa con diodi laser accordabili nell'infrarosso, *Il Nuovo Saggiatore* **7**, 5–16.

Baldacchini G., P.K. Chakraborti, F. D'Amato; 1992, Infrared diode laser absorption features of N_2O and CO_2 in a laval nozzle, *Appl Phys.* **B55**, 92–101.

Baldacchini G., P.K. Chakraborti, F. D'Amato; 1993, Lineshape in a laval molecular beam and pressure broadening of N_2O transition lines, *J. Quant. Spectrosc. Radiat. Transfer* **49**, 439–477.

Bergamaschi P., M. Schupp, G.W. Harris; 1994, *Appl. Opt.* **33**, 7704–16.

Bicanic D., H. Sauren, H. Jalink, M. Chirtoc, A. Miklós, K. van Asselt; 1991, Progress towards the development and application of laser based photothermal devices for environmental analysis, in: P. Borrell, P.M. Borrell, W. Seiler (eds) *Proc. EUROTRAC Symp. '90*, SPB Academic Publishing bv, The Hague, pp. 347–350.

Bicanic D., M. Franco, H. Sauren; 1992, The electric resonance tuning and intermodulation for selective and sensitive studies of small polar molecules,in: *Proc. SPIE Euroopto Int. Symp. on Environmental Monitoring*, Berlin.

Bicanic D., M. Franko, H. Sauren; 1993, On the potentiality of intermodulated electric resonance, photoacoustic and photothermal spectroscopies for ground based *in-situ* tracking of polar molecules in polluted air, in: H. Schiff, U. Platt (eds) *Proc. SPIE Conf. 1715*, Bellingham, Washington, pp. 222–234.

Bösenberg, J., D. Brassington, P.C. Simon (eds); 1997, *Instrument Development for Atmospheric Research and Monitoring*, Springer Verlag, Heidelberg.

Bone S.A., P.B. Davies, S.A. Johnson, N.A. Martin; 1991, Diode laser spectroscopy with optoacoustic detection, in: P. Borrell, P.M. Borrell, W. Seiler (eds) *Proc. EUROTRAC Symp. '90*, SPB Academic Publishing bv, The Hague, pp. 351–354.

Brassington D.J.; 1989, Measurements of Atmospheric HCl and NH_3 with a Mobile Tunable Diode Laser System, in: R.Grisar, G. Schmidke, M. Tacke, G. Restelli (eds), *Proc. Int. Symp. on Monitoring of Gaseous Pollutants by Tunable Diode Lasers*, Kluwer Academic Publishers, pp. 16–24.

Brassington D.J.; 1995, Tunable diode laser absorption spectroscopy for the measurement of atmospheric species; in: R.J.H. Clark, R.E. Hester (eds), *Spectroscopy in Environmental Science*, J. Wiley, pp. 85–148.

El-Sherbiny M., E.A. Ballik, J. Shewchun, B.K. Garside, J. Reid; 1979, *Appl. Opt.* **18**, 1198–1203

Fach M.A , H. Böttner, K. H. Schlereth, M. Tacke; 1994, Embossed-grating lead chalcogenide buried wave-guide distributed-feedback lasers, *IEEE J. Quantum Electron.* **30(2)**.

Favier J., A. Miklos, D. Bicanic; 1993, New and versatile cell for studies of powdered specimens across broad spectral range; *Acta Chim. Slovenica* **40**, 115–122.

Fischer H., H. Wolf, B. Halford, M. Tacke; 1991, Low-frequency amplitude-noise characteristics of lead-salt diode lasers fabricated by molecular-beam-epitaxy, *Infrared Phys.* **31**, 381–385.

Fischer, H., J. Bonifer, J.P. Borrows, D, Klemp, U. Parchatka, J. Roths, C. Schiller, T. Zenker, R. Zitzelsberger, G.W. Harris; 1997, *Instrument Development for Atmospheric Research and Monitoring*, Springer Verlag, Heidelberg, pp. 244–249.

Grisar R., J. Anders, M. Knothe, W.J. Riedel; 1992, Application of infrared fibers in diode laser trace gas analysis,in: *Proc. Conf. Infrared Fiber Optics III*, Boston, *Proc. SPIE* **1591** (1992) 201–205.

Harris G.W., D. Klemp, T. Zenker , J.P Burrows, B. Mathieu; 1992, *J. Atmos. Chem.* **15**, 315–326.

Harris G.W., D. Klemp, T. Zenker; 1992, *J. Atmos, Chem.* **15**, 327–332.

Harris G.W., J.P. Burrows, D. Klemp, T. Zenker; 1989, in: R.Grisar, G. Schmidke, M. Tacke, G. Restelli (eds), *Proc. Int. Symp. on Monitoring of Gaseous Pollutants by Tunable Diode Lasers,* Kluwer Academic Publishers, pp. 68–73.

Herman M., J. Vander Auwera; 1992, Spectroscopic analysis of the atmosphere, *Chimie Nouvelle* **10**, 1139–1145.

Herman M., T.R. Huet, Y. Kabbadj, J. Vander Auwera; 1991, *l*-type resonance in C_2H_2, *Mol. Phys.* **72**, 75–88.

Hinkley E.D.; 1976, *Opt. and Quantum Electron.* **8**, 155–167.

Hjorth J., F. Cappellani, C.J. Nielsen, and G. Restelli; 1989, Determination of the $NO_3 + NO_2 \rightarrow NO + O_2 + NO_2$ rate constant by infrared diode laser and Fourier transform spectroscopy, *J. Phys. Chem.* **93**, 5458.

Ku R.T., E.D.Hinkley, J.O.Sample; 1975, *Appl. Opt.* **14**, 854–861.

Lambrecht A., N. Herres, B. Spanger, S. Kuhn, H. Böttner, J. Evers, M. Tacke;1991, Molecular beam epitaxy of $Pb_{1-x}Sr_xSe$ for the use in IR-devices, *J. Cryst. Growth* **108**, 301–308.

Lambrecht A., H. Böttner, M. Agne, R. Kurbel, A. Fach, B. Halford, U. Schiessl, M. Tacke; 1993, Molecular beam epitaxy of laterally structured lead chalcogenides for the fabrication of buried heterostructure lasers, *Semicond. Sci. Technol.* **8**, 334–336.

Mac Leod H., I. Carrasco, H. Poncet, G. Mégie, F. Huard, W. Riedel; 1994, Development of an airborne tunable diode laser spectrometer for flux measurements of CO, O_3 and CH_4 by the eddy correlation technique, in: P.M. Borrell, P. Borrell, T. Cvitaš, W. Seiler (eds.) *Proc. EUROTRAC Symp. '94,* SPB Academic Publishing bv, The Hague, pp. 925–929.

Mélen F., F. Pokorni, M. Herman; 1992, Vibrational band analysis of N_2O_4, *Chem. Phys. Lett.* **194**, 181–186.

Mélen F., M. Herman; 1992a, Vibrational bands of $H_xN_yO_z$ molecules,*J. Phys. Chem. Ref. Data* **21**, 831–881.

Mélen F., M. Herman; 1992b, Fourier transform jet spectrum of the m_9 band of N_2O_4,*Chem. Phys. Lett.* **199**, 124–130.

Mücke R., P. Werle, F. Slemr, W. Prettl; 1991, Comparison of time and frequency multiplexing techniques in multicomponent FM spectroscopy, *Proc. SPIE* **1433**, 136–144.

Mücke R., J. Dietrich, B. Scheumann, F. Slemr, J. Slemr, P. Werle; 1994a, Application of tunable diode laser spectroscopy for QA/QC studies: A validation for formaldehyde measurement techniques, in: P.M. Borrell, P. Borrell, T. Cvitaš, W. Seiler (eds) *Proc. EUROTRAC Symp. '94,* SPB Academic Publishing bv, The Hague, pp. 910–914.

Mücke R., F. Slemr, P. Werle; 1994b, Measurement of atmospheric HO_2 radicals by tunable diode laser absorption spectroscopy: A feasibility study, in: P.M. Borrell, P. Borrell, T. Cvitaš, W. Seiler (eds.) *Proc. EUROTRAC Symp. '94,* SPB Academic Publishing bv, The Hague, pp. 915–919.

Reid J., J. Shewchun, B.S. Garside, A.E. Ballik; 1978a, *Appl. Opt.* **17**, 300–307.

Reid J., B.K. Garside, J. Shewchun, M. El-Sherbiny, E.A. Ballik; 1978b, *Appl. Opt.* **17**, 1806–1810.

Reid J., B.K. Garside, J.Shewchun; 1979, *Opt and Quantum Electron.* **11**, 385–391.

Reid J., M. El-Sherbiny, B.K. Garside, E.A. Ballik; 1980, *Appl. Opt.*, **19**, 3349–3354.

Sauren H., D. Bicanic, H. Jalink, J. Reuss; 1989, High sensitivity, interference free Stark tuned carbon dioxide laser photoacoustic sensing of urban ammonia, *J. Appl. Phys.* **66**, 5087.

Schiff H.I., D.R. Hastie, G.I. Mackay, T. Iguchi, B.A. Ridley; 1983, *Env. Sci. Tech.* **17**, 352–364.

Schiff H.I., G.I. Mackay, J. Bechara; 1994, The use of tunable diode laser absorption spectroscopy for atmospheric measurements; in: M.W. Sigrist (ed), *Air Monitoring by Spectroscopic Techniques*, J. Wiley, pp 239–333.

Schlereth K.-H., H. Böttner, M. Tacke; 1990a, "Mushroom" double channel double-heterostructure lead chalcogenide lasers made by chemical etching, *Appl. Phys. Lett.* **56**, 2169–2171.

Schlereth K.-H., H. Böttner, A. Lambrecht, U. Schießl, R. Grisar, M. Tacke; 1990b, Buried waveguide and mesa-DH-PbEuSe-lasers grown by MBE, in: P. Borrell, P.M. Borrell, W. Seiler (eds) *Proc. EUROTRAC Symp. '90*, SPB Academic Publishing bv, The Hague, pp.361–363.

Sigrist M.W.; 1994, Air monitoring by laser photoacoustic spectroscopy, in: M.W. Sigrist (ed), *Air Monitoring by Spectroscopic Techniques*, J. Wiley, pp.163–227.

Weinhold F.G., H. Fischer, G.W. Harris; 1996, *Infrared Phys. and Tech.* **37**, 67–74.

Werle P.; 1994a, Signal Processing Strategies for Tunable Diode Laser Spectroscopy, in: *Proc. Int. Symp. on Optical Sensing for Environmental Monitoring*, October 1993, *Proc. SPIE* **2112**.

Werle P.; 1994b, Atmospheric trace gas monitoring using high frequency modulation spectroscopy with semiconductor lasers, in: A.I. Nadezhdinskii, Y.N. Ponomarev, L. N. Sinitsa (eds), *11th Symp. and School on High-Resolution Molecular Spectroscopy*, *Proc. SPIE* **2205**, 83–94.

Werle P., F. Slemr, M. Gerhrtz, C. Brauchle; 1989a, *Appl Phys.* **B49**, 99–108

Werle P., F. Slemr, M. Gehrtz, C. Brauchle; 1989b, *Appl. Opt.* **28**, 1638–1642.

Werle P., R. Mucke and F. Slemr; 1992, in: R.Grisar, H. Bottner M. Tacke, G. Restelli (eds), *Proc. Int. Symp. on Monitoring of Gaseous Pollutants by Tunable Diode Lasers,* Kluwer Academic Publishers, pp. 169–182.

Werle P., R. Mucke, F. Slemr; 1993, *Appl. Phys.* **B57**, 131–139.

Werle P., B. Scheumann, J. Schandl;1994, Real time signal processing concepts for trace gas analysis by TDLAS, *Opt. Eng.* **33**, 3093–3105.

Wienhold F.G., H. Frahm, G.W. Harris; 1994, Measurements of N_2O fluxes from fertilised grassland using a fast response tunable diode laser spectrometer, *J. Geophys. Res.* **99D**, 16557.

12.2 TESLAS: Tropospheric Environmental Studies by Laser Sounding

Jens Bösenberg

Max-Planck-Institut für Meteorologie, Bundesstraße 55,
D-20146 Hamburg, Germany

12.2.1 Original goals

TESLAS, which stands for Tropospheric Environmental Studies by Laser Sounding, was formed in November 1987 as a subproject of EUROTRAC to enhance the measurement capabilities for vertical profiling of ozone in the troposphere by means of laser remote sensing. For studies of several atmospheric processes related to the formation and redistribution of photo-oxidants there was a clear need for measuring extended time series with appropriate vertical and temporal resolution. These could not be obtained by conventional *in-situ* techniques, at least not with affordable effort, so remote sensing appeared to be the best way to obtain the required information. At the beginning of the subproject, some Differential Absorption Lidar (DIAL) systems for measuring the vertical distribution of ozone already existed, but their use was restricted to very few laboratories and very few measurement campaigns, since the instruments were highly complex, rather unreliable, and required extensive efforts for maintenance and operation by skilled scientists. In addition, the accuracy of these measurements under a variety of meteorological conditions was not really well established. The main tasks within TESLAS therefore were to develop fully the DIAL methodology for remote sensing of tropospheric ozone, and to develop instruments which are accurate, reliable, easy to operate, and suitable for field deployment or airborne operation.

12.2.2 Organisation and activities of the subproject

A total of twelve groups from seven European countries were involved in TESLAS during it's lifetime:

* BAT: Battelle Institut e.V., Frankfurt, Germany

* ENEA: Ente per le Nuove Tecnologie, l'Energia e l'Ambiente, Frascati, Italy

* EPFL: Ecole Polytechnique Federale, Lausanne, Switzerland

* GKSS: GKSS Forschungszentrum, Geesthacht, Germany

* IAO: Institute for Atmospheric Optics, Tomsk, Russia

* IFU: Fraunhofer Institut für Atmosphärische Umweltforschung, Garmisch-Partenkirchen, Germany

* IROE: Institute di Ricerca sullo Onde Elettromagnetiche del CNR, Firenze, Italy

* LIT: Lund Institute for Technology, Lund, Sweden

* MPI: Max-Planck-Institut für Meteorologie, Hamburg, Germany

* NPL: National Physical Laboratory, Teddington, U.K.

* RIVM: Rijksinstituut voor Volksgezondheid en Milieu, Bilthoven, the Netherlands

* SA/CNRS: Service d'Aeronomie du CNRS, Paris, France

The TESLAS subproject started with nine groups from five European countries: IFU, GKSS, MPI, BAT, IROE, ENEA, SA/CNRS, LIT, NPL, co-ordinated by Jacques Pelon, SA/CNRS Paris. Over the years, some groups had to leave the project (BAT) or reduce their activities (IROE, LIT, ENEA, NPL) due to funding problems, but new groups were joining in: RIVM, EPFL, IAO. In 1990 the coordination was shifted from Jacques Pelon to Jens Bösenberg, MPI Hamburg. A regular exchange of information was established, formally through twelve mostly well attended workshops, which turned out to be very efficient for the dissemination of new results, and for the discussion of different approaches to the solutions of a large variety of technical and methodological problems. In addition, of course, close personal connections between the group members provided for additional informal exchange of information, which was at least as important as the official links.

In 1991 an intercomparison experiment was organised, TROLIX '91, in which four DIAL systems were compared to each other and to ground-level and helicopter-borne *in-situ* instruments as well as a special dual-path DOAS instrument. The results of this experiment are published in a detailed report (Bösenberg, 1993). This experiment clearly showed, that in spite of several shortcomings of the participating instruments, which were not fully developed at that time, the DIAL technique is capable of providing accurate measurements of the ozone distribution in the lower troposphere. Several smaller intercomparisons performed later by the individual groups showed that the initial problems were solved and that in fact high accuracy has been achieved.

In addition to the instrument intercomparison a corresponding exercise was organised for the intercomparison of retrieval algorithms. For this purpose four datasets originating from actual lidar measurements made for a range of atmospheric conditions in different parts of the troposphere were evaluated by the algorithms of four groups. The differences in the results were larger than expected, generally between 4 % and 8 %, but always smaller than the errors due to the signal statistics.

The observed differences were mainly attributed to differences in the smoothing schemes and hence are important only in cases were a considerable amount of smoothing had to be applied. It was concluded that the retrieval algorithms do not make a significant contribution to the overall error budget.

12.2.3 Principal scientific results

For studies of the role of ozone in the chemistry of the polluted atmosphere processes involving very different scales, both temporally and spatially, are important. In principle, the scales range from centimetre-sized eddies associated with turbulent transport, over mesoscale processes like convective transport, synoptic scale phenomena as tropopause folding events, to global distribution and secular trends. While the very small scales are still the domain of in situ measurements, the resolution of laser remote sensing has been pushed to several tens of meters in the vertical and several seconds temporally, so that a large variety of processes can actually be addressed with this technique. Also, the reliability, accuracy, and ease of operation which have been achieved provide for an excellent tool to address the processes associated with medium or large scales. The main advantage of using a remote sensing technique for such studies is, that they can be performed continuously (or at least quasi-continuously) over extended periods of time. The three-dimensional distribution of ozone can be derived using profiles from continuous measurements on board an aircraft.

Regarding the main methodology, two distinct approaches were taken. Most of the groups had decided to operate in the Hartley-Huggins band of ozone absorption, which is located in the UV part of the spectrum, but two groups were exploring the possibility of using absorption lines of the vibrational-rotational spectrum in the infrared around 10 μm, the emission region of CO_2 lasers (BAT, ENEA). Of these two, only ENEA was in a position to build and successfully test a system, Battelle had to give up due to lack of financial support for this project. For the UV systems, different approaches regarding the laser technology were used: three groups followed the traditional design using tunable dye lasers (EPFL, GKSS, LIT), four groups were using Nd:YAG lasers with subsequent Raman shifting (IROE, NPL, RIVM, SA/CNRS), and three groups employed KrF-excimer lasers with subsequent Raman shifting (IFU, LIT, MPI). The use of metal vapour lasers with frequency mixing was also explored (IAO). So a wealth of different techniques was investigated with good success, each of them having advantages for special applications. Of the twelve groups finally nine succeeded in setting up at least one complete system for use in atmospheric studies, compared to just one system which existed in Europe before TESLAS was started. One group (SA/CNRS) also finished the construction of an airborne system.

The development of DIAL systems within TESLAS was not performed as a pure engineering task, mainly because not all required properties of the instrument could be specified in the beginning. It was rather performed as a feedback process. The atmospheric studies carried out with prototype versions of the instruments yielded experience for the next step of improvement and adaptation to the specific needs of a large variety of applications. The feedback from actual experiments to the instrument design, and vice versa, was found to be essential in the development of a complex system to be applied for studies of processes which were not well known at the outset.

In addition to TROLIX '91, the intercomparison experiment organised by TESLAS, a number of smaller individual tests have been performed. Table 12.2.1 provides an overview over these intercomparison efforts and the level of performance which has been achieved for the different groups in different parts of the troposphere. It should be mentioned here explicitly, that the level of agreement with the reference instrument, which for practical reasons in most cases was an ECC sonde, not necessarily reflects the quality of the performance of the lidar, since the ECC sonde is not considered a true calibration standard.

In the following, a few examples of actual atmospheric studies involving TESLAS instruments shall be presented to highlight the range of applications and the degree of performance which has been achieved.

a.   Long term monitoring and seasonal cycles

The IFU lidar has been used for regular measurements of the ozone distribution up to about 10 km height, 580 individual soundings covering the complete seasonal cycle were performed in 1991. Fig. 12.2.1 shows the three-dimensional representation of these results. Throughout the year, the fair-weather number densities of ozone in the free troposphere are roughly 50 % of the peak values in the PBL. On average the density is roughly independent of the height which means an increase of the mixing ratio towards the stratosphere. As in the boundary layer, a clear annual cycle is also observed. Interestingly, such a behaviour is not seen near the coast where mostly clean marine air masses are analysed (Grabbe *et al.*, 1994). It is important to note that periodic diurnal density variations could not be found in the free troposphere, given our present accuracy.

As already indicated above the upward transport from the PBL is rather slow and cannot be discerned by measurements during a single day. Thus, the mechanisms governing the pronounced annual cycle of the ozone concentration in the free troposphere are not easily accessible. The observation of (Grabbe *et al.*, 1994), of a missing summer maximum in the coastal free troposphere might indicate that most of the excess O_3 in summer is produced over the European continent rather than caused by upward transport on a hemispherical basis. On the other hand, some import of ozone from North America may be expected, which would take place in the free troposphere (Jacob *et al.*, 1993; Fehsenfeld *et al.*, 1994).

Table 12.2.1: Summary of the results from the comparative measurements presented in this chapter. In the column 'Wavelength tuning', SRS indicates stimulated Raman scattering and Dye indicates the use of a dye laser. In the column 'Comparative instrument', ECC indicates an ECC sonde, BM indicates a Brewer-Mast sonde and UV indicates a UV-photometric ozone analyser. The operational mode, listed in the column 'Mode', was either horizontal (H) or Vertical (V). The values in the column 'Accuracy' give the absolute difference between the mean of the lidar measurements and the comparative measurements in the range interval and at the temporal resolution mentioned in the table. The precision given is the standard deviation of the lidar measurements. The CNRS lidar system was involved in several campaigns. This is indicated in the comparative instrument column.

Laser source	Institute	Wavelength tuning	Wavelengths (nm)	Comparative instrument	Mode (H/V)	Range interval (km)	Temporal resolution (min)	Accuracy ($\mu g/m^3$)	Precision (%)
KrF	MPI	SRS D_2	268/292	ECC	V	0.2–0.5	1	4–6	4–6
KrF	MPI	SRS D_2	268/292, 292/319	ECC	V	0.3–3.5	5–10	5	5
KrF	MPI	SRS H_2	277/313	ECC	V	1.0–2.5	5–10	3	3
KrF	IFU	SRS H_2	277/313	ECC/UV	V	0.4–10	12	6	–
KrF	LIT	SRS H_2	277/313	DOAS	H	0–2	5	15	5
Nd:YAG	LIT	Dye	278.7/ 286.4	DOAS	H	0–2	5	10	2
Nd:YAG	GKSS	Dye	280.91/ 282.72	UV	H	0.955– 1.105	26	4	2
Nd:YAG	CNRS	SRS D_2/H_2	289/299	ECC	V	6.5–9.5	30	8	–
Nd:YAG	CNRS	SRS D_2/H_2	289/299	BM	V	6.5–9.5	30	11	–
Nd:YAG	CNRS	SRS D_2/H_2	289/299	ECC/BM/ UV	V	4.5–8	60	4	5
CO_2	ENEA	Grating	9504/9569	UV	H	1.25–2.	2	4	–

Please see Bösenberg et al. (1997), volume 8 in this series, for more details.

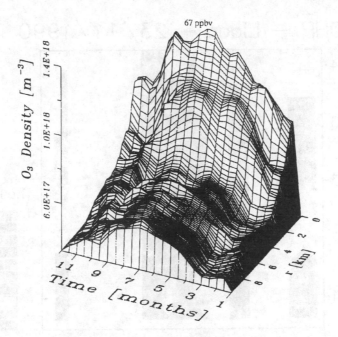

Fig. 12.2.1:   Three-dimensional representation of the 1991 annual series (note the shifted density scale).

We expect convective transport above mountains or in clouds to dominate the anthropogenic contribution to the ozone concentration in the free troposphere. However, also a more quantitative estimate of the *in-situ* chemical production has to be derived in the future. Unfortunately, there is a lack of regular measurements of ozone precursors in the free troposphere. The available numbers indicate rather low densities.

b.   Stratosphere-Troposphere exchange studies using ground based lidar

One DIAL system developed by the SA/CNRS within the frame of TESLAS was devoted to upper tropospheric studies, both for long term monitoring and case studies of stratosphere-troposphere exchange. The latter were conducted within the European program CEC/TOASTE (Transport of Ozone And Stratosphere Troposphere Exchange) during 1990/1991. During one of these campaigns (19th–30th Nov. 1990), a tropopause fold connected to a developing cut-off low system was observed over southern France (Ancellet *et al.*, 1994b). Quasi-continuous lidar measurements at the OHP on November 23rd represent a transverse cross section through an upper tropospheric frontal zone and make evident a tropopause fold with enhanced ozone concentrations (6–7×10^{10} mol cm^{-3} compared to about 4×10^{10} mol cm^{-3} outside of the fold) and sloping down to a 4.5 km altitude (Fig. 12.2.2).

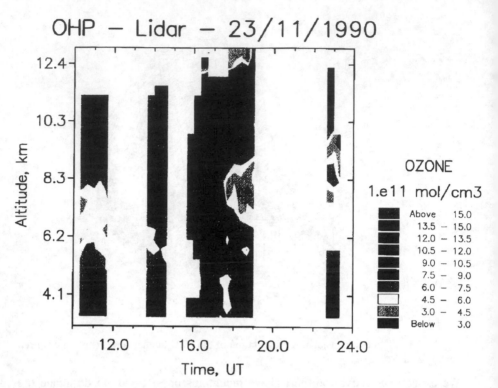

Fig. 12.2.2:   Lidar ozone measurements in 10^{11} mol cm^{-3}, at OHP on Nov. 23rd; a stratospheric intrusion from 12:00 to 18:00, descending down to 4.5 km, is visualised by high ozone values, in dark colour.

The horizontal and vertical extension of the fold was determined by using PV fields and by analysing radio soundings of the European network. Air-mass trajectory calculations show that the anticyclonic part of the fold is irreversibly transferred into the troposphere. Additionally, air masses on the cyclonic side of the fold are trapped by the circular motion around the cut-off low system and are irreversibly transferred into the troposphere in a second tropopause fold on the western flank of the cut-off low, 2 days later. The total air volume transferred to the troposphere is then approximately 1100 km × 500 km × 1.8 km and the ozone amount transferred to the troposphere about 6.5×10^{32} molecules. This estimation obtained during autumn, are at least three times lower than those obtained for spring (Danielsen, 1968; Viezee et al., 1983; Ancellet et al., 1991), when most of the tropopause folds have been investigated up to now.

At a later stage of the evolution of the same cut-off low, lidar measurements made in the low pressure while it was stationary above the OHP, show a decrease of ozone concentrations in the layer just above the tropopause (Ancellet et al., 1994). This corresponds to an ozone quantity transported into the troposphere of 6×10^{32} molecules, very similar to the value obtained for the tropopause fold observed at OHP some days before, showing the potential importance of cut-off lows for stratosphere-troposphere exchange processes. Although the exact mechanism of the observed cut-off low erosion is not clear (convective erosion, turbulent mixing), a vertical mixing coefficient K_z of 2.7 m^2 s^{-1} would explain most of the ozone change derived from the lidar measurements. Comparable values of K_z (1.5–2.2 m^2 s^{-1}) were obtained from ST-radar measurements in similar cut-off low systems, the erosion of the cut-off low being induced in one system by convective turbulence and in another by wave turbulence interaction (Bertin *et al.*, 1993).

c.   *Measurement of turbulent transport in the convective boundary layer*

The high resolution DIAL system of the MPI in combination with a Radar-RASS wind profiler for measuring the vertical wind component has been used to measure vertical profiles of turbulent ozone flux under convective conditions. The fluxes are determined directly from the highly resolved ozone density and vertical wind speed measurements by using the eddy correlation technique. The critical point for the application of the eddy correlation technique is the capability of the measurement systems to achieve high temporal and vertical resolution in combination with high accuracy. For the ozone flux measurements described here, the KrF laser based system was used with deuterium as the Raman shift gas, producing λ_{on} = 268 nm as on-line and λ_{off} = 292 nm as off-line wavelength. This choice offers a good compromise between high sensitivity to ozone absorption and attainable height range in the boundary layer. In addition, this is also an excellent choice for low sensitivity to aerosol interference, which is very important for high resolution measurements in the boundary layer. The repetition rate of the laser was 10 Hz, 100 shots were averaged on-line for subsequent evaluation. For the analysis, further averaging was performed, ending up with 60 sec temporal and 75 m vertical resolution.

The vertical wind measurements were obtained using the Radar-RASS system of the Meteorologisches Institut der Universität Hamburg, which consists of a FM-CW (frequency modulated - continuous wave) Radar operated at a centre frequency of 1235 MHz and a sound source emitting acoustic waves parallel to the Radar beam. The resolution of the Radar/RASS system is 10 sec temporally and 75 m vertically.

The height range of the combined DIAL/Radar-RASS system presently extends from 300 to a maximum of 800m above the ground. Below 300 m ozone data have not been retrieved from the DIAL signals due to an incomplete overlap between the receiver telescope's field of view and the laser beam. The upper limit of the height range is determined by a low signal-to-noise ratio of either the DIAL or the Radar-RASS system.

The ozone flux measurements with DIAL and Radar-RASS were complemented by ground level measurements of ozone density, ozone flux and sensible heat flux.

The mass budget of ozone for an air parcel in the atmosphere is described by the continuity equation, which in this particular application is approximated by stating that the sum of the local storage, the vertical flux divergence, and the horizontal advection equals the net production (or destruction) of ozone. Since with the lidar both the storage and the vertical flux divergence can be measured, in cases of horizontal homogeneity, where advection is negligible, the ozone production rate can be determined directly. Fig. 12.2.3 shows as an example the profiles of the storage term, the vertical flux divergence, and the resulting ozone production rate. It is obvious from this figure, that the errors for such a determination are rather large, but the results within their limits of error are very reasonable (this is confirmed by a more detailed study, Schaberl (1995), which will not be presented in the frame of this overview). The rather large errors are due to the fact that the main transport is caused by rather large eddies, of which only a few can be observed in a limited observation interval at a fixed station.

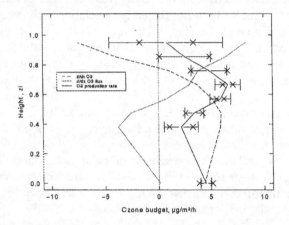

Czone budget, μg/m³/h

Fig. 12.2.3: Vertical profiles of the temporal derivative of ozone density, the vertical flux divergence, and of the ozone production rate as a function of normalised height for the measurement interval 11:34 to 13:14 UT on June 30th. The vertical bars indicate the rms error of the production rate, the crosses mark the errors due to uncertainties in the aerosol correction.

Better estimates of the ozone production rate profiles using this technique can be obtained by repeating these measurements several time under similar conditions, and by combining at least two DIAL instruments to get an estimate of the horizontal advection.

12.2.4 Conclusions

The few examples given above demonstrate clearly that the lidar technique has found its way to applications in tropospheric ozone research. The spectrum of applications is rather broad, ranging from long term monitoring via process studies of large scale phenomena to detailed investigations of the turbulent transport. The use of instruments developed within the frame of TESLAS in such applications is the best possible proof that the systems have reached a level of performance which make them valuable tools for atmospheric research.

Regarding the performance of the systems, the following can be stated.

* Within TESLAS most groups have finished the development of a lidar system for tropospheric ozone research.

* Most of these systems have demonstrated their capability of providing high accuracy measurements at least in some selected part of the troposphere.

* High temporal and spatial resolution has been achieved.

* Several systems are suitable and actually used for long term monitoring or routine operation.

* Operation by a single technician is possible.

* Reliability of the instruments has been greatly improved over previously existing systems.

* Several of these systems are transportable for use in field campaigns.

* One system has been developed for airborne operation.

From the results reported here it is obvious that the initial goals of the cooperation within TESLAS have been met, ozone lidars are now available for use in measurement campaigns addressing current problems in tropospheric ozone research.

12.1.5   Further information

The TESLAS contribution is shown in context in chapter 6 of this volume. A complete account of the subproject can be found in volume 8 of this series (Bösenberg *et al.*, 1997). It includes the following review articles.

Methodology   *G. Ancellet and J. Bösenberg*

Instruments   *A. Papayannis.*

Data processing   *E. Durieux and L. Fiorani*

Assessment of accuracy   *A. Apituley*

12.1.6   TESLAS: Steering group and principal investigators

Steering group

      Jens Bösenberg (Coordinator)   Hamburg
      Gerard Ancellet   Paris
      Roberto Barbini   Frascati
      Martin J. T. Milton   Teddington

with contributions from: *A. Apituley, E. Durieux, H. Edner, L. Fiorani, A. Papayannis, T. Trickl, C. Weitkamp and V. Zuev*

Further articles in the final report (Bösenberg *et al.*, 1997) from principal investigators (*) are listed below.

Applications

G. Ancellet*
The CNRS contribution

E. Durieux* and L. Fiorani
Application of a new shot per shot methodology to tropospheric ozone measurements with a shot per shot DIAL instrument

V.V. Zuev* and V.D. Burlakov
The multifrequency multichannel Siberian lidar station for sensing the tropospheric-stratospheric ozone and aerosol

V.V. Zuev* and B.S. Kostin
The aerosol correction method when determining the ozone concentration using a multifrequency lidar

W. Carnuth, U. Kempfer and T. Trickl*
Vertical soundings of tropospheric ozone with the IFU UV lidar

J. Bösenberg*, C. Senff and T. Schaberl
Remote sensing of turbulent vertical flux profiles and the budget of ozone in the convective boundary layer with DIAL and radar-RASS

12.2.7 References

Ancellet, G., J. Pelon, M. Beekmann, A. Papayannis; 1991, G. Mégie; Ground based lidar studies of ozone exchanges between the stratosphere and the troposphere, *J. Geophys.Res.* **96**, 22401–22421.

Ancellet, G., M. Beekmann, A. Papayannis; 1994b, Ozone transport in a tropopause fold / cut-off low event, *J. Geophys. Res.* **99D2**, 3451–3468.

Bertin F., A. Cremieu, R. Ney, A. Desautez; 1993, Tropospheric-stratospheric exchange coefficients in cut-off lows estimated with a high resolution UHF-radar, *TOASTE report.*

Bösenberg, J., G. Ancellet, A. Apituley, H. Bergwerff, G.V. Cossart, H. Edner, J. Fiedler, B. Galle, C.N. de Jonge, J. Mellquist, V. Mitev, T. Schaberl, G. Sonnemann, J. Spaakman, D.J.P. Swart, E. Wallinder; 1993, *Tropospheric Ozone lidar intercomparison experiment, Trolix '91: Field phase report*, Max-Planck-Institut für Meteorologie, Report No **101**, Hamburg.

Bösenberg, J., D. Brassington, P.C. Simon (eds); 1997, *Instrument Development for Atmospheric Research and Monitoring*, Springer Verlag, Heidelberg.

Danielsen, E.; 1968, Stratospheric-tropospheric exchange based on radioactivity, ozone and potential vorticity, *J. Atmos. Sci.* **25**, 502–518.

Fehsenfeld, F.C.; 1994, Transport of O_3 and O_3 Precursors from Anthropogenic Sources to the North Atlantic, in: P.M. Borrell, P. Borrell, T. Cvitaš, W. Seiler (eds) *Proc. EUROTRAC Symp. '94*, SPB Academic Publishing bv, The Hague 1994, pp. 57–64, and references therein.

Grabbe, G.C., J. Bösenberg, T. Schaberl; 1994, Ozone Distribution in the Lower Troposphere Investigated with the MPI Ozone Lidar, in: P.M. Borrell, P. Borrell, T. Cvitaš, W. Seiler (eds.) *Proc. EUROTRAC Symp. '94*, SPB Academic Publishing bv, The Hague,pp. 322–325.

Jacob, D.J, J.A. Logan, G.M. Gardner, R.M. Yevich, C.M. Spivakovsky, S.C. Wofsy, S. Sillman, M.J. Prather; 1993, *J. Geophys. Res.* **98**, 14817–14826.

Schaberl, T.; 1995, Messung des Ozonflusses in der unteren Troposphäre mit einem neuen Ozon-DIAL-System und einem Radar-RASS, PhD thesis, Universität Hamburg.

Viezee W., W.B. Johnson, H.B. Singh; 1983, Stratospheric ozone in the lower troposphere - II. assessment of downward flux and ground level impact, *Atmos. Environ.* **17**, 1979–1993.

12.3 TOPAS: Tropospheric Optical Absorption Spectroscopy

Paul C. Simon (Coordinator)

Institut d'Aeronomie Spatiale Belgique, 3 Avenue Circulaire
B-1180 Bruxelles, Belgium

12.3.1 Summary

The aim of the Tropospheric OPtical Absorption Spectroscopy (TOPAS) subproject of EUROTRAC was the development of high performance instruments based on Differential Optical Absorption Spectroscopy (DOAS) in order to measure minor tropospheric constituents using their absorption properties in the ultraviolet, visible and near infrared.

The target molecules detectable by this method are ozone (O_3), sulfur dioxide (SO_2), nitrogen dioxide (NO_2) and trioxide (NO_3). Other tropospheric constituents like HONO, HCHO and CS_2 as well as several hydrocarbons (toluene, benzene, xylene and naphthalene) are also measurable by DOAS.

Several DOAS systems were developed in the framework of this subproject. There are generally based on grating spectrometers with different detection systems. The Belgian groups used Fourier Transform Spectrometers (FTS), adapted for the first time to DOAS measurements.

New absorption long path systems were designed and extensively used, with the new concept of the combined transmitting/receiving telescope offering several advantages with respect to the classical double-ended system. Furthermore, a multipass cell was built and tested during campaigns.

The instruments were improved by using a variety of new type of detectors. Careful characterisation was carried out during the time-frame of this subproject. Several performances were improved *e.g.* in terms of measurement frequency. Dedicated software for instrument automation and data analysis were developed and improved, taking into account the specific problems imposed by each type of instrument and its associated detector.

Two intercomparison campaigns for DOAS instruments were organised, the first in 1992 in Brussels (Belgium) and the second in 1994 in Weybourne (UK).

The measurements made during the first campaign by eight different instruments were carefully analysed and yielded an absolute accuracy limit for NO_2, O_3 and SO_2 measurements made by the DOAS technique using an absorption path of a few hundred meters in a relatively non-polluted urban troposphere.

The second campaign compared seven different instruments through ten days of continuous measurement under field conditions. The comparisons concerned SO_2, NO_2, O_3, NO_3, HCHO and HONO. The DOAS were compared with commercial point monitors for NO_2, O_3 and SO_2 and a home-built monitor for HONO. Two other comparisons were also made during the campaign. Firstly the DOAS were tested using cells containing known amounts of NO_2 and SO_2. Secondly the group's software was tested using synthetic spectra including NO_2, O_3 and HCHO.

These two campaigns demonstrated the necessity for quality assurance and quality control procedures in order to obtain reliable tropospheric data relevant for high quality scientific studies of the troposphere.

On the other hand, intensive laboratory measurements were conducted to support the instrument developments and the aforementioned retrieval software.

Finally, scientific studies were performed with several of the TOPAS instruments during other field campaigns and are briefly presented.

One instrument has been fully commercialised, namely the "Système d'Analyse par Observations Actives" (SANOA) developed by the CNRS (France). For the first time, a Fourier transform spectrometer was used in the UV-visible range for DOAS and its performances demonstrated. This instrument is commercialised by Bruker (Germany). Furthermore, Hoffmann Messtechnik (Germany) has also developed an instrument for DOAS in the troposphere.

12.3.2 Aims of the TOPAS subproject

The Tropospheric OPtical Absorption Spectroscopy (TOPAS) subproject of EUROTRAC is one of the three subprojects concerned with instrument development and validation. It was officially accepted by the EUROTRAC SSC in 1988. Its objective was to develop high performance instruments based on Differential Optical Absorption Spectroscopy (DOAS) in order to measure minor tropospheric constituents using their absorption properties in the ultraviolet, visible and near infrared (wavelengths smaller than 1000 nm).

The target molecules detectable by this method are ozone (O_3), sulfur dioxide (SO_2), nitrogen dioxide (NO_2) and trioxide (NO_3). Other tropospheric constituents like HONO, HCHO and CS_2 have absorption bands in the UV range as well as several hydrocarbons (toluene, benzene, xylene and naphthalene) and are also measurable by DOAS.

In addition, new laboratory measurements of spectroscopic parameters used for the retrieval of trace species in the troposphere were needed to improve the performance of DOAS and were included in the TOPAS objectives.

During the last years the TOPAS objectives were focused on intercomparison campaigns in order to compare the performances of the available instruments and to fully characterise and validate this method. The priority was given first to O_3,

NO_2 and SO_2 measurements and second to NO_3, HONO and HCHO, without excluding the hydrocarbons.

12.3.3   Activities and principal results

a.   Instrument development

Several DOAS systems were developed in the framework of this subproject. There are generally based on grating spectrometers with different detection systems. The Belgian groups used Fourier Transform Spectrometers (FTS), adapted for the first time to DOAS measurements. They began first by using a fixed high resolution FTS Bruker 120 HR FTS located at the University of Brussels and therefore used in urban conditions. A mobile high resolution FTS (Bruker 120 M) was acquired in 1993 and adapted to field atmospheric measurements in the UV-visible range in 1994.

The main advantages of Fourier transform spectrometry compared to grating instruments are:

i)   the large spectral coverage achieved in one sampling, allowing to study many molecules simultaneously,

ii)   an internal high precision wave number calibration device (He-Ne laser line),

iii)   a constant, linear dispersion wave number scale,

iv)   a simple, well-known instrument line shape function when the instrument is perfectly tuned, and

v)   an available high resolution (0.01 cm^{-1} in the UV-visible) when required by sharp structured absorption features.

The grating spectrometers, usually commercially available, were adapted for tropospheric DOAS measurements by all groups. They have generally small focal lengths, from 125 to 500 mm, with large aperture (f/number from 2.9 to 6.9). The bandpass varies between 0.4 and 1.2 nm. The detectors are either photodiode arrays (PDA), NMOS type detectors, charged coupled devices (CCD), or photomultiplier tubes associated with a slotted disc system.

b.   Absorption device developments and improvements

DOAS measurements were originally performed using a double-ended system with a transmitting and a receiving telescope positioned some kilometres apart. In a number of applications however it is more convenient, and sometimes even necessary, to use single-ended systems. For this purpose a combined transmitting/receiving system was designed (Galle et al., 1990; Axelsson et al., 1990). The system uses different parts of the parabolic main mirror for transmitting and receiving, and uses retro-reflectors at the remote end of the path. The design is simple, easy to align, has good light efficiency and is compact. With this system a number of advantages can be gained. As the lamp is located close to the spectrometer, lamp reference spectra can easily be obtained and the lamp can be intermittently blocked out by an automatic shutter to record background noise and light levels. Multiple path measurements can easily be performed by realignment of the main mirror towards different retro-reflectors, opening up interesting measurement strategies. Finally, measurements can be made also on remote locations with the need for electric power and a stable platform at one end only.

This single-ended system was used by several groups. The Swedish group has implemented such a device for background trace monitoring at Draget, close to the EUROTRAC TOR Station at Rörvik. It is also used with a second instrument developed by the same group and optimised for urban air monitoring. A coaxial transmitting and receiving telescopes was also implemented by the British group at the Weybourne Atmospheric Observatory (WAO) in North Norfolk, providing an overall path of 5.5 km. The new long path DOAS system built by the German group is also based on this new coaxial design.

On the other hand, a long path system was set up on the campus of the University of Brussels. The light from Xenon or Tungsten lamps is collimated onto a 30 cm Cassegrain-type telescope, which projects the beam onto a slightly parabolic mirror, situated 394 m away. This mirror sends the beam back into a second 30 cm Cassegrain telescope, which collimates the light onto the entrance aperture or slit of the spectrometer.

Single path mobile system was designed by CNRS and used by both with the "Système d'ANalyse par Observations Actives" (SANOA) instrument developed by the French group (Laville *et al.*, 1991; Goutail *et al.*, 1993) and with the mobile FTS spectrometer acquired by the University of Brussels. On both cases, absorption measurements by several tropospheric constituents are performed over distances ranging from 150 to 400 m.

A totally different system, namely a multi-pass cell, was designed and tested in the framework of this sub-project by the German group. This multi-reflection cell is based on the original design of White (1976) and described by Ritz *et al.* (1993). The length of the cell was chosen to be 5 m allowing, with 144 reflections in the cell, light paths as long as 2 km. With the chosen mirror coating, only NO_2 and NO_3 were measured with the cell. The time between two manual readjustments was prolonged to 3 days, thus enabling the cell to perform automatic long term measurements. Another problem with multi-reflection cells is the stray light in the cell. Two different sources could be identified: solar light scattered in the cell that can easily be measured by taking spectra without the light of the lamp and the lamp light itself which produces stray light in the cell. The detection limits determined in field measurements in Heidelberg and Norwich are 1 ppb and 10 ppt for NO_2 and NO_3, respectively when using a 2160 m light path.

c.   DOAS intercomparison campaigns

Two intercomparison campaigns for DOAS instruments were organised in the framework of TOPAS, the first in 1992 in Brussels (Belgium) and the second in 1994 in Weybourne (UK).

The Brussels campaign

The campaign was organised under the leadership of the Laboratoire de Chimie Physique Moléculaire (CPM) of the Université Libre de Bruxelles (ULB).

Twenty two persons were present representing eight groups and five nations (Belgium, France, Germany, Sweden and UK). Four commercial companies, selling or developing DOAS instruments were invited to participate provided they brought one of their instruments. Hoffmann Messtechnik (Germany) participated with a prototype instrument. ATMOS Equipement (France) was represented by the CNRS French group who used one of the ATMOS instruments. Bruker Analytic (Germany), who is presently selling a high resolution portable Fourier transform spectrometer to be used in the UV-visible region, accepted, but was unfortunately not ready with its instrument in September 1992 and did not take part in the campaign. OPSIS (Sweden) declined the invitation.

The campaign took place from the 7th to the 18th of September 1992 on the campus of the Université Libre de Bruxelles. Seven instruments were installed in a laboratory situated at an altitude of ~ 20 m from ground level and ~ 120 m a.s.l. In addition to the 15m open White absorption cell and spectrograph was installed on the roof of the building (altitude 50m).

In order to reach a maximum efficiency in the comparisons of results, three molecules (NO_2, O_3 and SO_2) were chosen as imposed targets in spite of the fact that some instruments could not cover all three species. A common set of absorption cross sections for these three molecules was used.

Four optical path lengths were available from mirror, retro-reflectors and sources placed on several buildings within and outside the ULB campus. Path lengths of 230, 460, 780 and 2020 m were used. In several instances, the light from one optical path (780 m) was distributed simultaneously to several spectrographs.

Co-ordinated observation periods of 12 (night or day) and 24 consecutive hours were devoted to simultaneous measurements of the imposed species. Additional species (*e.g.* HCHO, HONO, hydrocarbons, *etc.*) were measured by some groups.

As expected in an urban site, the temporal variations of NO_2, O_3 and SO_2 were quite large and allowed to test the instruments in both high and low concentration conditions. The average concentrations for NO_2 and for O_3 were around 5×10^{11} molec cm^{-3}, with the classical anti-correlation in their diurnal variability. For SO_2, the average concentration was ~ 10^{11} molec cm^{-3} with a few spurious maxima values reaching 5 to 6×10^{11} molec cm^{-3}. For an urban troposphere, the level of pollution was reasonably low during the campaign period.

The results obtained during one of the 24 hour co-ordinated period have been extensively examined. A similar scatter of the points is observed for all three species and is of the same order of magnitude for high and low concentrations. Using exclusively the results of the fast instruments relative standard deviations are readily calculated for time intervals of 6 min. The relative standard deviations are small for high concentrations (± 5 %) and large for small concentrations (± 10 %); they are roughly proportional to the inverse of the concentration. A similar situation is observed for NO_2 and SO_2 but in the latter case the deviations are smaller by a factor of 2.

The comparison between the different data shows that rapid fluctuations in concentrations are not responsible for the discrepancies between instruments using almost the same paths. The absolute accuracy does not vary in spite of a fairly large concentration variation of the three species investigated during one co-ordinated period (by a factor of ~ 7 for NO_2, ~ 8 for O_3, and ~ 6 for SO_2). It may therefore be stated that the dispersion of the results are mainly due to the instruments themselves including the use of individual algorithms for the retrieval of the data. The dispersion (± 1σ) of the DOAS measurements on NO_2, O_3 and SO_2 performed by this large variety of UV-Visible instruments using a path length of a few hundred meters is for NO_2, $\pm 5 \times 10^{10}$ molec cm^{-3}, for O_3, $\pm 6 \times 10^{10}$ molec cm^{-3} and for SO_2, $\pm 1 \times 10^{10}$ molec cm^{-3}. The lower value for SO_2 was expected because the signature (cross section) of this molecule is stronger than that of the two others.

From the various comparisons made with the campaign data, it is clear that the discrepancies between the instruments are not random and are closely linked to each instruments. They are definitely connected to the different algorithmic treatments used be each group in the DOAS technique. In spite of many efforts on behalf of the participants to the campaign, it was impossible in a short time to conceive a uniform retrieval method (same algorithm) on account of the difference in conception of the various instruments (grating or Fourier, slotted disk, CCD or PDA). The technique used to adapt the resolution of the cross sections to the resolution of the instruments, the fact that several molecules are measured simultaneously or not, the type of regression procedure used, the fact that a dark current need to be measured or not, the filtering procedure used to eliminate the broad band component of the recorded spectrum, are a few examples of the necessity of different approaches to the treatment of the raw data.

Significant progress was made by all the participants due to the synergy of this type of campaign. After a first comparison of results made without any constraints, it was obvious that a common set of absorption cross sections had to be used by each group and also triggered improved measurements of these (Carleer et al., 1993). The campaign also served to exchange various instrument and software improvements between the groups.

The Weybourne campaign

The second TOPAS comparison campaign was held at the Weybourne Atmospheric Observatory between the 26th September and 7th October 1994.

The laboratory is situated on the north coast of Norfolk, UK (latitude 57° N, longitude 7° E) and is part of the TOR network (TOR station 11). This site was chosen because, as well as being the base for the UEA group's measurements it offered a location that was well equipped for the measurement of trace gases and was exposed to a wide range of conditions, from polluted European to very clean Arctic air.

The second exercise sought to extend the findings the first campaign held in Brussels in several ways: firstly, the seven DOAS instruments involved in the campaign were run continuously throughout the campaign to observe their performance in varied conditions and to test their reliability; secondly, the DOAS were compared with commercial point monitors for NO_2, SO_2, and O_3; thirdly, a quantitative comparison of cell measurements of both NO_2 and SO_2 was made; fourthly, trials of the participants analysis software were conducted; and lastly, as well as observing atmospheric concentrations of NO_2, O_3 and SO_2 the 1994 study also included measurements HONO.

Twenty four scientists, from seven different research institutions, participated in the Weybourne comparison exercise.

Other measurements were also made during the campaign using commercially available instrumentation and these data were compared with the DOAS measurements. Specifically, these were: an Ecophysics NO analyser and NO_2 photolytic converter; a Thermo-Electron SO_2 analyser and a Thermo-Electron O_3 analyser. Also, a home-built HONO instrument was run by the University of Birmingham.

As mentioned above, the experiment was split into three comparison exercises:

the first used synthetic spectra to test the participants analysis software; the second used a cell made up of known concentrations of NO_2 and SO_2; and in the third exercise the instruments were run alongside the commercial point monitors for 10 consecutive days in the field.

Five test spectra were made up to give typical atmospheric absorptions for a 5 km path length. Each contained NO_2, O_3 and HCHO in varying concentrations and increasing amounts of noise. Looking at the data as a whole the NO_2 and O_3 values were retrieved to within < 5 % in the majority of cases and the HCHO values to within 10 %.

A cell was made up with known amounts of NO_2 and SO_2 that gave typical atmospheric absorptions for a 5 km path. The results for the comparison are very encouraging. All the instruments, with the one exception agree with the cell make-up to within 7 % for NO_2 and all except two agree to within 7 % for SO_2.

Ambient atmospheric measurements were made by all seven instruments over the 10 day period between 26th September and 7th October 1994. The gases considered during this part of the exercise were NO_2, SO_2, O_3 and HONO. NO_3 was also measured by two instruments on three nights during this period and so there are gaps in these data. NO_3 was not strictly part of this comparison exercise.

The TOPAS community has amassed extensive comparison data during the Weybourne campaign. Analysis and interpretation of that data set is currently underway both within the respective groups and also in a series of discussion meetings. Although this process is some way from completion, it can clearly be seen that the results are both useful and encouraging both as a DOAS comparison exercise and as an atmospheric chemistry field measurement campaign.

d.   Other field campaigns and related scientific studies

In 1991, the latest version of the French instrument "Système d'Analyse par Observations Actives" (SANOA) was tested in urban (Paris) and rural (Vosges) sites in order to improve its performances. In Paris, the SANOA provided measurements of NO_2, SO_2, ozone, HONO, HCHO and toluene (Pommereau et al., 1993). They were compared with in-situ observations and with the UV-visible spectrometer made by OPSIS. The agreement with the OPSIS results is generally poor for all common detectable species.

The NUAC campaign in the frame of the EUROTRAC subproject ACE was performed in April–June 1991 on the west side of the Vosges, providing measurements of ozone, NO_2, SO_2 and HCHO. The data analysis (Goutail et al., 1993) brought important information on the uncertainties and precision of these observations, and on the crucial parameters and technical issues which were later addressed during the intercomparison campaigns in 1992 and 1994.

During 1991, the Swedish groups have further developed the large scale flux method based on the observation of the zenith sky-light by an airborne DOAS instrument, made at two different flight altitudes. The method was successfully tested for NO_2, during three field campaigns.

The dual-beam extinction measurement method developed by the Swedish group was tested in the TESLAS intercomparison campaign TROLIX '91 in Bilthoven, Netherlands, June 16th–27th, 1991 (Bösenberg et al., 1993). In this method two spectra are recorded, close in time, along two paths with the same direction but with different path lengths. If these two spectra are then divided, instrument factors, such as spectral structures from the light source and spectrometer anomalies, are to a high degree cancelled. This method was used routinely to improve the detection limit of the system when measuring background concentrations and made possible the measurement of absolute atmospheric extinction spectra. The idea was to compare the theoretical LIDAR corrections for particle attenuation and Rayleigh scattering with the measured values deduced for the dual-beam measurements. DOAS measurements were made over two paths, 885 m and 2 km respectively, and covering the wavelength range 270–315 nm, an interval containing most of the wavelengths used in the LIDAR ozone measurements. During some periods two of the Lidar systems were directed horizontally giving paths nearly collinear with the DOAS paths. Thus the different systems could be inter-calibrated. During the campaign the DOAS system was also measuring ground level concentrations of O_3 and SO_2. In connection with this campaign a NO_2 intercomparison between the DOAS and the RIVM NO_2 Lidar systems over a nearly collinear path of 1116 m was performed. The comparison was made without any preceding inter-calibration and showed good agreement between the two methods.

During Spring 1993 a measurement campaign was conducted at the Swedish field site at Draget close to the TOR station at Rörvik. The aim of this campaign was a comparative study of NO_2 measurements between Draget and Rörvik as well as a first attempt to measure NO_3. The NO_2 measurements show generally good agreement, however at some of high concentrations (20 $\mu g/m^3$) the DOAS concentrations are considerably higher than the TOR station data. This may be due to local contamination, different meteorological conditions or differences in the instruments. Significant concentrations of NO_3 (50 ng/cm^3) was detected during one night, the concentration being below the current detection limit of 20 ng/m^3 during the rest of the period.

A DOAS system optimised for operation in the wavelength region 215–245 nm was designed by the Swedish group for the detection of NO, NH_3 and SO_2. The system was tested in a field campaign at Lanna agricultural field station in Sweden in summer 1993. The aim of this campaign was to inter-compare different techniques for measurement of fluxes of NH_3 resulting from spreading of manure. The system was compared with a long path FTIR system as well as different gradient, denuder and field chamber techniques. The system showed good agreement with the other techniques, demonstrated linearity over three orders of magnitude, and had a detection limit for NH_3 of 15 ppb for 1 minute integration time. During the campaign the NO flux generated by microbial activity on nitrogen compounds was also measured.

Two other field campaigns at the Weybourne Atmospheric Observatory (WAO) were organised as part of the Land Ocean Interaction Study (LOIS). An important focus of LOIS is the chemistry of NO_3 in the marine boundary. This radical is a significant night-time oxidant of organic species, as well as providing a removal pathway for anthropogenic nitrogen oxides. The NO_3 lifetime in the marine boundary layer at Weybourne was found to be of the order of 10 minutes in the spring/summer, and appears to be controlled by dimethylsulfide (DMS) produced in the ocean by phytoplankton. It decreases to about 2 minutes in the autumn/winter. Since the DMS concentration is low at this time of the year, the most likely sink for NO_3 is the heterogeneous reaction on moist aerosol surfaces of N_2O_5 with which NO_3 is in equilibrium (Smith et al., 1995). Another highlight of these measurements in the marine boundary layer was the observation of nitrous acid (HONO) being formed during the night at sub-ppb levels.

The DOAS was also employed to study NO_3 at the Izaña Observatory in Tenerife in the Canary Islands. The Meteorological Station lies on a mountain platform at an altitude of 2376 m a.s.l. and 400 km west of Africa. The air at the observatory is usually representative of the free troposphere. During the campaign, the concentration of NO_3, NO_2 and O_3 were measured by the DOAS with a path length of 1.98 km. The measurements of NO_2 and O_3 were compared with point sampling instruments at the Meteorological Station, yielding excellent correlation.

Measurements of NO_3 were made for 20 full nights during the campaign. The NO_3 was seen to be strongly correlated with O_3, and anti-correlated with the relative humidity (RH) during the night. The reason for the inverse relationship with RH, which has been reported previously (Smith *et al.*, 1995; Platt *et al.*, 1984) is most likely that the surfaces of aerosols (and the ground) must be deliquesced in order to provide effective sinks for gas-phase NO_3 or N_2O_5. However, the RH at Izaña at night-time was often very low. Since the measured concentrations of reactive hydrocarbons such as isoprene were very low, one possible mechanism for removing NO_3 is the homogeneous gas-phase reaction $N_2O_5 + H_2O \rightarrow 2HNO_3$. The field data indicate that this reaction would require a rate coefficient of about 9×10^{-21} cm^3 molecule^{-1} s^{-1} which is in sensible accord with the upper limit to this reaction of 2×10^{-21} cm^3 molecule^{-1} s^{-1} at 298 K measured in the laboratory.

e.   Laboratory measurements and wavelength range definition for DOAS

New measurements of absorption cross sections in the UV and visible range were performed for SO_2, NO_2 and CS_2 at different resolutions by means of the FTS at the University of Brussels. High resolution spectra (0.02 nm at 300 nm) were obtained at room temperature for SO_2, NO_2 and CS_2 and also at 250 and 273K for NO_2. Comparison with previous measurements shows good agreement, better than 5%. SO_2 results were published by Vandaele *et al.* (1994).

12.3.4 Achievements

The basic objective of TOPAS, namely the development and improvement of instruments for the detection of tropospheric constituent by DOAS were fulfilled and intercomparison campaigns were carried out to evaluate their performances.

New absorption long path systems were designed and extensively used, with the new concept of the combined transmitting/receiving telescope offering several advantages with respect to the classical double-ended system. Furthermore, a multipass cell was built and tested during campaigns.

The instruments were improved by using a variety of new type of detectors. Careful characterisation was carried out during the time-frame of the subproject. The Fourier transform spectrometers were used for the first time for DOAS measurements. Several performances were improved *e.g.* in terms of measurement frequency. Dedicated software for instrument automation and data analysis were developed and improved, taking into account the specific problems imposed by each type of instrument and its associated detector.

The major achievements were the two intercomparison campaigns conducted in 1992 and 1994, respectively in an urban (Brussels, Belgium) and in a pristine site (Weybourne, UK).

This first intercomparison campaign of DOAS instruments operating in the visible and UV ranges brought to light several fundamental issues which have contributed to improve this technique in its aim to study the composition of the troposphere. Significant progress has been made by all the participants due to the synergy of this type of campaign. After a first comparison of results made without any constraints, it was obvious that a common set of absorption cross sections had to be used by each group and also triggered improved measurements of these (Vandaele *et al.*, 1994). The campaign also served to exchange various instrument and software improvements between the groups.

The measurements made during the campaign by eight different instruments were carefully analysed and yielded an absolute accuracy limit for NO_2, O_3 and SO_2 measurements made by the DOAS technique using an absorption path of a few hundred meters in a relatively non-polluted urban troposphere.

There remain nevertheless differences between the results obtained from the various instruments. These are most certainly linked to the various retrieval methods used although it was concluded that these retrieval methods are well optimised and well adapted to each instrument.

This campaign and its results are described in detail in Camy-Peyret *et al.* (1996)

In order to improve our knowledge of the accuracy of the presently available type of instruments. In particular it appeared interesting to compare results obtained by the DOAS technique to those obtained by other techniques. In order to do so, it was felt useful to conduct a campaign in a more stable environment, where concentrations vary less than in an urban site. This would reduce the difficulties encountered in the first campaign related to the various integration times used by the different instruments. Also, cell measurements should be performed by all instruments. Although cell measurements poorly represent the difficulties associated with real atmospheric measurements, they would probably reveal more clearly differences in results due to the retrieval methods used. Finally, the atmospheric pollution conditions encountered in the first campaign did not allow the detection limits of each experimental set-up to be reached.

All these points were addressed in a second campaign which took place in September 1994 at the Weybourne Atmospheric Observatory situated in a remote area on the coast of East Anglia (UK). The comparisons concerned SO_2, NO_2, O_3, NO_3, HCHO and HONO.

The campaign compared seven different instruments through ten days of continuous measurement under field conditions. The DOAS were compared with commercial point monitors for NO_2, O_3 and SO_2 and a home-built monitor for HONO. The preliminary results are very encouraging although the full analysis of the data has not yet been completed. Two other comparisons were also made during the campaign. Firstly the DOAS were tested using cells containing known amounts of NO_2 and SO_2. The agreement was to within about 7%. Secondly the group's software was tested using synthetic spectra including NO_2, O_3 and

HCHO. Agreement was found within about a 5% error. This represents a considerable improvement on the results of the first comparison exercise.

In the framework of the TOPAS subproject, tropospheric measurements of trace gases by using DOAS were significantly improved from the instrument and retrieval method point of views. They were fully validated, mainly during the two intercomparison campaigns conducted in 1992 and 1994. This kind of exercise demonstrated the necessity for quality assurance and quality control procedures in order to obtain reliable tropospheric data for high quality scientific studies of the troposphere. It is also obvious that further improvements are required and can only be obtained by means of instrument-oriented projects in future tropospheric research programmes (*e.g.* EUROTRAC-2).

One instrument has been fully commercialised, namely the "Système d'Analyse par Observations Actives" (SANOA) developed by the CNRS (France) . For the first time, a Fourier transform spectrometer has been used in the UV-visible range for DOAS and its performances demonstrated. This instrument is commercialised by Bruker (Germany). Furthermore, Hoffmann Messtechnik (Germany) has also developed an instrument for DOAS in the troposphere. It is very unfortunate that OPSIS declined to participate in the intercomparison campaign in 1992. Consequently, the performances of this commercial instrument were not evaluated.

12.1.5 Further information

The TOPAS contribution is shown in context in chapter 6 of this volume. A complete account of the subproject can be found in volume 8 of this series (Bösenberg *et al.*, 1997) which includes a general scientific report by the subproject coordinator.

12.1.6 TOPAS: Steering group and principal investigators

Steering group

Paul C. Simon (Coordinator)	Brussels
Reginald Colin	Brussels
Bo Galle	Göteborg
John Plane	Norwich
Ulrich Platt	Heidelberg
Jean-Pierre Pommereau	Verrières-le-Buisson

Individual contributions from principal investigators (*) to the final report (Bösenberg *et al.*, 1997) are listed below.

R. Colin*, M. Carleer, J.M. Guilmot, P.C. Simon, A.C. Vandaele, C. Hermans,
   P. Dufour and C. Fayt
   Belgian contribution to differential optical absorption studies of the
   troposphere between 1989 and 1994

B. Galle*, H. Axelsson, B. Bergqvist, A. Eilard, J. Mellqvist and L. Zetterberg
Development of DOAS for atmospheric trace species monitoring

N. Smith, H. Coe, B. Allan and J. Plane*
Differential optical absorption studies at East Anglia

J. Stutz and U. Platt*
A new generation of DOAS instruments

J.-P. Pommerea*u, F. Goutail, P. Laville and M. Nunes-Pinharanda
Development of a long path UV-visible spectrometer for atmospheric
composition monitoring

12.3.7 References

Axelsson, H., B. Galle, K. Gustavsson, P. Ragnarsson M. Rudin; 1991,
A transmitting/Receiving telescope for DOAS measurements using retroreflector
technique, in: P. Borrell, P.M. Borrell, W. Seiler (eds.) *Proc. EUROTRAC Symp. '90.*
SPB Academic Publishing bv, The Hague.

Bösenberg, J., B. Galle, J. Mellqvist *et al.;* 1993, Tropospheric ozone Lidar experiment,
TROLIX '91 field phase report, Max-Planck-Institut für Meteorologie, report No. 102.

Bösenberg, J., D. Brassington, P.C. Simon (eds); 1997, *Instrument Development for
Atmospheric Research and Monitoring*, Springer Verlag, Heidelberg.

Camy-Peyret, C., B. Bergquist, B. Galle, M. Carleer, C. Clerbaux, R. Colin, C. Fayt,
F. Goutail, M. Nunes-Pinharanda, J.P. Pommereau, M. Hausmann, F. Heinz, U. Platt,
I. Pundt, T. Rudoph, C. Hermans, P.C. Simon, A.-C. Vandaele, J. Plane, N. Smith;
1996, Intercomparison of instruments for tropospheric measurements using differential
optical absorption spectroscopy, *J. Atmos. Chem.* **23**, 51–80.

Carleer M., R. Colin, A.C. Vandaele, P.C. Simon; 1994, Measurement of NO_2 and SO_2
absorption cross sections, in: P.M. Borrell, P. Borrell, T. Cvitaš, W. Seiler (eds) *Proc.
EUROTRAC Symp. '92.* SPB Academic Publishing bv, The Hague, pp. 419–422.

Galle, B., H. Axelsson, P. Ragnarsson, M. Rudin; 1990, A transmitting/receiving
telescope for DOAS measurements using retroreflector technique, *Proc. OSA Conf.
Optical Remote Sensing of the Atmosphere*, Nevada.

Goutail, F., J.P. Pommereau, M. Nunes-Pinharanda; 1993, Ambient Air Monitoring by
Differential Absorption Spectroscopy in the Ultraviolet and the Visible : the SANOA
Instrument, *Proc. Symp. Atmos. Spectros. Applications*, Reims.

Goutail, F., J.P. Pommereau, M. Nunes-Pinharanda; 1993, Surveillance de la composition
de l'air par spectrométrie uv-visible, *Proc. Coll. Horizons de l'Optique,* Limoges.

Laville, P., J.P. Pommereau, F. Goutail; 1991, Evaluation of a long path diode array
spectrometer for O_3, NO_2, SO_2 and water vapour monitoring, in: P. Borrell,
P.M. Borrell, W. Seiler (eds) *Proc. EUROTRAC Symp. '90.* SPB Academic Publishing
bv, The Hague, p. 479.

Plane, J.M.C., N. Smith; 1995, Atmospheric monitoring by differential optical absorption
spectroscopy, in: R.E Hester, R.J.H. Clark (eds), *Spectroscopy in Environmental
Sciences*, John Wiley, London, pp. 223–262.

Platt, U., A.M. Winer, H.W. Biermann, R. Atkinson, J.N.Jr Pitts; 1984, Measurement of
nitrate radical concentrations in continental air, *Environ. Sci. Technol.* **18**, 365–369.

Pommereau, J.P., F. Goutail, M. Nunes-Pinharanda; 1993, Application de la spectroscopie UV-Visible à la mesure de la pollution urbaine, *Coll. Pollution Atm. à l'échelle locale et régionale*, Ademe, Paris.

Ritz, D., M. Hausmann, U. Platt; 1992, An improved open path multi-reflection cell for the measurement of NO$_2$ and NO$_3$, in: H.I. Schiff, U. Platt (eds), *Proc. Int. Symp. on Environmental Sensing in Optical Methods in Atmospheric Chemistry, Proc. SPIE* **1715**, 200–211.

Smith, N., J.M.C. Plane, C.-F. Nien, P.A. Solomon; 1995, Night-time radical chemistry in the San Joaquin Valley, *Atmos. Environ.* **29**, 2887–2897.

Vandaele, A.-C., P.C. Simon, J.M. Guilmot, M. Carleer, R. Colin; 1994, SO$_2$ absorption cross section measurement in the UV using a Fourier transform spectrometer, *J. Geophys. Res.* **99**,) 25599–25605.

White, J.U.; 1976, Very long optical paths in air, *J. Opt. Soc. Am.* **66,5**, 411-416.

Appendix A: Table 1

EUROTRAC Subprojects and Subproject Coordinators 1987 to 1995

ACE	Acidity in Cloud Experiments	Dr. Anthony R. Marsh	1988 to 1992
ALPTRAC	High Alpine Aerosol and Snow Chemistry Study	Prof. Hans Puxbaum	1988 to 1992
		Prof. Dietmar Wagenbach	1993 to 1995
ASE	Air-Sea Exchange	Dr. Peter Liss	1988 to 1990
		Dr. Søren E. Larsen	1990 to 1995
BIATEX	Biosphere-Atmosphere Exchange of Pollutants	Prof. J. Slanina	1987 to 1995
EUMAC	European Modeling of Atmospheric Constituents	Prof. Adolf Ebel	1987 to 1995
GCE	Ground-based Cloud Experiments	Dr. John Ogren	1987 to 1990
		Prof. Sandro Fuzzi	1991 to 1995
GENEMIS	Generation of European Emission Data	Prof. Rainer Friedrich	1992 to 1995
GLOMAC	Global Modeling of Atmospheric Chemistry	Prof. Henning Rodhe	1988 to 1995
HALIPP	Heterogeneous and Liquid Phase Processes	Prof. Peter Warneck	1988 to 1995
JETDLAG	Joint European Development of Tunable Diode Laser Absorption Spectroscopy for the Measurement of Atmospheric Trace Gases	Dr. Franz Slemr	1987 to 1993
		Dr. David Brassington	1993 to 1995
LACTOZ	Laboratory Studies of Chemistry Related toTropospheric Ozone	Dr. R. Anthony Cox	1987 to 1990
		Prof. Karl-Heinz Becker	1990 to 1992
		Prof. Georges Le Bras	1993 to 1995
TESLAS	Joint European Programme for the Tropospheric Environmental Studies by Laser Sounding	Dr. J. Pelon	1987 to 1991
		Prof. Jens Bösenberg	1991 to 1995

EUROTRAC Subprojects and Subproject Coordinators (continued)

TOPAS	Tropospheric Optical Absorption Spectroscopy	Prof. Paul Simon	1987 to 1995
TOR	Tropospheric Ozone Research	Dr. Dieter Kley	1987 to 1995
TRACT	Transport of Pollutants over Complex Terrain	Prof. Franz Fiedler	1989 to 1995
Working Groups			
AP	Application Project	Dr. Peter Borrell	1992 to 1995
CMWG	Chemical Mechanism Working Group	Dr. Dirk Poppe	1993 to 1995
CG	Cloud Group	Prof. Peter Warneck	1994 to 1995

Appendix A: Table 2

EUROTRAC International Executive Committee
Members 1987 to 1996

Austria	Dr. Peter Tuschl	1987 to 1993
	Dr. Rudolf Orthofer	1993 to 1996
Belgium	Mr. C. de Wispelaere	1987
	Mr. Peter Vanhaecke	1987 to 1988
	Mrs Martine Vanderstraeten	1989 to 1996
Denmark	Dr. Eigil Praestgaard	1987 to 1994
	Dr. Ole John Nielsen	1994 to 1996
European Commission	Dr. Anver Ghazi	1986
	Dr. Heinz Ott	1986 to 1995
	Dr. Giovanni Angeletti, *Member*	1995 to 1996
	Observer	1987 to 1995
Finland	Prof. Antti Kulmala	1986 to 1989
	Dr. Sylvain Joffre	1990 to 1996
	Dr. Pekka Kostamo *Observer*	1987 to 1991
France	Dr. Lucien Chabason	1987 to 1989
	Dr. Jean-Claude Oppenau	1989 to 1996
	Mr. Jacques Dubois *Observer*	1989 to 1996
Germany	Dr. Wolf van Osten	1986 to 1987
	Mr. H.J. Block	1987
	Dr. Harmut Keune	1987 to 1990
	Vice Chairman	1987 to 1990
	Dr. Bernhard Rami	1990 to 1993
	Vice Chairman	1990 to 1993
	Dr. Gerhard Hahn	1993 to 1996
	Vice Chairman	1993 to 1996
	Dr. Helmut Bauer *Observer*	1987 to 1996
	Dr. Eric Weber *Observer*	1987 to 1991

EUROTRAC International Executive Committee (continued)

Greece	Dr. Nicolas Moussiopoulos	1991 to 1996
Ireland	Dr. Tom O'Connor	1988 to 1996
Italy	Dr. Ivo Allegrini	1988 to 1996
Netherlands	Dr. Roul van Aalst	1986
	Mr. Said Zwerver	1987 to 1988
	Dr. M. Bovenkerk	1989 to 1991
	Mr. Volkert G. Keizer	1991 to 1996
	Mr. A.R. Manuel *Observer*	1987
	Dr. Erik Schenk *Observer*	1989 to 1996
Norway	Dr. Armund Barstad	1986
	Dr. Kari Kveseth	1987 to 1996
Portugal	Dr. Renato Carvalho	1986
	Prof. Carlos A. Borrego	1987 to 1996
Spain	Prof. Juan Albaiges	1986 to 1990
	Prof.. Millán Millán	1994 to 1996
Sweden	Dr. Göran Persson	1987 to 1991
	Chairman	1987 to 1991
	Dr. Erik Fellenius	1991 to 1996
	Chairman	1993 to 1996
Switzerland	Prof. Hans Richner	1988 to 1996
Turkey	Dr. Saim Öskar	1987
	Dr. Altan Acara	1989 to 1996
	Dr. Önder H. Ösbelge *Observer*	1992
United Kingdom	Dr. Geoffrey Jenkins	1986 to 1989
	Mr. David Warrilow	1989 to 1993
	Chairman	1991 to 1992
	Dr. Martin L. Williams	1993 to 1996
Observers		
EUREKA Secretariat	Mr. Wolfgang Locker	1993 to 1995
	Dr. Benno Schmidt-Küntzel	1995 to 1996
European Science Foundation	Mr. Josip Hendekovic	1987

Appendix A: Table 3

Countries Participating in EUROTRAC

The following countries were represented on the IEC

Austria	Netherlands
Belgium	Norway
Denmark	Portugal
Finland	Spain
France	Sweden
Germany	Switzerland
Ireland	Turkey
Greece	United Kingdom
Italy	

together with the European Commission (EC)

There were also principal investigators from

Croatia	Romania
German Democratic Republic (until 1991)	Poland
Hungary	Russia (also from the USSR)
Lithuania	Yugoslavia
Slovenia	

Appendix A: Table 4

EUROTRAC Scientific Steering Committee Members 1986 to 1995

Dr. Roul M. van Aalst	RIVM, Bilthoven	1994 to 1995
Dr. Lennart Bengtsson	Max Plank Institute, Hamburg	1993
Dr. P. Buat-Ménard	University of Bordeaux	1992 to 1993
Prof. Reginald Colin	Université Libre, Brussels	1994 to 1995
Dr. R. Anthony Cox	University of Cambridge	1986 to 1995
Prof. Paul Crutzen	Max Plank Institut, Mainz	1986 to 1989
Dr. Robert Delmas	Lab. Glaciologie et Geophys., Grenoble	1986 to 1992
Prof. Adolf Ebel	Universität zu Köln	1994 to 1995
Dr. Dieter Ehhalt	Forschungszentrum, Jülich	1986 to 1989
Prof. Franz Fiedler	Forschungszentrum, Karlsruhe	1986 to 1993
Prof. Sandro Fuzzi *Vice Chairman*	CNR, Bologna	1993 to 1995 1993 to 1995
Prof. Heinz Gäggeler	Paul Scherrer Institute, Villigen	1992 to 1995
Dr. Peringe Grennfelt	IVL, Göteborg	1994 to 1995
Dr. Robert Guicherit *Chairman*	TNO, Delft	1986 to 1993 1991 to 1993
Dr. Jost Heintzenberg	University of Stockholm	1986 to 1994
Prof. Ivar Isaksen *Chairman*	University of Oslo	1986 to 1991 1986 to 1991
Dr. S. Gerard Jennings	University College, Galway	1994 to 1995
Dr. Niels-Otto Jensen	Risö National Laboratory, Roskilde	1994 to 1995
Dr. Dieter Kley	Forschungszentrum, Jülich	1989 to 1995
Prof. Alain Marenco	Université Paul Sabatier, Toulouse	1993 to 1994

EUROTRAC Scientific Steering Committee (continued)

Dr. Gérard Mégie	CNRS, Service de Aeronomie, Paris	1986 to 1991
Dr. Anthony R. Marsh *Chairman*	Imperial College, London	1989 to 1995 1993 to 1995
Prof. Stuart A. Penkett *Vice Chairman*	University of East Anglia, Norwich	1986 to 1995 1986 to 1991
Prof. Casimiro Pio	Universidade de Aveiro	1992 to 1995
Prof. Hans Puxbaum	Technical University, Vienna	1986 to 1992
Dr. Wolfgang Seiler	IFU, Garmisch-Partenkirchen	1986 to 1995
Prof. Han van Dop	KNMI, de Bilt, *later* WMO, Geneva	1986 to 1993
Dr. Bruno Versino *Vice Chairman*	JRC, Ispra	1986 to 1993 1991 to 1993
Prof. Rudi Zander	University of Liége	1989 to 1993

Appendix A: Table 5

EUROTRAC International Scientific Secretariat
1986 to 1996

Dr. Wolfgang Seiler	Director	1986 to 1996
Dr. Peter Borrell	Scientific Secretary	1989 to 1996
Prof. Tomislav Cvitaš	Scientist	1990 to 1995
Ms. Kerry Kelly	Scientist	1995 to 1996
Dr. Patricia Borrell	Contract Scientist	1990 to 1996
Frau Irmgard Réthoré	Secretary	1988 to 1996

Appendix A: Table 6

Chemical Mechanism Working Group Members

Prof. Karl-Heinz Becker	Universität, Wuppertal	LACTOZ
Dr. Peter Borrell	IFU, Garmisch-Partenkirchen	ISS
Prof. Peter Builtjes	TNO, Apeldoorn	TOR
Dr. R.G. Derwent	UK Meteorological Office	–
Prof. Adolf Ebel	Universität zu Köln	EUMAC
Dr. Gary Hayman	Harwell	LACTOZ
Prof. Dieter Kley	Forschungszentrum, Jülich	TOR
Prof. Mike Pilling	Leeds University	LACTOZ
Prof. Dirk Poppe, *Chairman*	Forschungszentrum, Jülich	EUMAC
Dr. David Simpson	Meteorological Inst. Oslo	EMEP
Dr. W.R. Stockwell	IfU, Garmisch-Partenkirchen	TOR
Prof. P. Warneck	Max Plank Inst., Mainz	HALIPP

Appendix A: Table 7

The Cloud Group Members

Dr. Urs Baltensperger	Paul Scherer Inst. Villingen	ALPTRAC
Dr. Erik Berge	Meteorological Inst., Oslo	EMEP
Dr. Keith Bower	UMIST, Manchester	GCE
Dr. George Buxton	Leeds University, Cookridge	HALIPP
Dr. Sylvia Cautenet	Université Blaise Pascal	EUMAC
Dr. Nadine Chaumerliac	Université Blaise Pascal	EUMAC
Prof. Tom Choularton	UMIST, Manchester	GCE
Dr. Frank Dentener	Max Plank Inst. Mainz	GLOMAC
Prof. Adolf Ebel	Universität zu Köln	EUMAC
Prof. Andrea Flossmann	Université Blaise Pascal	EUMAC
Dr. Renate Forkel	IFU, Garmisch-Partenkirchen	EUMAC
Prof. Sandro Fuzzi *Chairman*	CNR, Bologna	GCE
Prof. H.C. Hansson	Stockholm University	GCE
Prof. J. Hoigné	Universität, Bern	HALIPP
Prof. Wolfgang Jaeschke	J.W. Goethe Universität, Frankfurt	GCE
Dr. Anne Kasper	Technische Universität, Wien	ALPTRAC
Dr. Manfred Laube	Universität zu Köln	EUMAC
Prof. Jos Lelieveld	IMAU, Utrecht	GLOMAC
Prof. Detlev Möller	Brandenburg Univ. Cottbus	GCE
Dr. Arthur Salmon	Leeds University, Cookridge	HALIPP
Dr. Rolf Sander	Max Plank Inst. Mainz	GLOMAC
Dr. Winfried Seidl	IFU, Garmisch-Partenkirchen	EUMAC
Prof. Peter Warneck	Max Plank Inst. Mainz	HALIPP
Dr. Wolfram Wobrock	J.W. Goethe Universität, Frankfurt	GCE

Appendix A: Table 8

Application Project Members

Dr. Roel van Aalst	RIVM Bilthoven
Dr. Peter Borrell *Coordinator*	IFU, Garmisch-Partenkirchen
Prof. Peter Builtjes *Convener (tools)*	TNO, Apeldoorn
Dr. David Fowler	ITE, Penicuik
Dr. Peringe Grennfelt *Convener (acidity)*	IVL, Göteborg
Prof. Øystein Hov *Convener (photo-oxidants)*	NILU Kjeller
Prof. Gérard Mégie	CNRS Service d'Aeronomie, Paris
Prof. Nicolas Moussiopoulos	Aristotle University, Thessaloniki
Dr. Andreas Volz-Thomas	Forschungzentrum Jülich
Prof. Peter Warneck	Max-Planck-Inst. für Chemie, Mainz
Prof. Richard Wayne	Physical Chemistry Lab., Oxford

Appendix B

EUROTRAC Publications

EUROTRAC Project Descriptions

A two volume project description and project descriptions for each of the fifteen subprojects were published between 1988 and 1991.

EUROTRAC Newsletters

Issues 1 to 16.

EUROTRAC Brochures

The EUROTRAC Information Folder, published in 1989

A brief description of the project published in June 1994, 32 pp.

EUROTRAC-1 Annual Reports

1989 in nine parts;
1990 in nine parts;
1991 in nine parts;
1992 in ten parts;
1993 in ten parts.

The final report replaced the annual reports for 1994 and 1995.

EUROTRAC Final Report

The final report from EUROTRAC-1 is published in 10 volumes by the Springer Verlag, Berlin, Heidelberg and New York

Volume 1:   **Transport and Chemical Transformation of Pollutants in the Troposphere** (editors: Peter Borrell and Patricia M. Borrell) 2000.

Volume 2:   **Heterogeneous and Liquid Phase Processes** (editor: Peter Warneck) 1996. ISBN 3-540-60792-7.

Volume 3:   **Chemical Processes in Atmospheric Oxidation** (editor: Georges Le Bras) 1997. ISBN 3-540-60998-9.

Volume 4:   **Biosphere-Atmosphere Exchange of Pollutants and Trace Substances** (editor: J. Slanina) 19976. ISBN 3-540-61711-6.

Volume 5:	**Cloud Multi-phase Processes and High Alpine Air and Snow Chemistry**
(editors: Sandro Fuzzi and Dietmar Wagenbach) 1997.
ISBN 3-540-62496-1.

Volume 6:	**Tropospheric Ozone Research** (editor: Øystein Hov) 1997.
ISBN 3-540-63359-6.

Volume 7:	**Tropospheric Modelling and Emission Estimation**
(editors: Adolf Ebel, Rainer Friedrich and Henning Rodhe), 1997.
ISBN 3-540-63169-0.

Volume 8:	**Instrument Development for Atmospheric Research and Monitoring**
(editors: Jens Bösenberg, David Brassington and Paul Simon), 1997.
ISBN 3-540-62516-X.

Volume 9:	**Exchange and Transport of Air Pollutants over Complex Terrain and the
Sea** (editors: Søren Larsen, Franz Fiedler and Peter Borrell), 2000.

Volume 10:	**Photo-oxidants, Acidification and Tools: Policy Applications of EUROTRAC
Results** (editors: Peter Borrell, Øystein Hov, Peringe Grennfelt and
Peter Builtjes), 1997. ISBN 3-540-61783-3.

EUROTRAC Symposium Proceedings

Proceedings of the EUROTRAC Symposium '90; (editors: P. Borrell, P.M. Borrell and
W. Seiler) SPB Academic Publishing bv, The Hague 1991, 586 pp.
ISBN 90-5103-058-4-

Proceedings of the EUROTRAC Symposium '92; (editors: P.M. Borrell, P. Borrell,
T. Cvitaš and W. Seiler), SPB Academic Publishing bv, The Hague 1993, 830 pp.
ISBN 90-5103-082-7.

Proceedings of the EUROTRAC Symposium '94; (editors: P.M. Borrell, P. Borrell,
T. Cvitaš and W. Seiler), SPB Academic Publishing bv, The Hague 1994, 1283 pp.
ISBN 90-5103-095-9.

Proceedings of the EUROTRAC Symposium '96 (2 volumes); (editors: P.M. Borrell,
P. Borrell, T. Cvitaš, K. Kelly and W. Seiler), Computational Mechanics,
Southampton 1996, pp 1057(vol. 1); 822(vol. 2). ISBN 1-85312-495-8.

EUROTRAC Special Publications

ALPTRAC Data Catalogue; H. Kromp-Kolb, W. Schöner and P. Seibert
EUROTRAC ISS, Garmisch-Partenkirchen 1993.

The Tor Network; T. Cvitaš and D. Kley (eds)
EUROTRAC ISS, Garmisch-Partenkirchen 1994.

LACTOZ Re-evaluation of the EMEP MSC-W Photo-oxidant Model;
K. Wirtz, C. Roehl, G.D. Hayman and M.E. Jenkin
EUROTRAC ISS, Garmisch-Partenkirchen 1994.

The TRACT Field Measurement Campaign (Field Phase Report); H. Zimmermann
EUROTRAC ISS, Garmisch-Partenkirchen 1995.

Clouds: Models and Mechanisms; A. Flossmann and T. Cvitaš (eds)
EUROTRAC ISS, Garmisch-Partenkirchen 1995.

Comparison of Photo-oxidant Dispersion Model Results;
H. Haas, P.H.J. Builtjes, D. Simpson, R. Stern, together with H.J. Jakobs,
M. Memmersheimer, G. Piekorz, M. Roemer, P. Esser and E. Reimer.
EUROTRAC ISS, Garmisch-Partenkirchen 1996.

Gas-phase Reactions in Atmospheric Chemistry and Transport Models: A Model Intercomparison;
D. Poppe, Y. Andersson-Sköld, A. Baart, P.H.J. Builtjes, M. Das, F. Fiedler, Ø. Hov,
F. Kirchner, M. Kuhn, P.A. Makar, J.B. Milford, M.G.M. Roemer, R.Ruhnke,
D. Simpson, W.R. Stockwell, A. Strand, B. Vogel and H. Vogel.
EUROTRAC ISS, Garmisch-Partenkirchen 1996.

EUROTRAC Data handbook; Jürgen Hahn
EUROTRAC ISS, Garmisch-Partenkirchen 1997.

Copies of some of these publications are available from the EUROTRAC-2 ISS, GSF-Forschungszentrum für Umwelt und Gesundheit, Kühbachstraße 11 D-81543 München, Germany. See the web page: www.gsf.de/eurotrac for further information

Subject Index

A

T

Printing: Mercedes-Druck, Berlin
Binding: Buchbinderei Lüderitz & Bauer, Berlin